Nanocomposites
Ionic Conducting Materials and Structural Spectroscopies

Electronic Materials: Science and Technology

Series Editor: Harry L. Tuller
Professor of Materials Science and Engineering
Massachusetts Institute of Technology
Cambridge, Massachusetts
tuller@mit.edu

Nanocomposites: Ionic Conducting Materials and Structural Spectroscopies
Philippe Knauth and Joop Schoonman, eds.
ISBN 978-0-387-33202-4, 2008

Electroceramic-Based MEMS: Fabrication-Technology and Applications
N. Setter
ISBN 978-0-387-23310-9, 2005

Nanostructured Materials: Selected Synthesis Methods, Properties and Applications
Philippe Knauth and Joop Schoonman, eds.
ISBN 978-1-4020-7241-3, 2002

Nanocrystalline Metals and Oxides: Selected Properties and Applications
Philippe Knauth and Joop Schoonman, eds.
ISBN 978-0-7923-7627-9, 2002

High-Temperature Superconductors: Materials, Properties, and Applications
Rainer Wesche
ISBN 978-0-7923-8386-4, 1999

Amorphous and Microcrystalline Silicon Solar Cells: Modeling, Materials and Device Technology
Ruud E.I. Schropp and Miro Zeman, eds.
ISBN 978-0-7923-8317-8, 1998

Microactuators: Electrical, Magnetic, Thermal, Optical, Mechanical, Chemical and Smart Structures
Massood Tabib-Azar,
ISBN 978-0-7923-8089-4, 1998

Thin Film Ferroelectric Materials and Devices
R. Ramesh, ed.
ISBN 978-0-7923-9993-3, 1997

Wide-Gap Luminescent Materials: Theory and Applications
Stanley R. Rotman, ed.
ISBN 978-0-7923-9837-0, 1997

Piezoelectric Actuators and Ultrasonic Motors
Kenji Uchino
ISBN 978-0-7923-9811-0, 1996

Philippe Knauth • Joop Schoonman
Editors

Nanocomposites

Ionic Conducting Materials and Structural Spectroscopies

Springer

Editors
Philippe Knauth
Université de Provence
Centre St Jérôme
F-13397 Marseille Cedex 20
France

Joop Schoonman
Delft University of Technology
Department Delft Chem Tech-Energy
Julianalaan 136
2628 BL Delft
The Netherlands

Series Editor
Harry L. Tuller
Professor of Materials Science and Engineering
Massachusetts Institute of Technology
Cambridge, MA 02139

ISBN 978-0-387-33202-4 e-ISBN 978-0-387-68907-4

Library of Congress Control Number: 2007932056

Printed on acid-free paper.

9 8 7 6 5 4 3 2

springer.com

Preface

Nanocomposite organic/inorganic materials are a fast expanding area of research, part of the growing field of nanotechnology. The term "nanocomposite" encompasses a wide range of materials mixed at the nanometer scale, combining the best properties of each of the components or giving novel and unique properties, unknown in the constituent materials, with great expectations in terms of advanced applications. Significant effort is focused on the ability to control the nanoscale structures via innovative synthetic approaches. The properties of nanocomposite materials depend not only on the properties of their individual components but also on their morphology and interfacial characteristics.

Experimental work has shown that virtually all types and classes of nanocomposite materials lead to new and improved properties, when compared to their macrocomposite counterparts: they tend to drastically improve the electrical conductivity, specifically the ionic conductivity, and thermal conductivity of the original material as well as the mechanical properties, e. g., strength, modulus, and dimensional stability. Other properties that might undergo substantial improvements include decreased permeability to gases, water and hydrocarbons, thermal stability and chemical resistance, surface appearance and optical clarity. Therefore, nanocomposites promise new applications in many fields such as mechanically reinforced lightweight components, nonlinear optics, battery cathodes and solid state ionics, nanowires, sensors, and many others. Much effort is going on to develop more efficient combinations of materials and to impart multifunctionalities to the nanocomposites.

In previous volumes of this book series, we have reported on the properties and applications of materials with characteristic dimensions in the nanometer scale. These reports were however mainly devoted to inorganic single phase materials. Recent years have seen a widespread development of polymer and hybrid organic/inorganic systems, which improve the variability of the materials properties and give supplementary freedom for realization of hitherto unattainable combination of properties.

In this book, we want to review some recent advances in composite materials with domain sizes in the nanometer range, emphasizing polymeric and hybrid systems, which have advanced spectacularly. Given the scientific background of the editors, most of the chapters are devoted to ionic conducting materials; some emphasis is put on materials potentially useful in fuel cells and lithium ion batteries, including polymer nanocomposites with clay and other plate-shaped particles as second phase.

The first chapter sets the general frame for the investigation of composite polymeric electrolytes. This chapter is especially devoted to lithium ion conducting solid electrolytes for lithium batteries, with polyethylene-oxide-based systems playing a central role. But theoretical models of electrical properties and ionic conduction in polymers are discussed in depth and various scenarios for conductivity enhancement effects are outlined, including space charge and Lewis acid–base model. These concepts are useful for any kind of polymer electrolyte.

On the basis of these foundations, the second chapter addresses proton-conducting nanocomposite and hybrid polymers used as electrolyte membranes in proton exchange membrane fuel cells. An overview of recent literature in this domain is given: besides traditional Nafion, polyaromatic polymers play an increasingly prominent role in the field. Models used for description of the structure, stability, and transport properties of proton-conducting polymer nanocomposites are also outlined.

Thin-film metal–polymer and metal oxide–polymer nanocomposites are the subject of the third chapter: they are prepared by vacuum phase codeposition of metal and polymer, using *para*-xylylene as monomeric unit, and subsequent oxidation. The vacuum deposition technique might be applicable for related materials; relevant properties are reported, including adhesion and electrical resistance. These composite films can be used in different domains, such as microelectronics and Li-ion battery electrodes.

The mechanical properties of polymer nanocomposites with rod- and plate-shaped nanoparticles are described in the fourth chapter. The anisotropy plays a central role for improvement of mechanical properties. The materials preparation and analysis are described, including specific techniques such as dynamical mechanical analysis or moisture diffusion measurements. Modeling of mechanical properties is also treated.

The fifth chapter presents a small outlook on the vast and rapidly growing domain of computer simulation of materials. Relevant methods, quantum mechanics, Monte-Carlo simulations, and molecular dynamics, are briefly introduced. Cationic and anionic clay–polymer nanocomposite materials attract great attention as they offer enhanced mechanical, thermal, and catalytic properties as compared to conventional materials. They are also studied as possible solid electrolytes for batteries and fuel cells. Simulations of structure and dynamics of clay–polymer nanocomposites are presented, including Li-ion conduction and catalytic properties.

The last three chapters present specific structural spectroscopies, which were extensively applied to the domain of nanocomposites and have brought significant advances in the understanding of these systems. In all these chapters, the specific techniques are first introduced briefly; the advantages of the technique, the information available and significant examples are then presented again with particular emphasis on ionic conducting systems.

X-ray absorption spectroscopy studies of nanocomposites can provide information on the local environment and oxidation state of an atom, the technique being element specific, usable at low concentrations of target atom, and not restricted to crystalline systems. The case of nanoparticles dispersed in an inorganic matrix or a polymer matrix and nanoparticle/nanoparticle composites are examined.

Nuclear magnetic resonance (NMR) is a valuable tool for studying ionic diffusion in materials and dynamical aspects of nanocrystalline ceramics and composites. This technique is particularly useful to differentiate alternative transport mechanisms, like fast interfacial vs. slower bulk diffusion, via the NMR relaxation rates. In this chapter, F- and Li-ion conducting nanostructured materials are particularly discussed.

Mössbauer spectroscopy is another nonconventional technique, which has proven to be of great relevance for the investigation of electrode materials for Li-ion batteries. The mechanism of lithium insertion/deinsertion during cycling of the battery can be followed in situ during the cycles. The local structure and oxidation state of ions can also be deduced from Mössbauer spectra.

The sum of these eight contributions should give a broad range of readers from solid state chemistry, solid state physics, and materials science an outlook on the status of research and development in the domain of nanostructured composites. The emphasis on ionic conducting materials makes this book particularly attractive for the solid state ionics and electrochemistry community. Given the particular impact of these materials for environmental and energy applications, readers interested in these topics should also profit from this book.

Marseille — Philippe Knauth
Delft — Joop Schoonman

Contents

Preface v

Contributors xi

Composite Polymeric Electrolytes 1
Władysław Wieczorek and Maciej Siekierski

Proton-Conducting Nanocomposites and Hybrid Polymers 71
Y.D. Premchand, M.L. Di Vona, and P. Knauth

Hybrid Metal Oxide–Polymer Nanostructured Composites: Structure and Properties 119
Alla Pivkina, Sergey Zavyalov, and Joop Schoonman

Structure and Mechanical Properties of Nanocomposites with Rod- and Plate-Shaped Nanoparticles 143
S.J. Picken, D.P.N. Vlasveld, H.E.N. Bersee, C. Özdilek, and E. Mendes

Gaining Insight into the Structure and Dynamics of Clay–Polymer Nanocomposite Systems Through Computer Simulation 175
Pascal Boulet, H. Christopher Greenwell, Rebecca M. Jarvis, William Jones, Peter V. Coveney, and Stephen Stackhouse

X-ray Absorption Studies of Nanocomposites 205
Alan V. Chadwick and Shelley L.P. Savin

Dynamical Aspects of Nanocrystalline Ion Conductors Studied by NMR 227
P. Heitjans, Sylvio Indris, and M. Wilkening

Mössbauer Spectroscopy and New Composite Electrodes for Li-ion batteries **247**
Pierre-Emmanuel Lippens and Jean-Claude Jumas

Index **273**

Contributors

H.E.N. Bersee
Design and Production of Composite Structures, Faculty of Aerospace Engineering, Delft University of Technology, Kluyverweg 3, 2629 HS Delft, The Netherlands

Pascal Boulet
UMR 6121 CNRS, Université de Provence - Aix Marseille I
Centre Saint Jérôme, 13397 MARSEILLE Cedex 20, France

Alan V. Chadwick
Functional Materials Group, School of Physical Sciences, University of Kent, Canterbury, Kent CT2 7NH, UK.

Peter V. Coveney
Centre for Computational Science and Department of Chemistry,
University College of London, 20 Gordon Street, London WC1H 0AJ, UK

H. Christopher Greenwell
Centre for Applied Marine Sciences, School of Ocean Sciences,
University of Wales, Bangor, Menai Bridge Anglesey LL59 5Ab, UK

P. Heitjans
Institute of Physical Chemistry and Electrochemistry, and Center for Solid State Chemistry and New Materials, Leibniz University Hannover, Callinstr. 3a, 30167 Hannover, Germany

S. Indris
Institute of Physical Chemistry and Electrochemistry, and Center for Solid State Chemistry and New Materials, Leibniz University Hannover, Callinstr. 3a, 30167 Hannover, Germany
Forschungszentrum Karlsruhe, Institute of Nanotechnology, 76201 Karlsruhe, Germany

Rebecca M. Jarvis
Centre for Applied Marine Sciences, School of Ocean Sciences,
University of Wales, Bangor, Menai Bridge Anglesey LL59 5Ab, UK

William Jones
Department of Chemistry, University of Cambridge, Lensfield Road, Cambridge, CB2 1WE, UK

Jean-Claude Jumas
AIME, Institut Charles Gerhardt, UMR 5253 CNRS, Université Montpellier II, CC15, Place Eugène Bataillon, F-34095 Montpellier Cedex 5, France

P. Knauth
Université de Provence, UMR 6121 CNRS, Centre St Jérôme, F- 13397 Marseille Cedex 20, France

Pierre-Emmanuel Lippens
AIME, Institut Charles Gerhardt, UMR 5253 CNRS, Université Montpellier II, CC15, Place Eugène Bataillon, F-34095 Montpellier Cedex 5, France

E. Mendes
Nanostructured Materials, Delft University of Technology, Julianalaan 136, 2826 BL Delft, The Netherlands

C. Özdilek
Fundamentals of Advanced Materials, Faculty of Aerospace Engineering, Delft University of Technology, Kluyverweg 3, 2629 HS Delft, The Netherlands

S.J. Picken
Nanostructured Materials, Delft University of Technology, Julianalaan 136, 2826 BL Delft, The Netherlands

Alla Pivkina
Semenov Institute of Physical Chemistry, Russian Academy of Science, Kosygin st. 4, 119991, Moscow, Russia

Y.D. Premchand
Université de Provence, UMR 6121 CNRS, Centre St Jérôme, F-13397 Marseille Cedex 20, France

Shelley L.P. Savin
Functional Materials Group, School of Physical Sciences, University of Kent, Canterbury, Kent CT2 7NH, UK

Joop Schoonman
Delft University of Technology, Delft Institute for Sustainable Energy, P.O. Box 5045, 2600 GA Delft, The Netherlands

Maciej Siekierski
Polymer lonics Research Group, Warsaw University of Technology, Chemical Faculty, ul. Noakowskiego 3, 00-664 Warszawa, Poland

Stephen Stackhouse
Department of Earth Sciences, University College of London, Gower Street, London WC1E 6BT, UK

D.P.N. Vlasveld
Promolding BV, Laan van Ypenburg 100, 2497GB Den Haag, The Netherlands

M.L. Di Vona
Università di Roma Tor Vergata, Dip. Scienze e Tecnologie Chimiche, I-00133 Roma, Italy

Władysław Wieczorek
Polymer lonics Research Group, Warsaw University of Technology, Chemical Faculty, ul. Noakowskiego 3, 00-664 Warszawa, Poland

M. Wilkening
Institute of Physical Chemistry and Electrochemistry, and Center for Solid State Chemistry and New Materials, Leibniz University Hannover, Callinstr. 3a, 30167 Hannover, Germany

Sergey Zavyalov
Karpov Institute of Physical Chemistry, Vorontsovo Pole, 10, 103064 Moscow, Russia

Composite Polymeric Electrolytes

Władysław Wieczorek and Maciej Siekierski

1 Introduction: Early Steps and Ideas

Polymer electrolytes are "complexes" of electrodonor polymers with various inorganic or organic salts or acids [1]. The main requirements for a polymer to be used as a matrix in polymer electrolyte systems are the following:

- The presence of an heteroatom (usually O, N, S) with lone electron pairs of a donor power sufficient to complex cations
- Appropriate distances between the coordinating centers to insure the hopping of charge carriers
- Sufficient flexibility of polymer chain segments to facilitate movements of ionic carriers

In polymer electrolytes, ionic transport occurs in a highly amorphous, viscoelastic (solid) state. The most intensively studied polymer electrolytes are based on poly(oxa alkanes), poly(aza alkanes), or poly(thia alkanes). The present work deals with polymer electrolytes based on poly(oxa alkanes)–polyethers and particularly on alkali metal salt complexes with poly(ethylene oxide) (PEO).

The recent intense interest in polymer solid electrolytes results from a varicty of possible applications of these materials [2, 3].

Among them, the following seem to be of particular importance:

- Solid state primary and secondary microbatteries with a lithium, lithium alloy, or intercalated graphite anode and a composite cathode based on intercalated materials,
- Electrochemical sensors
- Electrochromic devices (windows or displays)
- Fuel cells with proton polymeric electrolyte membranes

Polymer Ionics Research Group, Warsaw University of Technology, Chemical Faculty, ul. Noakowskiego 3, 00-664 Warszawa, Poland

P. Knauth and J. Schoonman (eds.), *Nanocomposites: Ionic Conducting Materials and Structural Spectroscopies.*

Despite almost 30 years of history of polymeric electrolytes and their application in ion storage devices, still some fundamental problems and questions from the initial times of development of these solid ionic conductors remain unsolved [1–3]. Among them low ambient temperature ionic conductivity, lithium transference numbers much lower than unity and the formation of resistive layers on the electrode–polymer electrolyte interface are crucial ones for further development of lithium or (and) lithium-ion batteries containing polymer solid electrolytes, in particular those having polyether as a polymeric matrix [1–2]. Various types of modification of poly(ethylene oxide)-based polymeric electrolytes have been applied but the successes are limited [3]. These various modifications of PEO-based electrolytes can be divided into three main categories.

First is the preparation of amorphous polymer matrices in which the ether segments consist of 4–15 ethylene oxide monomeric units. These are long enough to effectively complex alkali metal cations but too short to show a tendency towards crystallization. Examples are polymer networks, random, and block ethylene oxide copolymers and comb-like systems with short chain ethylene oxide sequences [1–3].

Second is the utilization of an appropriate ionic dopant, one which tends to form complexes having low temperature eutectics with the pristine PEO phase. These are the so-called plasticizing salts [1–3].

Third is the addition of substances that reduce the crystallizing ability of the polyether hosts. One of the most promising approaches was to prepare composite polymeric electrolytes [4, 5]. This is due to their higher conductivity, improved cation transport numbers, and enhanced electrolyte–lithium electrode stability compared to standard polyether-based electrolytes. Composite electrolytes usually consist of three components: polymer matrix, dopant salt, and filler (see Fig. 1). The role of the latter is to modify polymer–ion and ion–ion interactions leading to an improvement in the ion transport. The area of composite polymeric electrolytes began in the early 1980s

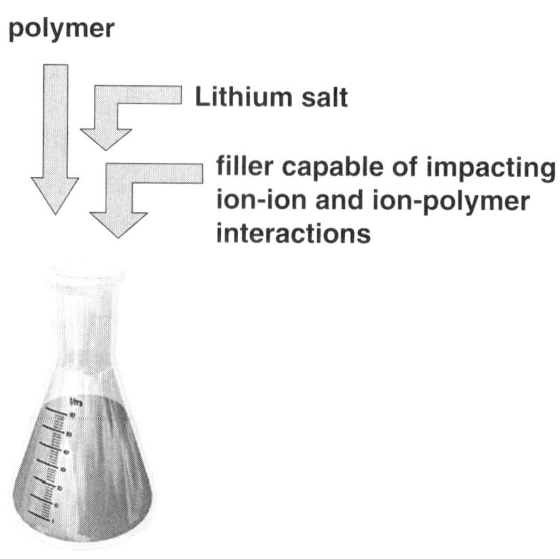

Fig. 1 Preparation of the composite polymeric electrolyte

with the studies of composite systems containing conducting fillers. Soon the number of approaches including the use of commercially available micro- and nanosize silica or alumina powders, polar polymers, supramolecular receptors, etc. became widespread. A variety of models have been designed and used to describe ion transport phenomena in composite polymeric electrolytes with particular attention paid to the role of the filler [6–8]. These models are similar in many respects but some contradictions can also be found.

The aim of this chapter is not to present a comprehensive review of the work carried out on polymer ionic composites. This would be a rather difficult task due to the very large number of papers dealing with basic and applied studies of the wide range of composite polymeric electrolytes. We rather intend to present the most important ideas, which mark the milestones in the development of novel generations of ion-conducting polymer composites. The chapter is divided into several sections dealing with ideas leading to the increase in ionic conductivity of polyether electrolytes via formation of composite structure; enhancement in the cation transport number obtained with the help of specially designed inorganic and organic fillers; and the effect of additives on the stabilization of composite polymeric electrolyte–lithium electrode interface. In the following sections we aimed to generalize the phenomena observed in composite polymeric electrolytes using the previously developed models as well as design a new approach that would be helpful in describing changes in the conductivity and lithium ion transference numbers occurring upon addition of fillers to polymeric electrolytes. The chapter ends with the presentation of variety of semiempirical and quantum mechanics models used to describe conduction phenomena in composite polymeric electrolytes. The entire discussion will be illustrated by a variety of electrochemical and structural data obtained for composite electrolytes containing specially designed inorganic and organic fillers.

2 Towards an Increase in Ionic Conductivity

2.1 Synthesis of the Systems with Conducting and Nonconducting Inorganic Fillers

Composite polymeric electrolytes are prepared according to the procedures described in numerous papers [6, 9–11]. These procedures varied depending on the type of filler used and the form of polymer matrix. For high molecular weight PEO, the typical approach is based on a casting technique with acetonitrile used as the solvent. Samples are later cast on the glass or Teflon substrate followed by the evaporation of the solvent. The second path that is also frequently used in the preparation of high molecular weight composites is based on a grinding–melting technique without using polar volatile solvents [2, 5]. Electrolytes based on low molecular weight PEO oligomers were prepared by the direct dissolution of the salt in the polymer. Fillers were added to the electrolytic solution [9]. In all the cases, polymers, salts, and fillers are dried under vacuum prior to use in electrolyte synthesis. Solvents and low molecular weigh PEO are also dried according to widely described procedures [9].

2.2 Mixed Phase Systems: Inert Matrices + Conducting Fillers

Originally, the aim to use composite polymeric electrolytes was to increase the ionic conductivity of the PEO-based system in ambient temperature range. The initial idea of the mixed phase systems comprising polymeric electrolytes with fillers composed of ceramic fast ionic conductors was based on the expectation to get percolation pathways composed of inorganic filler grains through the polymeric matrix. Such a phenomenon could lead to an increase in ionic conductivity followed, possibly, by an enhancement of the cation transport number, while preserving mechanical properties and flexibility of the composite electrolyte prepared in the thin film configuration. This concept was explored by several research groups. Our previous studies on mixed-phase polymeric electrolytes containing conductive fillers such as NASICON [12], β-alumina [13, 14], and glassy fillers [13, 14] have shown that these fillers do not contribute to ionic conductivity of the mixed phase systems. Figure 2 presents changes in ionic conductivity for composite polymeric electrolytes from the PEO–NaI–NASICON system. Reproducible ambient temperature conductivities were obtained only for composite systems with a low amount of the filler (ca. 5% by volume). For higher filler concentration, impedance spectra of composite polymeric electrolytes obtained in the symmetrical cell with stainless steel blocking electrodes reveal the presence of an additional relaxation phenomena attributed to the formation of polymer filler interfaces. As shown in Fig. 3, the resistivity at these interfaces is considerably higher than bulk resistance of the electrolyte and, therefore, is a serious limitation toward the application of mixed-phase systems in electrochemical devices.

Scrosati and co-workers [15–18] found similar behavior in polymeric electrolytes containing β- and β''-aluminas. However, Skaarup et al. [19, 20] reported that for composite systems containing high amounts of conducting

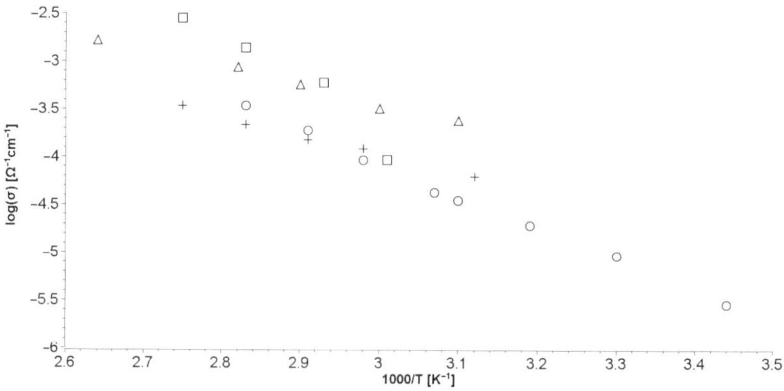

Fig. 2 Bulk conductivity versus reciprocal temperature for $(PEO)_xNaI(NASICON)_y$ composite electrolyte: (*open circle*) $x = 10$, $y = 0.5$; (*open triangle*) $x = 10$, $y = 1$; (*plus*) $x = 10$, $y = 2$; and (*open square*) $x = 10$, $y = 10$

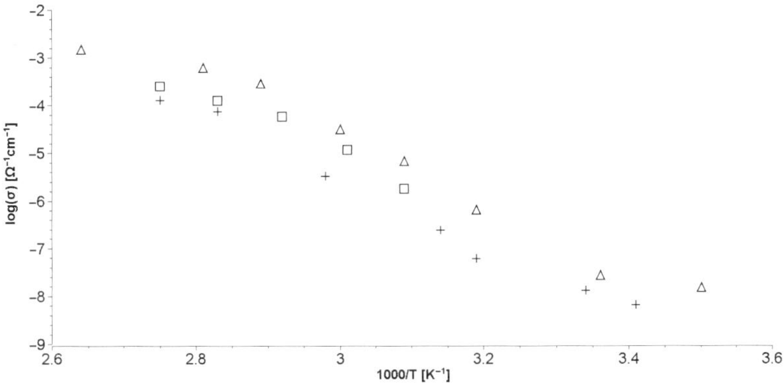

Fig. 3 Phase boundary conductivity versus reciprocal temperature for $(PEO)_xNaI(NASICON)_y$ composite electrolyte: (*open circle*) $x = 10$, $y = 0.5$; (*open triangle*) $x = 10$, $y = 1$; (*plus*) $x = 10$, $y = 2$; and (*open square*) $x = 10$, $y = 10$

fillers (exceeding 85 vol%) conductivity occurs via the dispersed phase and the polymers act as binders for ceramic grains. The decrease in conductivity in comparison with pristine ceramic electrolytes is due to the dilution effect of the polymer host. Similar results demonstrating the contribution of the conducting filler to the conductivity of the mixed-phase electrolytes have been obtained by Stevens and Mellander [21] for systems containing PEO and $RbAg_4I_5$ or KAg_4I_5 as conductive ceramic additives.

Based upon the above observations as well as the fact that even for very high concentrations of the conducting filler the conductivities are much lower than the ones characteristic for the dispersed additive, a different explanation for the increase in the conductivity must be found. The enhancement observed for composite electrolytes is too weak compared to the pristine PEO-based systems to be explained by phenomena directly related to the percolation through the conductive filler phase. NMR experiments that will be discussed in details later in this chapter reveal the reduction in the fraction of crystalline phases, which on the basis of the concept developed by Berthier and co-workers [22] should lead to the observed increase in conductivity.

2.3 Composite Systems Containing Nonconducting Ceramic Additives: The Effect of Filler Type and Size, Type of Polymer Matrices, and Type and Concentration of the Salt Used

Because of the unsuccessful development of the idea of the mixed phase systems, the initial idea of composite solid electrolytes introduced by Liang [23] who had improved the electrical properties of a LiI solid electrolyte by the addition of finely grained α-Al_2O_3 was explored by several research teams. Weston and Steele [24] used α-Al_2O_3 particles (grain size 40 µm) to improve the mechanical stability of a PEO–$LiClO_4$ electrolyte. Later, it was recognized that the addition of fine inorganic

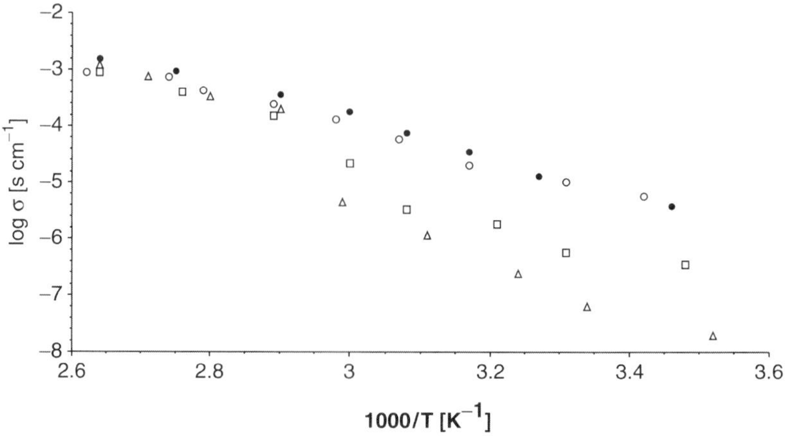

Fig. 4 Bulk conductivity versus reciprocal temperature for (PEO) 10-NaI–Θ-Al_2O_3 (10% by weight) composite electrolyte: (*filled circle*) grain sizes 2 μm, measured two weeks after preparation; (*open circle*) grain sizes 2 μm, measured a month after preparation; (*open square*) grain sizes 4 μm; and (*open triangle*) grain sizes 7 μm

fillers (grain size 1–3 μm) led to an improvement in the mechanical properties and an increase in ambient temperature conductivity of the electrolytes studied [25–31].

Figure 4 presents changes in ionic conductivity for $(PEO)_{10}NaI–Θ-Al_2O_3$ (10% by weight) composite polymeric electrolytes [6]. Three fractions of different grain size distribution were separated and used for preparation of polymeric electrolytes. It can be clearly seen that conductivity strongly depends on the size of the particle used and is highest for composite electrolytes containing the fraction of fillers characterized by the lowest grain size. The conductivity of this composite electrolyte is stable over the time as shown by the values measured at various times after the electrolyte preparation. This increase results from a decrease in electrolyte crystallinity as has been shown by NMR [32], DSC [33, 34], Raman Spectroscopy [35], and X-ray investigations [6, 27] (see Fig. 5). As shown in Fig. 5, the degree of polymer crystallinity decreases with an increase in the filler concentration. The fraction of crystalline phase remains unchanged over the time.

The addition of small inorganic particles often decreases the crystallinity of the system, however, stiffening simultaneously the electrolyte host [32–34]. The effect of grain size distribution, particle concentration, and surface area on conductivity and the phase structure of the composite electrolytes have been discussed [29–31]. Significant improvements in the conductivity of polymeric systems are usually obtained using fine-grained (1–2 μm) powders, with large effective surface areas and concentrations of 10–20 wt%. Even higher conductivities can be achieved [4, 5] when using nanosize fillers, which, however, are much more prone to agglomeration than the microsize ones during the electrolyte synthesis. For higher concentrations, formation of non-conducting particle aggregate regions, which lower the bulk conductivity of the electrolytes, are observed.

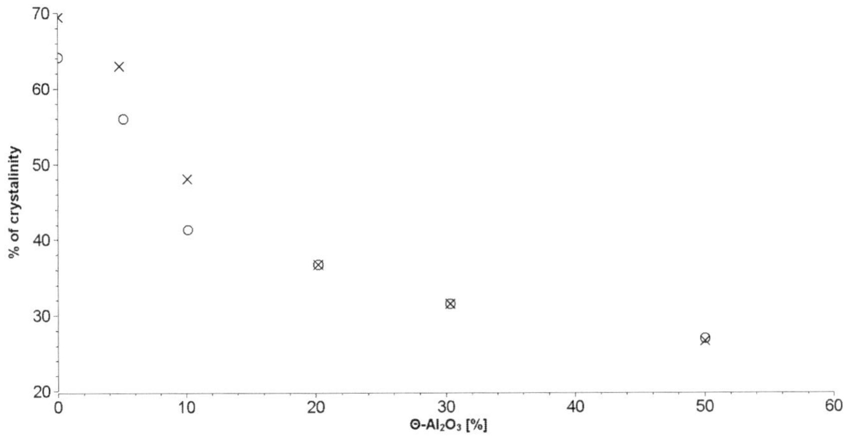

Fig. 5 Degree of crystallinity of PEO–Θ-Al_2O_3 composite electrolyte measured by means of X-ray diffraction: (*open circle*) just after preparation and (*times*) two weeks after preparation

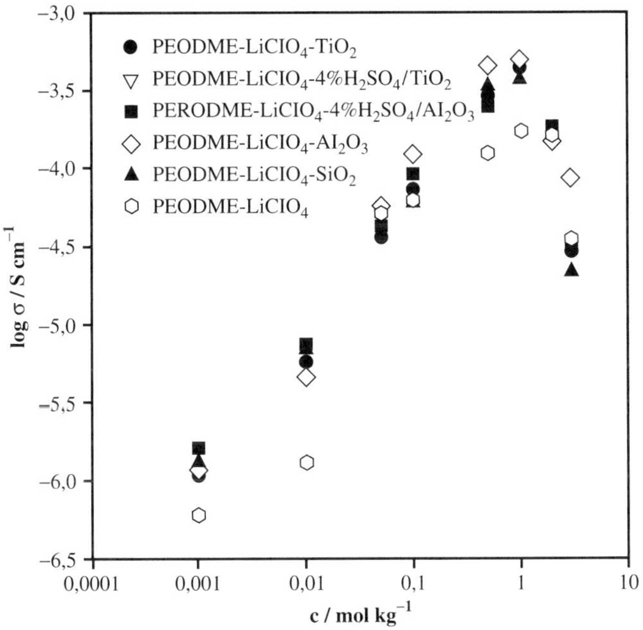

Fig. 6 Changes in ionic conductivity as a function of salt concentration obtained at 25°C for composite PEODME–$LiClO_4$-based electrolytes containing various inorganic additives. Comparison is made with the data obtained for the pristine PEODME–$LiClO_4$ electrolyte

Following the positive effect of inorganic fillers on the conductivity of high molecular polyether-based electrolytes the concept was extended to low and medium weight analogues of PEO. Figure 6 presents changes in the bulk conductivity of poly(ethylene oxide dimethyl ether) polymeric electrolytes measured at 25°C as a function of $LiClO_4$ concentration and type of the filler used. At low salt concentrations (up to 10^{-2} mol kg^{-1}) as well as in the concentration range from 0.5 to 1 mol kg^{-1} conductivities measured for composite electrolytes are slightly higher than for pure PEODME–$LiClO_4$ system. For other salt concentrations studied ionic conductivities for pure and modified electrolytes are similar. Generally, for almost all salt concentrations the highest conductivity has been achieved for composite electrolyte containing nanosize Al_2O_3.

Figure 7 presents changes in molar conductivity as a function of the square root of salt concentration for poly(ethylene glycol methyl ether) (PEGME)–$LiClO_4$–Al_2O_3 composite electrolytes [9]. Filler surfaces are modified to obtain neutral, Lewis acid, or Lewis base type surface groups. The conductivity data are compared to those obtained for unmodified PEGME–$LiClO_4$ electrolyte. Only for salt concentrations higher than 1 mol kg^{-1}, PEGME conductivities of composite electrolytes are higher than for model system. For lower salt concentrations, conductivities are similar for all systems studied and do not differ by more than the experimental

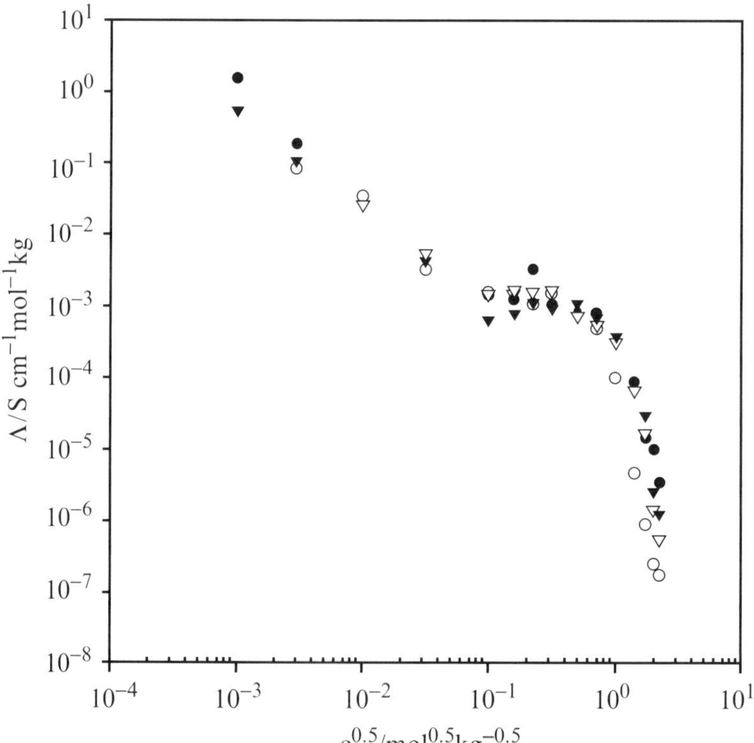

Fig. 7 Changes in molar conductivity as a function of the square root of molar concentration for: (*filled circle*) PEGME–$LiClO_4$–Al_2O_3 neutral surface groups; (*filled triangle*) PEGME–$LiClO_4$–Al_2O_3 basic surface groups; (*open triangle*) PEGME–$LiClO_4$–Al_2O_3 acidic surface groups; and (*open circle*) PEGME–$LiClO_4$

error of impedance measurements. It is evident that the increase in conductivity is related to the salt concentration region in which a high degree of ionic associations is expected. Therefore, we have assumed that the increase in conductivity should result from changes in ionic association due to ion–ion and ion–polymer interaction involving inorganic or organic additive. These interactions led to the lowering of the electrolyte viscosity and changes in the fraction of ionic associates. On the basis of these observations, the Lewis acid–base model of the polymer–salt–filler interactions was developed [7]. This model assumes that the final conductivity of composite electrolytes depends on the equilibrium of the Lewis acid–base reactions involving an additive, a matrix polymer, and salt (with cations acting as Lewis acids and anions as Lewis bases). This model was successfully used to explain the changes of conductivity in a variety of polymeric electrolytes based on high and low molecular weight amorphous polymeric matrices [7, 9].

Some groups try to solidify composite polymeric electrolytes based on low molecular weight polyethers. [36–38] The simplest approach, originated from the work of Fedkiv and co-workers [37], rely on the use of fumed nanosize silica fillers. An addition of a small fraction of fumed silica (ca. 5–10% by weight) results in formation of gel-type electrolytes with conductivities comparable to those measured for liquid electrolytes incorporated into the silica framework.

Several groups used layered ceramic precursors from silicate, hectorite, and montmorillonite families into which molten high molecular weight PEO or low molecular weight organic solvents were incorporated together with the dopant salt [36, 39–41]. Such systems exhibit good mechanical stability and conductivities comparable to polymeric electrolytes used for their formation. Also the improvement in the cation transference number in these systems compared to pure polymeric electrolytes has been reported [36]. Computer simulations of clay–polymer nanocomposites are presented in the last chapter of this book.

2.4 *The Use of Organic Additives: Polymers and Supramolecular Compounds*

Over the past few years, a number of papers have been published, which describe the application of polymer blends as matrices in polymer solid electrolytes [42–53]. Blends of two macromolecules behaving as a real solid system are the main subject here. Le Mehaute and co-workers [47] stabilized the amorphous structure of polymer networks by blending styrene-terminated PEO with a butadiene–styrene copolymer. The mixture was subjected to X-ray and thermal treatment leading to polymerization and crosslinking of the system. The conductivity of 2.5×10^{-4} S cm^{-1} at ambient conditions was achieved and retained for a long time.

A similar conductivity range was obtained by Kaplan et al. [54] who prepared a blend comprising a high molecular weight analogue of propylene carbonate mixed with crown ether and a salt. The author suggests that crown ethers are hindered in the polymer matrix and the ion transport occurs by the exchange of cations between nearest-neighbour crown ether molecules. Tsuchida et al. [43] studied the PEO–$LiClO_4$ system supported by poly(methacrylic acid). Because of the formation of hydrogen bonds between both polymers, rigid polymeric electrolytes were synthesized. The ionic conductivities of these electrolytes were poor, because of strong interactions between polar groups hindering ionic motion. A similar system consisting of a PEO–$LiClO_4$ electrolyte mixed with poly(acrylic acid) was described [55]. The ambient temperature conductivities measured are significantly higher and approach 10^{-5} S cm^{-1}.

The PEO–polystyrene–LiX system was intensively studied by Gray et al. [46]. The blends were obtained by thermal polymerization of styrene in the presence of high molecular weight PEO. The mechanical properties of these electrolytes are far better than for the pristine PEO-based system. Conductivities are similar to those measured for a PEO–LiX electrolyte for polystyrene concentration lower than 40% by weight. For higher concentrations conductivities dropped. This was attributed by the authors to the formation of nonconductive polystyrene clusters in an electrolyte structure. At temperatures exceeding the melting point of the crystalline PEO phase, a fall in the conductivity was observed for electrolytes containing more than 60% by weight of polystyrene.

Stevens and co-workers [56, 57] have studied poly(propylene glycol)–(PPG)–poly(methyl methacrylate)–PMMA–LiX amorphous blend based electrolytes in order to apply them as polymeric membranes in electrochromic windows. The electrolytes were obtained by the mixing of the methyl methacrylate (MMA) with PPG followed by free-radical polymerization of MMA. A low molecular weight PPG solution of the lithium salt was trapped in a higher molecular weight PMMA matrix. The authors found that the miscibility range of PMMA and PPG increases after the addition of the lithium salt. On the basis of FTIR studies, this effect was attributed to a weak coupling of the ether oxygen atoms with the ester groups of the PMMA via lithium cations. The miscibility range was extended up to 20% by weight of PMMA without significant loses in conductivity. However, more recent studies [58] have shown a microseparation of the blend components. The other problem is poor mechanical stability and lack of formation of a laminate structure desirable for applications in electrochromic windows.

Substantial increase in the electrolyte conductivity can be obtained by the addition of polyacrylamide (PAAM) [59, 60]. Results from DSC, FTIR, and EDX indicate that the increase in conductivity is due to enhanced segmental flexibility at the interface of the filler with the polymer matrix in the presence of alkali metal salts. This is due to the formation of complexes involving the filler and salt that results in a reduction of the number of transient crosslinks between polyether oxygens and alkali metal cations. The addition of PAAM to the semicrystalline PEO-based electrolytes results in a decrease in crystallinity that also enhances the ionic conductivity.

It has been proposed [61] that polymer–ion coordination phenomena occurring in these composite systems are governed by equilibria between the various Lewis acid–base reactions occurring between Lewis base centers of the polyether host, Lewis base centers on the PAAM chain, and alkali metal cations that can be treated as Lewis acids. The interpretation of these phenomena in the polyether–PAAM–alkali metal salt systems is, however, complicated by the possibility of formation of hydrogen bonds between protons of NH_2 amide groups and oxyanions (such as ClO_4^-) or polyether oxygens. The competition between Lewis acid–base complex formation and hydrogen bonding leads in some cases to the lowering of the electrolyte conductivity [60]. To eliminate the possibility of hydrogen bonding the organic filler was changed from PAAM to poly(*N*,*N*-dimethylacrylamide) (NNPAAM) [62] because NNPAAM lacks an amidic hydrogen. The reason for using polyacrylamides comes from the high donicity of the Lewis base centers (carbonyl oxygens and amide nitrogens) of the polyamides, which for low molecular weight analogues is comparable or even slightly higher than that of low molecular weight polyether analogues [63, 64]. It has been shown [65] that there are strong interactions between NNPAAM and $LiClO_4$ leading to changes in the ultrastructure and conductivity of oxymethylene linked PEO (OMPEO)–NNPAAM–$LiClO_4$ electrolytes.

Figure 8 presents the conductivity isotherms obtained for the OMPEO–NNPAAM–$LiClO_4$ composite electrolytes at −20°C, 0°C, 25°C, and 100°C as a function of vol% of NNPAAM. At −20°C and 0°C conductivities measured for all of the composite systems are higher than those measured for the OMPEO-$LiClO_4$ electrolyte. A conductivity maximum is obtained for the sample containing 25 Vol% of NNPAAM and is over one order of magnitude higher than the conductivity of the OMPEO-$LiClO_4$ electrolyte at −20°C and 0°C. At 25°C only the conductivities measured for samples containing 20 and 25 vol% of NNPAAM are higher than those measured for the pure OMPEO–$LiClO_4$ system, whereas at 100°C the conductivity obtained for the OMPEO–$LiClO_4$ system is the highest (see Fig. 8). At this temperature a decrease in the conductivity with an increase in the NNPAAM concentration is observed. At 25°C the conductivity decreases slightly for samples containing up to 15 Vol% of NNPAAM, then reaches a maximum at 20–25 Vol% of NNPAAM followed by a decrease in the conductivity for higher NNPAAM concentrations.

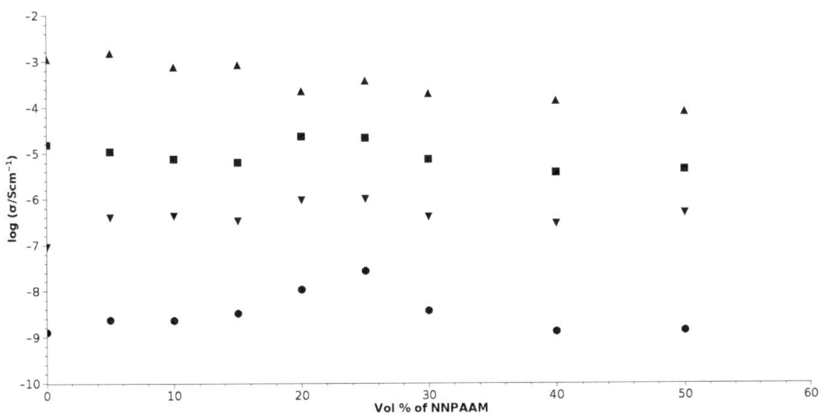

Fig. 8 Isotherms of ionic conductivity of OMPEO–NNPAAM–$LiClO_4$ versus volume fraction of NNPAAM with 10 mol% $LiClO_2$: (*filled circle*) −20°C, (*filled down triangle*) 0°C, (*filled square*) 25°C, and (*filled up triangle*) 100°C

Considering the success of the use of organic additives as conductivity enhancing agents in polymeric electrolytes this idea was further extended aiming not only an improvement in the conductivity but also in lithium transference numbers. New supramolecular compounds have been synthesized and used as anion receptors in polyether-based composite polymeric electrolytes [66]. The properties of the systems studied do not depend on molecular weight of the polymer matrix used and are sensitive only to the concentration of the supramolecular compound used as an additive. The changes in the electrolyte conductivity were associated only to the complexation of anions by supramolecular additives, without considerable modification of the ion transport mechanism. A considerable increase in lithium transference numbers has been noticed especially for electrolytes with higher salt concentration [67]. Addition of supramolecular compounds has been found to suppress formation or growth of secondary passive layers at the lithium electrodes. The last two effects will be more extensively discussed in the forthcoming sections.

3 Addition of Specially Designed Fillers as a Method Toward an Increase in Lithium Transference Numbers

3.1 Inorganic Fillers with Specially Design Surface Groups

The Lewis acid–base model is also useful for designing fillers, which might act as anionic receptors, thus, possibly increasing the cation transport number. Both anions and cations are generally mobile in most polymer electrolytes, whereas restricting the mobility of the anions without adversely affecting the lithium cations is desirable for battery applications. The use of inorganic fillers proved to be one of the most effective as demonstrated by Scrosati's group [4–5]. However, despite an increase in the cation transport numbers, the values obtained were still much below unity. Recently this group as well as others has developed a new generation of inorganic fillers based on a superacid concept having its roots in the catalytic chemistry [11, 68–71]. Surface-modified superacid fillers consisted of particles of oxide grafted with "SO_4^{2-}" groups characterized by high acidity ($H_0 \approx -15$ on the Hammett scale) [72]. These systems seem to be more efficient in the complexation of anions.

Table 1 present the values of lithium transference numbers measured for PEO–$LiClO_4$ and PEO–$LiBF_4$ electrolytes and composite electrolytes based on these model systems with surface-modified superacid Al_2O_3 and ZrO_2 additives [71]. The lithium transference number increases with the addition of alumina filler and a further increase is observed for electrolytes with surface-modified additives. It should be noticed that the higher the acidic groups concentration the higher is lithium transference number. Similar observations can be made for composite electrolytes containing ZrO_2 [71].

Table 1 Lithium transference numbers for $(PEO)_{20}LiClO_4$-based composite electrolytes containing 10% by weight of inorganic filler additives

Type of the electrolyte	Type of the filler	Temperature (°C)	Lithium transference number
$(PEO)_{20}LiClO_4$	Filler-free sample	40	0.31
$(PEO)_{20}LiClO_4$	Al_2O_3	40	0.61
$(PEO)_{20}LiClO_4$	Al_2O_3 (1% ASG)	40	0.66
$(PEO)_{20}LiClO_4$	Al_2O_3 (4% ASG)	40	0.72
$(PEO)_{20}LiClO_4$	Al_2O_3 (8% ASG)	40	0.77
$(PEO)_{20}LiBF_4$	0	70	0.32
$(PEO)_{20}LiBF_4$	Surface-modified ZrO_2	70	0.81

ASG-acidic surface groups. Data for ZrO_2 were obtained from [72]

3.2 Boron Family Receptors

Anion receptors based on boron compounds were applied to the solutions of lithium salts in aprotic (inert) electrolytes based on low molecular weight solvents [73, 74] as well as in gel polyelectrolytes [75]. Boron-based aza ether compounds (borane, borate complexes) have been studied by McBreen and co-workers [76–79] using mainly Near edge X-ray absorption fine structure spectroscopy (NEXAFS). These studies showed that the degree of complexation of Cl^- or I^- anions strongly depends on the structure of boron compounds. Also a significant enhancement in ionic conductivity upon the addition of boron compounds has been noticed in these electrolytes.

Several types of boroorganic compounds exhibiting Lewis acid properties can be used as anion-complexing additives for polymer electrolytes. For example, various compounds, such as $(CF_3)_2CHO_3B$, $(C_6F_5O)_3$, or cyclic boronate compounds were applied by McBreen in EC/DMC-based electrolytes [80, 81], giving an increase in conductivity of almost one order of magnitude. A series of the cyclic receptors – fluorinated 1,3,2-benzodioxaboroles and 1,3,2-dioxaborolanes – were studied by the same authors in dimethoxyethane-based solutions [82]. Trisarylboranes were used as additives to PEO–LiF and PEO–$LiCF_3SO_3$ solid electrolytes, leading to an increase in cation transference numbers [83]. Lewis acid properties of six different receptors with the formula $(C_6F_5)_x(C_6F_5O)_{(3-x)}B$ were compared using Guttman's and Child's method [84]. It was shown that fluorinated compounds are a stronger acid than non-fluorinated analogues. Acidic properties increase in the order $(C_6F_5)_3B < (C_6F_5)_2BOC_6F_5 < C_6F_5B(OC_6F_5)_2 < B(OC_6F_5)_3$.

Triphenylborane [85] and pentafluorophenylboronic acid esters were also applied in PEODME-based electrolytes. The formation of complex with different anions $CF_3SO_3^-$, ClO_4^-, and BF_4^- were confirmed by FTIR experiments. Conductivity of these systems was slightly lower for the electrolyte doped with BPh_3 and slightly higher for other studied additives comparing to the reference PEODME–$LiCF_3SO_3$ system. Both BPh_3 and esters of pentafluoroboronic acid interact strongly with polar solvents, e. g. glymes or DMF.

3.3 The Role of Supramolecular Additives

Another successful approach leading to enhancement in lithium transference number is to use supramolecular additives as anionic receptors. Several different additives such as linear or cyclic aza-ether compounds (with electron-withdrawing groups) [86] or calix[4]arene derivatives with various types of active groups in the lower rim were shown to be very effective in complexing anions, thus giving lithium transport numbers close to unity [66, 67]. However, this effect was quite frequently observed for relatively large fractions of the supramolecular additive that act as a steric hindrance, thus lowering electrolyte conductivity.

Table 2 presents the effect of type and concentration of calyx[4]arene additives on the transference number obtained for PEO–LiI electrolytes. Two types of supramolecular receptors with different anion complexing capability have been used as additives. The structure of calyx[4]arene receptors is shown in Fig. 9. Figure 10 shows respectively the calyx[6]pyrrole-type compound complexed with the I^- anion. For small salt concentrations corresponding to 0.25 mol of LiI kg^{-1} of polyether, the addition of calixarenes has a negligible effect on the enhancement of lithium transference numbers. For higher salt concentrations, the lithium transference number increase with an increase in the calix[4]arene concentration and for calixarene 2 derivative equals 1 for the electrolyte containing roughly 1 mol of LiI kg^{-1} of polyether. It can also be seen that lithium transference number decreases with an increase in temperature.

Table 3 presents values of lithium transference numbers obtained for PEO-based electrolytes doped with various type of lithium salts and containing various amounts of calixpyrrole type supramolecular additive (for calyx[6]pyrrole structure see Fig. 10). The addition of an even small molar fraction (~0.125) of calix[6]pyrrole results in a considerable increase in the lithium transference numbers. (For this supramolecular additive concentration all composite electrolytes seem to be homog-

Table 2 Lithium transference numbers for PEO–LiX–Calix-4-arene electrolytes obtained by means of dc–ac electrochemical experiment [66, 67]

Additive in electrolyte	Salt in electrolyte	t (°C)	O:Li ratio[a]	x ratio[b]	t^+
Calixarene **5a**	LiI	90	100	0	0.14
		90	100	0.25	0.15
		90	100	0.50	0.18
		50	20	0	0.35
		50	20	0.30	0.59
Calixarene **5c**	LiI	90	7	0.30	0.69
		75	7	0.30	0.74
		90	20	1.00	0.80
		75	20	1.00	0.93
		50	20	1.00	1.00

[a]Oxirane monomeric units in respect to salt molar concentration.
[b]Calixarene content in respect to LiI salt molar concentration.

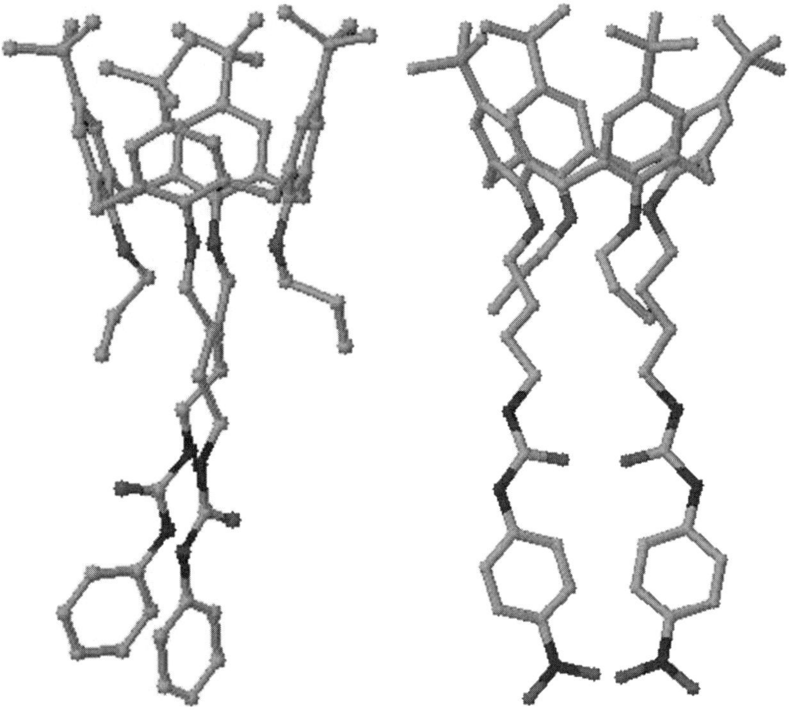

Fig. 9 Chemical structure of (**a**) 5,11,17,23-tetra-*p*-tert-butyl-25,27-bis(((*N*-phenylureido)butyl) oxy)-26,28-dipropoxycalix[4]arene and (**b**) 5,11,17,23-tetra-*p*-tert-butyl-25,27-bis(((*N*-*p*-nitrophenylureido)butyl) oxy)-26,28-dipropoxycalix[4]arene

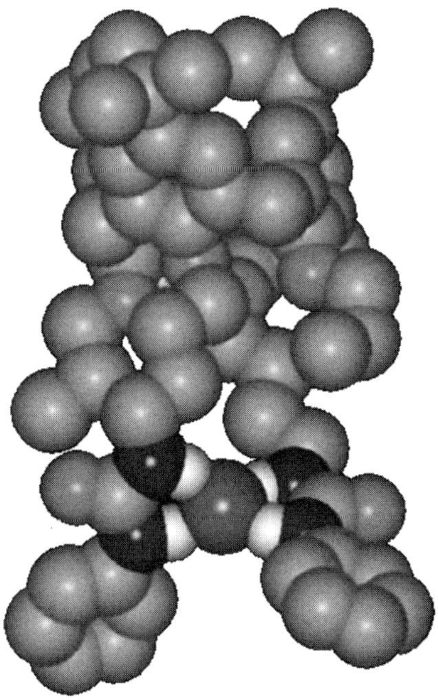

Fig. 10 Chemical structure of the iodide anion complex with 1,1,3,3,5,5-*meso*-hexaphenyl-2,2,4,4,6,6-*meso*-hexamethyl-calix[6]pyrrole

Table 3 Lithium transference numbers for PEO–LiX–Calix-6-pyrrole electrolytes obtained by means of the dc–ac electrochemical experiment [10]

Type of the electrolyte	Molar fraction of calix-6-pyrrole	Temperature (°C)	Lithium transference number
$(PEO)_{20}LiI$	0	70	0.25
$(PEO)_{20}LiI$	0.125	70	0.56
$(PEO)_{20}LiAsF_6$	0	75	0.44
$(PEO)_{20}LiAsF_6$	0.5	75	0.84
$(PEO)_{20}LiBF_4$	0	70	0.32
$(PEO)_{20}LiBF_4$	0.125	70	0.78
$(PEO)_{20}LiBF_4$	0.25	70	0.81
$(PEO)_{20}LiBF_4$	0.5	70	0.85
$(PEO)_{100}LiBF_4$	0.25	70	0.95
$(PEO)_{100}LiBF_4$	1	70	0.92
$(PEO)_{20}LiCF_3SO_3$	0	75	0.45
$(PEO)_{20}LiCF_3SO_3$	0.125	75	0.68

Table 4 Self-diffusion coefficients D and t^+ at 363 K [10]

	$D_{polymer}$ 10^{-8} $cm^2\ s^{-1}$	D^- 10^{-8} $cm^2\ s^{-1}$	D^+ 10^{-8} $cm^2\ s^{-1}$	t^+
PEO–$LiBF_4$–calixpyrrole	6.51	27.5	24.6	0.47
PEO–$LiBF_4$	3.37	36.1	20.0	0.36

enous.) The increase in lithium transference number is particularly well seen for PEO–$LiAsF_6$ and PEO–$LiBF_4$ electrolytes. A smaller enhancement has been achieved for PEO–$LiCF_3SO_3$ system and the smallest one for PEO–LiI electrolytes. These observations are in good correlation with computational calculations [87] showing the following preference of calix[6]pyrrole in the coordination of anions BF_4 $ASF_6 > ClO_4 > CF_3SO_3^- > PF_6^- > I$. A further increase in the fraction of calixpyrrole results in only a small increase in the lithium transference number. For the 0.5 molar fraction of calixpyrrole the values of lithium transference numbers are very similar independently of the type of the dopant salt. A further increase in the calix[6]pyrrole concentration as well as a rise in temperature results in a decrease in the lithium transference number.

To confirm these observations PFG NMR experiments were performed for PEO–$LiBF_4$ and PEO–$LiBF_4$–CP electrolytes showing the diffusion coefficient for protons belonging to the polymeric chain, cations, and anions [10]. The diffusion data collected in Table 4 show that (a) the anion (i. e. BF_4^-) has a larger diffusion coefficient than the cation in both samples, (b) the cation (i. e. Li^+) in the sample with calixpyrrole has a larger diffusion than in the sample without calixpyrrole, (c) the polymer diffusion (i. e. H) in the sample with calixpyrrole has a larger diffusion than in the sample without calixpyrrole, and (d) the anion (i. e. BF_4^-) in the sample with calixpyrrole has a smaller diffusion than the anion in the sample without the agent.

Please note that the effect of the calixpyrrole seems to be a mild plasticizing one (i. e. increasing chain mobility). This result is consistent with a lower glass transition temperature observed for the material containing calixpyrrole.

4 The Effect of Additives on Electrode–Electrolyte Interfacial Behavior

The commercialization of batteries with polymeric electrolytes is a still very challenging task due also to the formation of resistive passive layers at the electrode–electrolyte interface characterized with resistance increasing in time [88, 89]. The growth of the passive layers is often connected with the predominant anionic conductivity of polymeric electrolytes. Basically, the high cation transference number is equally important for the practical application as the high conductivity value of the system [88]. The overall efficiency of the lithium cell is dependent both on the rate of charge transported by the electrodically active ion and on the resistance of the passive layer forming on the electrode surface as these phenomena lead to the internal voltage drops in the system. A good influence of high cation transference number can be observed for both above-mentioned parameters.

The nature of the formation and growth of resistive passive layers at the electrolyte–lithium electrode interface has been recognized and widely studied for liquid electrolytes used in lithium and lithium ion batteries [90, 91]. There are also number of studies performed for gel and solid polymer electrolytes. Peled and co-workers proposed a model of the Solid Electrolyte Interface applicable for polymeric electrolytes when in contact with a lithium electrode [92, 93]. The studies of the polymer electrolyte–lithium electrode interface were summarized by Scrosati in his excellent review paper [89].

Scrosati and co-workers were the first to recognize also the positive effect of inorganic additives on the stabilization of polymer electrolyte–lithium electrode interface [4, 5]. The suppression of formation and growth of the resistance of interfacial layers has been demonstrated. Similar observations were afterwards reported by number of research teams, which used different type of organic and inorganic additives [9, 89, 94, 95]. Most of the studies were, however, performed in the limited salt concentration range corresponding to that practically used in lithium batteries.

In most of the studies commercially available micro- and nanosize fillers were used. Recently several research groups have developed fillers with specially designed superacid surface groups. It has been shown above that these fillers have a very positive effect on the enhancement of lithium transference numbers. The superacid fillers also have a considerable effect on the improvement of the stability of lithium electrode–composite polymer electrolytes interface [11, 96]. Figs. 11a–c show time evolution of the electrolyte resistance (11a), native passivating layer resistance (11b), and interfacial resistance coupled with the charge transfer resistance (11c). Electrolyte resistances are stable in time with the filler-free electrolyte resistance being twofold less than resistances of composite polymeric electrolytes. The addition of fillers has a very pronounced effect on the reduction in the passive layer and interfacial resistances. These resistances for composite electrolytes are at least one order of magnitude lower than those for filler-free electrolyte. The origin of these changes is still unclear and requires some additional studies using combination of in situ spectro-electrochemical techniques.

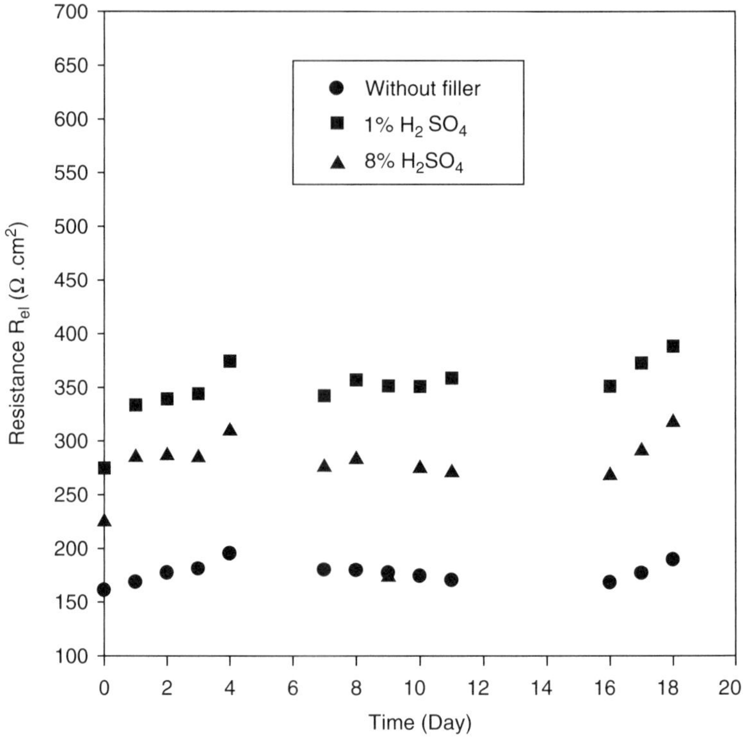

Without filler
1% H2 SO4
8% H2SO4
Resistance Rel (Ω .cm2)
Time (Day)

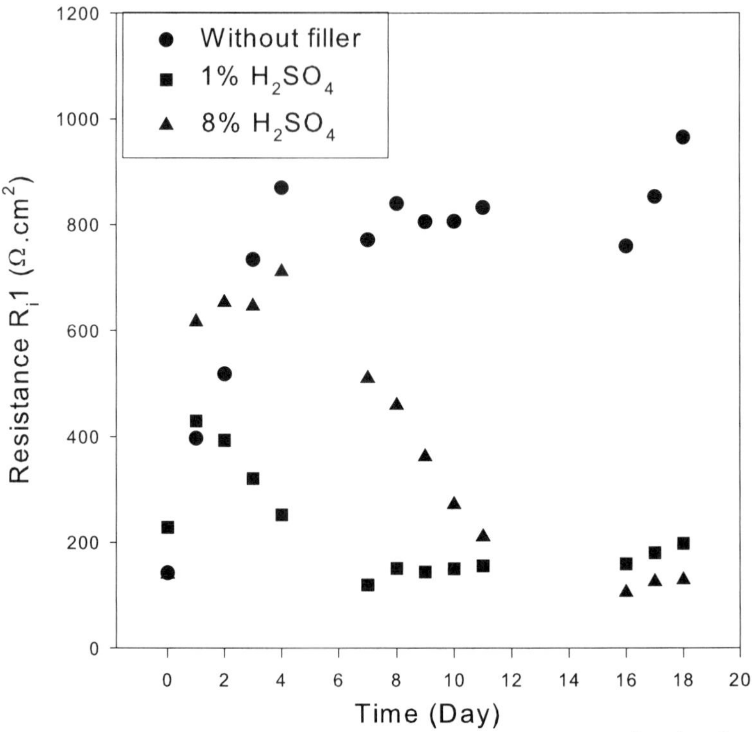

Without filler
1% H2SO4
8% H2SO4
Resistance Ri1 (Ω.cm2)
Time (Day)

(continued)

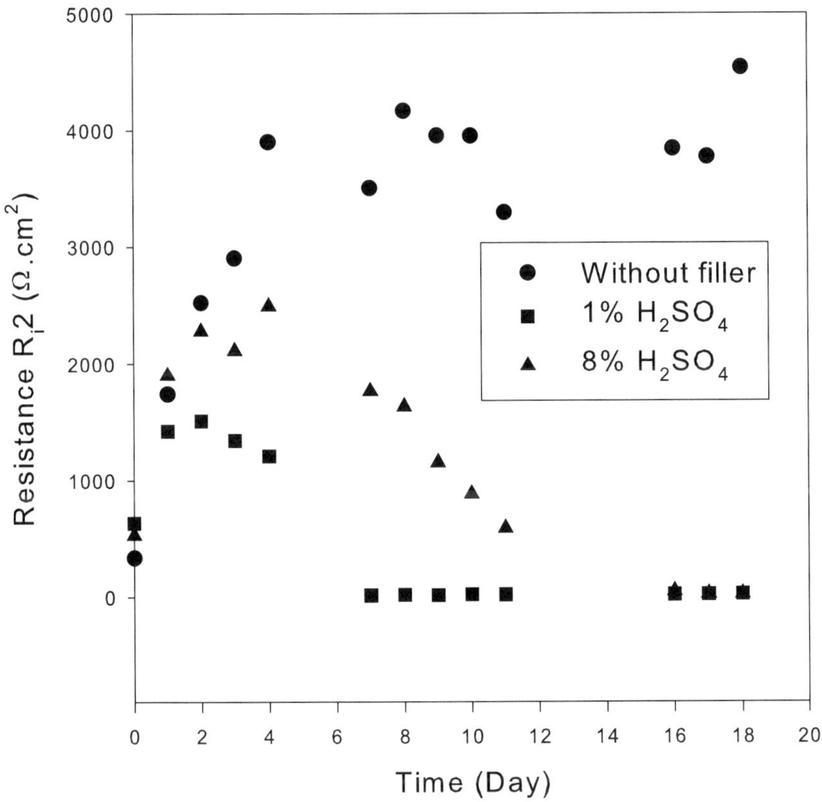

Fig. 11 Time evolution of electrolyte resistance (**a**), native passivating layer resistance (**b**), and interfacial resistance coupled with charge transfer resistance (**c**) for composite polymeric electrolytes of various composition

5 Towards Understanding Ionic Transport Phenomena in Composite Polymeric Electrolytes

Through the history of studies of composite polymeric electrolytes, a variety of models have been developed to describe ion transport phenomena in these systems. These models have been mentioned previously in this chapter. The following sections summarized the idea of the models most commonly used and useful for the description of ionic transport in composite electrolytes.

5.1 An Amorphous Phase Model and its Use for Partially Crystalline Polymer Matrices

As mentioned before for a variety of composite electrolytes based on high molecular weight PEO matrix, a decrease in the degree of crystallinity was identified and related to an increase in the ionic conductivity measured in composite systems compared to PEO-based electrolytes. On the basis of the obtained results an "amorphous phase" model was developed, which explains the increase in the conductivity in composite electrolytes compared to the pure PEO-based systems [6]. In the crystalline PEO-based systems filler particles (e. g. α-Al_2O_3) act as nucleation centers and most probably are attached to PEO segments via acid Al surface centers. Since there is a large number of these nucleation centers, the crystallization process goes faster due to the higher nucleation rate and, in consequence, a bigger level of disorder typical for the liquid state is frozen during the solidification of the polymeric matrix observed in the cooling process or solvent evaporation.

The idea of the amorphous phase model is schematically drawn on Figs. 12a and b [97]. Figure 12a shows the structure of the so called composite grain, e. g., the vicinity of the isolated filler grain. The structure of the filler polymer interface is composed of three separate components: dispersed filler grain (denoted as phase 2), a highly conductive layer covering the surface of the grain (phase 1), and the bulk polymeric electrolyte (phase 3). Figure 11b shows the result of theoretical RRN calculations indicating that the highest current density is obtained at phase 2. The increase in the conductivity for the entire composite polymeric electrolyte is possible due to the formation of conductive pathways composed of surface layers throughout the bulk of polymer electrolyte (as shown in Fig. 12b).

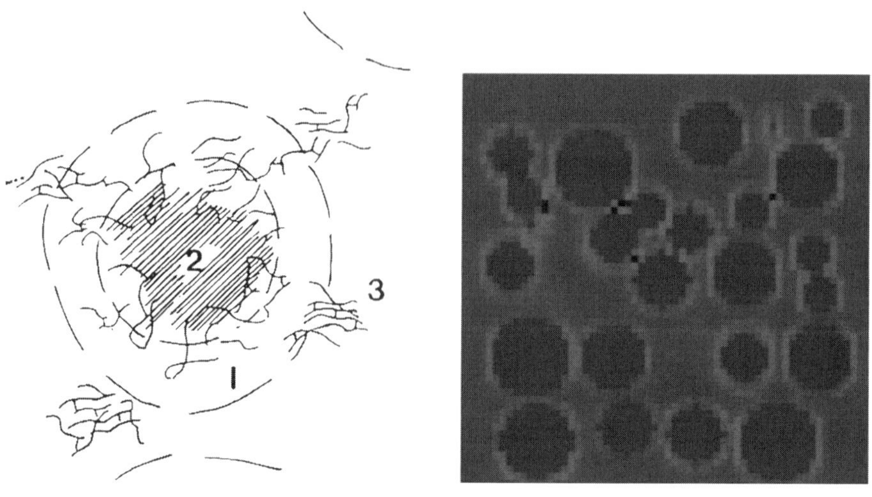

Fig. 12 (**a**) Schematic drawing of morphology of composite polyether–nonconductive filler electrolytes. Numbers are attributed to (1) highly conductive interface layers coating the surfaces of grains, (2) dispersed insulating grains, (3) matrix polymer ionic conductor. (**b**) RRN modeling of electrical properties of the composite system

5.2 *Space Charge Models*

The previously presented models resemble in the physical approaches to phenomena occurring in composite solid electrolyte the approach developed by Maier for a liquid nonaqueous electrolyte [8] and known as the space charge model. In the case of the liquid system, a composite can also be formed but only with high inorganic particle content. In this situation an enhancement of conductivity (compared to the pure solutions of identical composition) is observed. This type of material can be described as a viscous grain ensemble wetted by the liquid or "soggy sand" like system. Because of interfacial interactions a synergetic effect is observed yielding about one order of magnitude increase of the conductivity value. "Soggy sand" systems show some similarities with the properties of composite polymeric electrolytes. In both cases a covalent organic matrix can produce a ground state of the present charge carriers in the form of undissociated salt particles (contact ion pairs). Thus, the conductivity effect would consist of absorption of one of the pair's constituents, resulting in a breakup of the ion pair and generating a mobile counter ion. In all these cases, a percolation type of behavior is observed, which is typical for the enhancement of the interfacial conductivity. Additionally, the increase is higher for the acidic type of filler (SiO_2) when comparing to analogous system with the basic oxide (Al_2O_3). This suggests the existence of a mechanism related to anion absorption on the grain surfaces. This, in turn, leads to an increase of the number of the Li^+ cations in the space charge layer surrounding the filler particle. The relative enhancement of the conductivity value is higher for less polar solvents (THF ε=7.4) in comparison with MeOH (ε=32.6), resulting finally in lower absolute values for the first of the systems studied. This observation confirms that the absorption mechanism as the salt dissociation constant is significantly lower for the less polar system.

5.3 *The Lewis Acid–Base Approach*

This model assumes that final conductivity of composite electrolytes depends on the equilibrium of the Lewis acid–base reactions involving an additive, a matrix polymer, and a salt (with cations acting as Lewis acid and anions as Lewis bases). This model was successfully used to explain changes of conductivity in a variety of polymeric electrolytes based on both high and low molecular weight amorphous polymeric matrices [7, 9]. Figure 13 shows examples of possible polymer–salt–filler interaction occurring in polymeric electrolytes [98]. Both Lewis acid and Lewis base type fillers are considered. The presented approach also accounts for the possibility of interactions of the filler with the side chain OH groups of the polyether.

It is possible to explain the influence of the filler on the properties of poly(ethylene glycol) methyl ether (PEGME)-based electrolytes using the

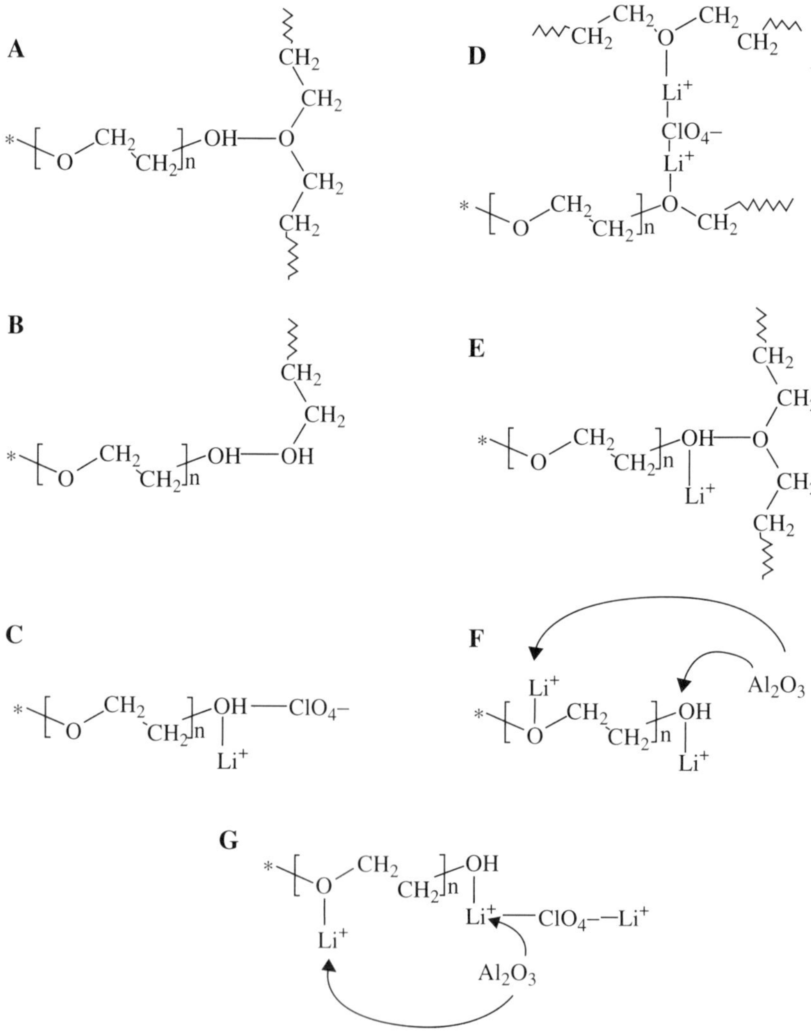

Fig. 13 Types of interactions between the polymer matrix, salt, and inorganic filler. Types **A** and **B** are without salt and apply to both polyethers. Types **C–E** are for PEGME–$LiClO_4$ systems, and types F (acidic surface groups) and G (basic surface groups) are for PEGME–$LiClO_4$–Al_2O_3 systems

interactions **F** (acidic surface groups) and **G** (basic surface groups), which are expected to occur (see Fig. 13). For the neutral filler these two effects can act in opposition. **F** type interactions are typical for the acidic surface modified Al_2O_3, which competes with the Li^+ cation in oxygen coordination. Al_2O_3 with the basic groups can interact with Li^+ (strong Lewis acid) and, thus, form an additional steric effect, hindering the wrapping of the polymer chain around the cation.

On the basis of the FTIR and FT-Raman analysis, it has been shown that the **C**, **D**, and **E** type interactions can come into play with increased salt concentration [98]; however, for electrolytes with very low salt concentration **C** type interactions are dominant. With a higher salt concentration there is a possibility of existence of **E** type complexes. As shown this type of **E** systems can also interact with the polyether oxygen.

Finally, the highest salt concentrations provide proper conditions for the creation of **D** type complexes. The **D** complex is both cation and anion dependent. Some of the **D** type inter crosslinks may involve the OH oxygen instead of the polyether oxygen in the PEGME-based electrolytes. Increases in conductivity are observed in the salt concentration ranges where complexes of the **D** type exist. Such crosslinks may be less probable with the addition of the filler via the **F** and **G** types of interactions.

The **D**, **F**, and **G** complexes are of greatest importance in the composite electrolytes. The **D** type of interaction is responsible for the formation of intertransient crosslinks involving the positively charged triplet; this results in the stiffening of the polymer host. The addition of the filler results in breaking of these transient crosslinks following **E**, **F**, and **G** (acidic and basic fillers); Li^+ can be part of a triplet or exist as a free cation. These effects are observed as a decrease in the viscosity of the electrolyte and weakening of the polymer–salt interactions, which can be observed from FTIR spectra. In the case of monomethyl-capped systems in which there is a participation of OH-end groups in the formation of intertransient crosslinks, the filler particles have easier access to these crosslinks and more interchain connections can be weakened or eliminated. From this we conclude that the increase in conductivity observed is larger than in the case of the dicapped systems. For the dicapped system, the increase in ionic mobility does not compensate the reduction in the number of mobile anions resulting from the **F** and **G** type interactions.

6 Novel Approaches Toward Understanding Ionic Transport Phenomena in Polymeric Electrolytes

As has been shown earlier, several models have been applied to describe ion transport phenomena in polymeric electrolytes. The aim of the present chapter is to generalize on ion transport phenomena observed in composite polymeric electrolytes using a new approach, which would be helpful in describing changes in the conductivity and lithium ion transference numbers upon addition of fillers to polymeric electrolytes. The concept is based on the observation of changes in ionic associations in the polymeric electrolytes studied in a wide salt concentration range.

It is demonstrated that the addition of calix[6]pyrrole or superacid-type fillers to polyether –LiX (X = I^-, BF_4^-, ClO_4^-, AsF_6^-, $CF_3SO_3^-$) electrolytes results in a considerable increase in lithium cation transference numbers as ac–dc polarization experiments have revealed [99]. In the case of the supramolecular receptor, the highest increase in lithium transference number is received for low molar fraction of calix[6]pyrrole (0.125 with respect to the molar concentration of anions) used and is followed by only a slight increase for higher receptor concentrations [10]. It is also known that for high concentrations of calix[6]pyrrole, the self association of the receptor samples occurs, which in turn limits the amount of calix[6]pyrrole capable of complexing anions. In the case of anion–receptor complexation, the concentration of the anion–receptor complex depends linearly on the free-receptor concentration, $[AnRec]=[An][Rec]/K_{complex}$, while concentration of receptor agglomerates depends as follows: $[Rec_n]=[Rec]^n/K^n_{self\text{-}ass}$, where $n\geq 2$. For this reason self-association could be privileged in samples with higher calix[6]pyrrole concentration. It is also evident that in this concentration ranges the dissolution of calixpyrrole in polar solvents decreases.

Because of the low dielectric constant of polyethers (which can be assumed as a highly viscous or solid solvent) their complexes with lithium salts can be considered as weak electrolytes. Therefore, for different salt concentration ranges, the charge carriers exist in the form of single ions (low salt concentration range), ion-pairs (medium to high salt concentration range), or higher aggregates (high salt concentration range). Ion pairs and triplets formation constants (see Table 5 [100]) were calculated for polyether-based electrolytes using the Fuoss-Kraus procedure [101]. The calculations are based on our own experimental results and literature data ($LiCF_3SO_{3-}$ and $LiAsF_6$-doped samples). Figure 14 shows the schematic interaction of supramolecular compounds with anions. A similar scheme can be drawn to describe the interaction between anions and superacid fillers surface groups. The three different scenarios showing various relations between the complexation constant of the reaction of the inorganic or organic receptor and anions with respect to the values of ion-pairs and ionic triplets constant are considered here (see caption of Fig. 14). The ionic triplets formation constant is lower than that of ion pair formation and usually lower than the receptor–anion complex formation constants, which, according to the literature data, are within range 2–7 × 10^3 [10].

Table 5 Ion pairs (K_A) and Ionic Triplets (K_T) formation constants calculated for PEO-based electrolytes doped with various lithium salts

Salt	K_A	K_T
LiI	3.87×10^4	130
$LiCF_3SO_3$	3.18×10^4	72
$LiBF_4$	1.75×10^5	78
$LiClO_4$	5×10^2	54
LiSCN	1.2×10^3	124
$LiAsF_6$	1×10^5	28

Data for $LiCF_3SO_3$ and $LiAsF_6$ obtained from reference [100].

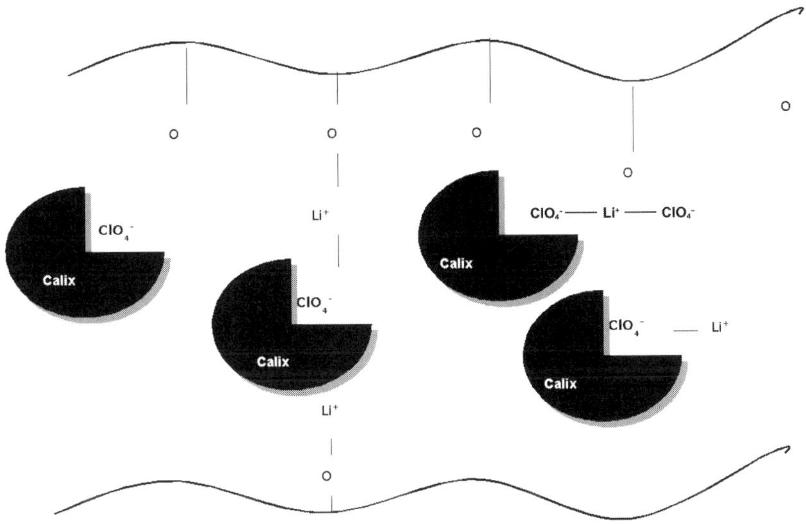

Fig. 14 A diagram illustrating various types of activity of the active filler against the anions of the polymeric electrolyte. The validity of the particular schemes is dependent on the equilibriums present in the sample. From left to right (1) $K_I > K_T > K_{cal}$ – only free anions are complexed; (2) $K_I > K_{cal} > K_T$ – free anions and ones belonging to ionic triplets can form complexes – transient crosslink breaking; (3) as above but for noncrosslinking negative ionic triplet; and (4) $K_{cal} > K_I > K_T$ – ionic pairs can also be affected. Typical values for PEO-based systems: $K_I = 10^5$–10^4 kg mol^{-1} – ion pairs formation constant, $K_T = 10^1$–10^2 kg mol^{-1} – ionic triplets formation, and $K_{cal} = 10^2$–10^3 kg mol^{-1} – calix-anion complex constant

The increase in the lithium transference numbers, which is observed upon the addition of calix[6]pyrrole, is usually associated with a decrease in the conductivity of the samples studied. The simplest explanation of this phenomenon relies on the fact of immobilization of mobile anions and, thus, a decrease in conductivity. This effect is particularly evident for low concentrations of the lithium salt for which a high concentration of "free anions "can be expected (see Fig. 14). However, the behavior observed for high salt concentration for the various types of salt used cannot be explained by such a simple assumption. For the high salt concentration range (which is the case for the most systems under study), the presence of a large fraction of ion pairs and higher aggregates should be expected. Therefore, it can be assumed that it is easier for the receptor to release an anion from a triplet than from an ion-pair.

As has already been mentioned earlier, also in this salt concentration range, in spite of the increase in the cation transport number, ionic conductivity of composite electrolytes is often much lower than for pure polyether-based systems. Besides the explanation based on the reduction in a fraction of mobile anions the additional phenomenon of steric hindrances creation must be taken into consideration. This assumption is related to the fact that receptor molecules exhibit high molecular volume and can negatively influence the ionic transport. On the other hand, the addition of calixpyrrole results in breaking inter- and intramolecular crosslinks (due to the release of the anion) and, therefore increases the flexibility of polyether chains (see Table 4 containing PFG NMR data), thus enhancing the anion transport. The final decrease or increase in conductivity results from the above-described contradictory effects. For LiI-doped electrolytes, the fraction of triplets is higher than the maximum fraction of the reacted calixpyrrole. Therefore, the stiffening of the polyether chain due to the presence of crosslinks still affects the polymer flexibility, which is additionally influenced by the presence of the bulky receptor. For $LiCF_3SO_3$-doped samples the fraction of calixpyrrole available is higher than the maximum concentration of crosslinks, which means an increase in the flexibility of polyether chains and thus an increase in conductivity.

Similar rules can easily be applied to composite polymeric electrolytes with inorganic receptors. A nice illustration comes from the early days of studies of the PEO–NaI–Nasicon composite polymeric electrolyte. Figure 15 presents a comparison of the fraction of protons belonging to various polymer electrolyte phases for filler-free and composite $(PEO)_{10}$ NaI-based polymer electrolyte [32]. Three phases exhibiting different proton mobility, crystalline PEO phase, amorphous PEO phase, and PEO–NaI crystalline complex phase, can be distinguished. The fraction of protons belonging to the amorphous phase of PEO in composite electrolyte is 0.2 higher (below the melting temperature of crystalline PEO) compared with unmodified $(PEO)_{10}$ NaI electrolyte. Even at temperatures higher than the melting temperature of the PEO, the fraction of amorphous phase of composite system is still higher. The increase in the fraction of the amorphous phase results from the decrease in the fraction of both crystalline PEO phase and crystalline PEO-NaI crystalline complex phases.

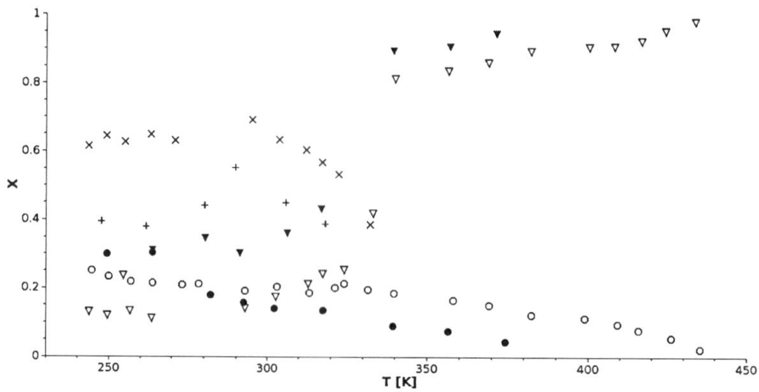

Fig. 15 Fraction of protons belonging to different phases of PEO mixed-phase electrolyte compared with unmodified $(PEO)_{10}$ NaI electrolyte. Fraction of protons belonging to (*open down triangle*) amorphous phase, (*open circle*) crystalline complex, (*times*) pure crystalline PEO of $(PEO)_{10}$NaI electrolyte, (*filled down triangle*) amorphous phase, (*filled circle*) crystalline complex, (*plus*) pure crystalline PEO of $(PEO)_{10}$NaI(NASICON) mixed-phase solid electrolyte

These results are a confirmation of both the amorphous phase model and Lewis acid–base model. A decrease in the fraction of the complex phase might results from breaking the transient crosslinks formed by the positively charged triplet (see Fig. 14) or the filler taking the lithium place in the coordination sphere of polyether oxygen. Because of the higher glass transition temperature of the composite the latter effect is more probable.

The whole above-presented analysis shows clearly that the issue of composite polymer electrolyte optimization and property tailoring is complicated and lacks a single, simple answer. A variety of presented modification paths lead to the further complication of the system studied introducing, in consequence, additional phenomena related to charge carrier transportation. Despite the great amount of information that can be gathered by application of the various experimental techniques, a more theoretical approach is needed when thinking about both tailoring of properties of the known systems and new material designing. This approach can be applied by both theoretical studies and computer modeling.

From the point of view of theoretical thermodynamics, an effort aimed at building a theoretical model able to predict conductivity and other properties of composite electrolytes is possible but with limited practical meaning. Models of conductivity in solid polymeric electrolytes such as Free Volume Theory [102], Configurational Entropy Theory [103], and Dynamic Bond Percolation Theory [104] were widely described in the literature, but unfortunately are valid only for simple, single-phase, and fully amorphous systems and, in consequence, cannot be directly applied to the description of the whole composite system not loosing their importance for the description of the amorphous part of the polymer matrix.

When thinking about a theoretical approach to the conductivity of the composite polymeric electrolytes, one must take into consideration that they are multiphase systems containing at least two different crystalline phases (pure polymer and polymer–salt complex), a phase of the dispersed grains and an amorphous phase of nonhomogenous properties. Additionally, both the phase composition of the system and the properties of the particular phases change with temperature. To make the overall image of the studied composites more complicated, one must also consider the surface state of filler particles as it influences the filler–host interactions in both strength and mechanism terms [19, 20, 105]. Models applicable to the systems described can be divided mainly into three groups:

- Models based on the studies of the dielectric relaxations observed in the sample.
- Mesoscale models of conductivity based on predicting global material parameters on the basis of its phasc composition and properties of the individual phases.
- Molecular scale models based on both classical mechanics (molecular dynamics) and quantum mechanics approach (ab initio and semiempirical QM).

7 Dielectric Relaxation Studies of Ionic Transport in Composites

As the presented approaches are arranged with the macroscopic ones at the beginning, one should concentrate on electrochemical impedance spectroscopy that appears here to be one of high importance experimental techniques allowing for acquiring multiple material parameters from a single measurement. Different experimental setups are used with the application of both blocking and transporting electrodes, two and three electrode cells, a wide temperature and frequency range and active probe shielding. The data obtained can be fitted to the equivalent circuit where different phenomena observed in the sample tested are modeled as a various electrical elements (resistors, capacitors, constant phase elements, and others) [106, 107]. This allows to find not only the DC bulk conductivity value but also to determine the existence of grain boundary conductivity. The diffusion process in the electrolyte, transport properties of the electrolyte–electrode boundary in correlation with passive layer formation and transference numbers of the charge carriers are also a matter of interest.

7.1 Theory of Mismatch and Disorder

A novel approach was introduced by Funke and co-workers [108–110] to various type of composites. The authors tested the introduced method on various well-defined inorganic systems such as single crystals, conductive glasses, and molten salts declaring its applicability to polymeric systems. Conductivity data were collected in a very wide frequency range combining classical impedance spectroscopy measurements, microwave spectroscopy,, and far infrared. The spectra obtained in this manner were reviewed and discussed, with particular emphasis on their high frequency plateaus and their low frequency scaling properties. A concept of mismatch and relaxation (CMR) was introduced to correlate the spectral data with the mobile ion dynamics in the samples. To explain the spectra, a jump relaxation model was built over the CMR basis. In amorphous materials such as conducting glasses, ions encounter different kinds of site and the model must be modified accordingly. The proposed model was applied to crystalline conductors such as $RbAg_4I_5$, where the order–disorder transition can be observed below 298 K by a change of conductivity mechanism from nonrandom to random hopping. In supercooled molten salts the ion dynamics can be explained by yielding a new equation for the thermal dependence of the DC conductivity. Finally, the introduction of a structure-sensitive parameter allows the CMR approach to fit experimental conductivity spectra that display significant differences in shape characteristic for various types of composite materials.

7.2 *Universal Power Law of Dielectric Response*

High frequency dielectric response of the electrolyte can be also analyzed by combining Jonscher's universal power law of dielectric response [111]. In this purpose a log(σ)–log(ω) representation is used. Figure 16 shows the exemplary experimental data for the composite polymeric electrolyte measured in various temperatures while the Fig. 17 shows the theoretical prediction calculated for various n values. The universal power law allows computing the hopping frequency of the charge carrier (ω_p) on the basis of high frequency impedance by means of (1) and (2).

$$\sigma(\omega) = \sigma_{DC} + A\omega^n, \tag{1}$$

$$\omega_p = (\sigma_{DC} / A)^{(1/n)}, \tag{2}$$

where σ_{DC} is the direct current conductivity of the sample, *A* and *n* are the material parameters, and the alternate current frequency.

The material parameter *n* varies for different materials in range (0,1). When *n* is in range (0.5,1) the real part of the electrical susceptibility is higher than the imaginary one. For the range (0,0.5) the reverse is true. For $n=0.5$ both parts are equal.

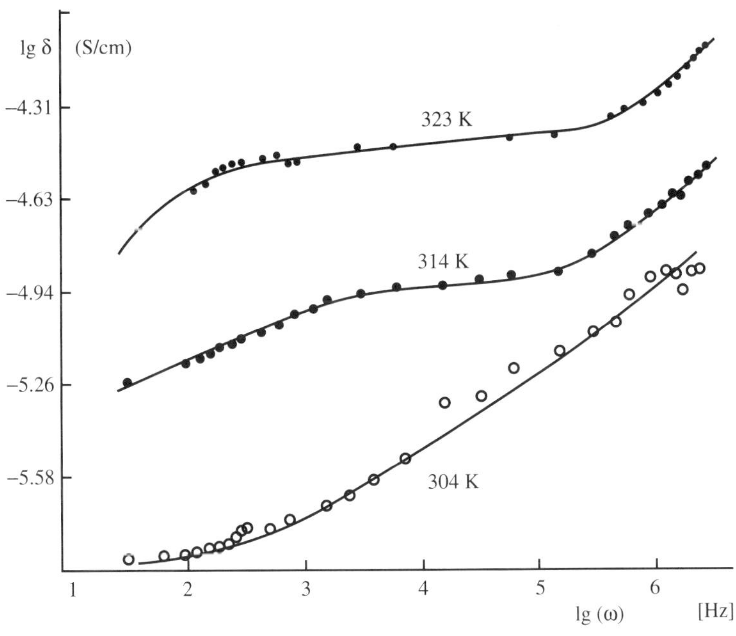

Fig. 16 Impedance spectra of polymer electrolyte samples in log(σ)–log(ω) representation (used for Jonsher's Universal Power Law – experimental data for $(PEO)_{10}NaI$–20% Θ-Al_2O_3

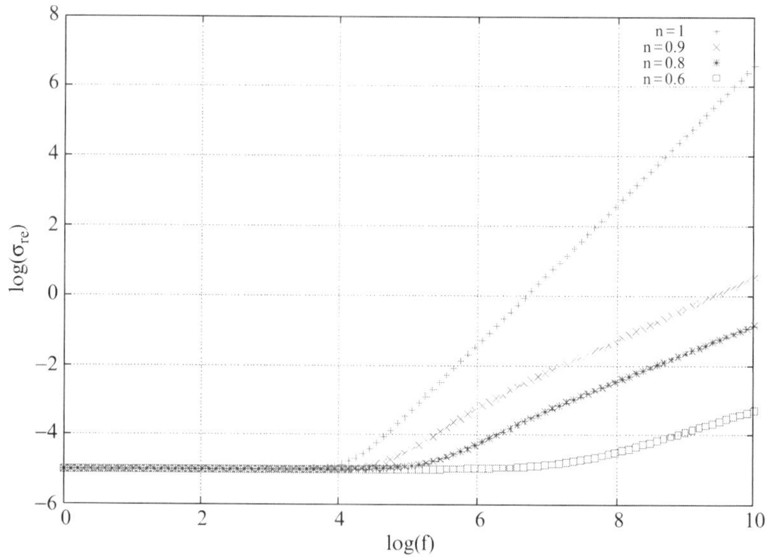

Fig. 17 Theoretical prediction calculated for various n values of the impedance spectra of polymer electrolyte samples in $\log(\sigma)$–$\log(\omega)$ representation (used for Jonsher's Universal Power Law)

The case of $n = 1.0$ is impossible for the real materials (ideal dielectric) and can be reached only as an experimental error for low loss materials. In general a higher n value is related to a better correlation of the charge carriers movements in the sample. Typical observed values for the ionically conductive solids lay in range 0.6–0.8.

The real concentration of charge carriers taking part in the charge transport process can be calculated from (3).

$$K = \sigma_{\mathrm{DC}} T / \omega_{\mathrm{p}}. \tag{3}$$

The obtained value compared with the theoretical charge carrier concentration found from sample stoichiometry on the basis of full dissociation assumption gives information on the real fraction of the doping salt converted into the current transporting species.

7.3 The Almond–West Formalism

It is generally accepted that one can divide the overall ionic conductivity of a material (σ) into two independent factors: one dependent on charge carrier mobility (μ) and the other on charge carrier concentration (c) (ez: factor represents the average charge carrier charge in electron units).

$$\sigma(T) = \mu(T)c(T)ez. \tag{4}$$

For most of the systems studied it can be also assumed that conductivity together with charge carrier mobility and concentration show an Arrhenius-type temperature dependency below the melting point of the polymer crystalline phase. In this temperature range, the value of the activation energy of conduction E_a can be easily found from the Arrhenius equation:

$$\sigma(T) = \sigma_o \exp(-E_a / kT). \tag{5}$$

Taking into consideration the two previously mentioned factors, it should be possible to divide the E_a value into the terms related to charge carrier migration E_m and creation E_c [112, 113]. For calculations based on the Almond–West formalism, the impedance data must be collected in a temperature range above the glass transition temperature of the sample and below its melting point. The calculations must be performed for all collected data sets. A quasi-Arrhenius temperature dependency of ω_p can be later used to calculate the E_m

$$\omega_p = \omega_o \exp(-G_m / kT) = \omega_e \exp(-E_m / kT), \tag{6}$$

where ω_o is the ions oscillation frequency,G_m is the free energy of migration of the charge carrier, and ω_e is the effective vibrational frequency for mobile charge carriers.

The energetic charge carrier creation factor E_c is obtained by a simple subtraction of E_m from E_a. The values of the two fractional activation energies provide information on the type of the process, which predominantly limits the conductivity.

The above-described approach was used to study impedance data gathered for a wide range of polymeric electrolytes such as blends [114], thermoplastic polyurethanes [115], and ethylene oxide–propylene oxide copolymers [116]. For composites with a nonconductive filler [117], it is clearly shown that conductivity changes follow the migrational term changes. Generally, the lower is the activation energy of conduction and the E_m value, the higher is the ambient temperature conductivity of the sample. This observation in part related to the E_a value was previously described in [118] for a wide range of composite electrolytes. For samples of low concentration of the filler added, the value of the concentration term E_c is almost constant and lies within the experimental error equal to the value characteristic for the pristine system of analogous composition that contains no filler. Table 6 gathers the data for different composite systems that were analyzed. For higher concentration of the filler, the samples containing θ-alumina exhibit higher E_c values, whereas an addition of α-alumina leads to the decrease of this factor. On the basis of the results presented it can be assumed that the mobility of charge carriers plays the dominant role in conductivity in comparison with the concentration of the charged species. The behavior observed is in good agreement with the theory

Table 6 Parameters calculated from the Almond–West Formalism and the Universal Power Law of Dielectric Response for a set of composite polymeric electrolytes

Type of the electrolyte	σ_{rt} (S cm^{-1})	E_a (kJ mol^{-1})	E_m (kJ mol^{-1})	E_c (kJ mol^{-1})	n	K
$(PEO)_{10}NaI$	1.1×10^{-8}	86.2	419	44.3	0.70	1.9×10^{-9}
					0.93	2.5×10^{-8}
$(PEO)_{10}NaI$ + 10% Θ-Al_2O_3 (grain size 4 μm)	3.2×10^{-7}	53.7	17.8	35.9	0.78	3.5×10^{-9}
					1.00	1.9×10^{-8}
$(PEO)_{10}NaI$ + 10% Θ-Al_2O_3 (grain size 7 μm)	2.0×10^{-8}	92.1	54.6	37.5	0.63	2.0×10^{-9}
					1.00	2.4×10^{-8}
$(PEO)_{10}NaI$ + 5% Θ-Al_2O_3 (grain size 5 μm)	1.5×10^{-7}	57.9	19.4	38.5	0.65	4.6×10^{-10}
					1.00	1.2×10^{-8}
$(PEO)_{10}NaI$ + 10% Θ-Al_2O_3 (grain size 5 μm)	9.5×10^{-8}	59.4	19.4	40.0	0.75	1.4×10^{-9}
					1.00	1.4×10^{-8}
$(PEO)_{10}NaI$ + 20% Θ-Al_2O_3 (grain size 5 μm)	1.5×10^{-7}	69.6	16.6	53.0	0.48	1.3×10^{-9}
					0.88	3.1×10^{-7}
$(PEO)_{10}NaI$ + 30% Θ-Al_2O_3 (grain size 5 μm)	1.8×10^{-7}	73.9	9.4	64.5	0.81	1.9×10^{-9}
					1.00	2.5×10^{-7}
$(PEO)_{10}NaI$ + 50% Θ-Al_2O_3 (grain size 5 μm)	1.2×10^{-8}	147.2	79.0	68.2	0.77	2.3×10^{-9}
					0.85	5.0×10^{-8}
$(PEO)_{10}NaI$ + 10% α-Al_2O_3 (grain size 5 μm)	2.0×10^{-7}	75.0	18.4	56.6	0.77	8.8×10^{-10}
					1.00	5.8×10^{-9}
$(PEO)_{10}NaI$ + 20% α-Al_2O_3 (grain size 5 μm)	1.5×10^{-8}	176.8	124.0	52.8	0.74	3.5×10^{-9}
					0.91	2.9×10^{-8}
$(PEO)_{10}NaI$ + 30% α-Al_2O_3 (grain size 5 μm)	1.6×10^{-8}	131.9	117.2	14.7	0.80	2.4×10^{-9}
					0.85	6.1×10^{-8}

(continued)

Table 6 (continued)

Type of the electrolyte	σ_{rt} (S cm^{-1})	E_a (kJ mol^{-1})	E_m (kJ mol^{-1})	E_c (kJ mol^{-1})	n	K
$(PEO)_{10}NaI$ + 50% α-Al_2O_3 (grain size 5 μm)	5.3×10^{-8}	140.9	130.6	10.3	0.68	2.7×10^{-9}
					1.00	4.5×10^{-8}
$(PEO)_{10}NaI$ + 5% α-Al_2O_3 (grain size 4 μm)	5.5×10^{-8}	124.4	86.4	38.0	0.61	6.1×10^{-9}
					1.00	3.3×10^{-8}
$(PEO)_{10}NaI$ + 10% α-Al_2O_3 (grain size 4 μm)	2.3×10^{-7}	52.1	10.0	42.1	0.65	1.8×10^{-9}
					0.84	1.7×10^{-8}

For n and K the upper value corresponds to the ambient temperature and the lower to the one above the melting point of the crystalline phase [117].

explaining the matrix–filler interactions. The formation of highly conductive amorphous shells connected to the percolation paths eases the movements of charge carriers, while their quantity is determined by salt dissociation, which, in turn, is correlated with the dielectric constant of the polymer matrix independent of the filler addition. At low concentrations of the dispersoid, the volume of the polymer converted into amorphous material is relatively low. Thus, the majority of the charged species is created in the bulk of the polymer matrix where the interaction of the added grains can be neglected. In consequence, the values of the E_c do not vary significantly from the one characteristic for the pristine system. In contrast, for the high concentrations of the fillers added, the stiffening of polymer chains leads to a partial hindrance of segmental movements. This leads to an increase of the activation energy of ion migration. Additionally, for high concentrations of dispersoids, the creation of ions also occurs in the regions distorted by the polymer–grain interaction, which is evidenced by the changes of the E_c values. Therefore, in this case the final character of the E_c change depends on the filler type. This observation can be attributed to the contrary influence of two factors (amorphization and stiffening) on the salt dissociation. Their relative strength varies with filler type and microstructure influencing finally on the sign of the E_c value change.

Table 6 also presents the values of the calculated charge carriers density K and the frequency exponent n. The typical temperature dependence of K shows an increase starting from the ambient temperature to about 340–350 K. Above this temperature the value of the K factor remains almost constant. An exemplary thermal dependence is presented in Fig. 18. The highest increase is observed in the temperature range where melting of the crystalline phase is observed. This can be explained by a sudden increase of the amorphous phase amount in the sample. The constant increase observed in the lower temperature range can be attributed to changes of the material dielectric constant leading to salt

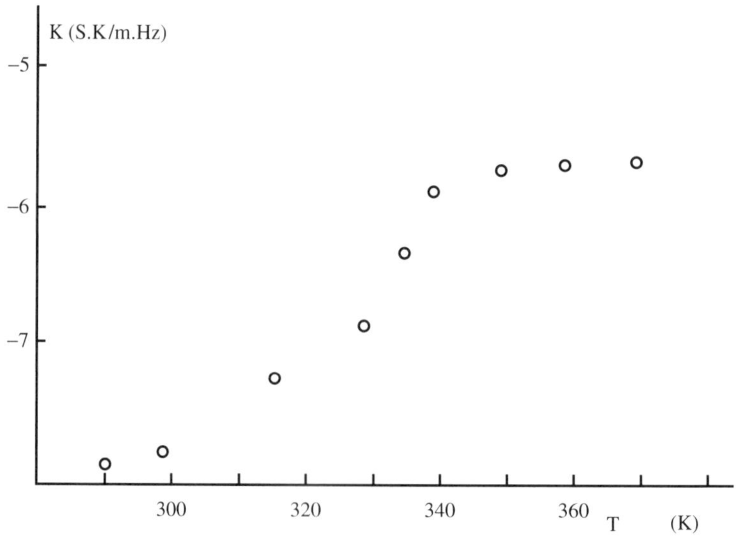

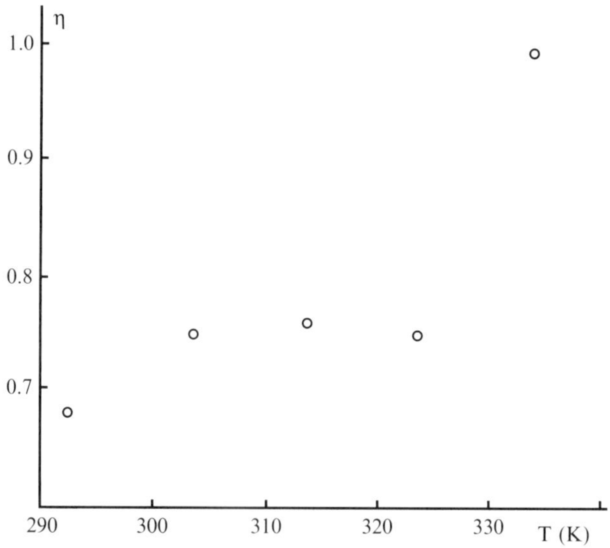

Fig. 18 Variation of *K* (**a**) and *n* (**b**) parameters as a function of temperature for $(PEO)_{10}NaI$–20% Θ-Al_2O_3 composite polymeric electrolyte

dissociation together with thermal activation of the charge carrier creation process. Also an abrupt increase of the n value is observed when the order–disorder transition occurs in the sample (see Fig. 18b). For higher temperatures the value of the *n* factor is almost constant and equal to unity. This implies that AC conductivity of the composites above the crystalline phase melting temperature is a linear function of frequency.

8 Phase Scale Models of Conductivity

The second group of semiempirical models gives the possibility of the conductivity value calculation for a set of composite systems on the basis of the measurement performed only for some of them. Taking into the consideration the assumption of the filler–matrix interaction mechanism, one can obtain an oversimplified dependence of the composite conductivity on filler concentration consisting of three different regions. In the small filler concentration range, almost no change of the conductivity in comparison with the pristine system is observed as the point of the percolation threshold is not yet achieved. At some point the conductivity of the system changes dramatically (few orders of magnitude) and achieves values characteristic for the pure amorphous phase of the electrolyte. In the second range the conductivity value drops slowly due to the dilution effect of the filler. An abrupt decay is observed at the point, in which the polymer matrix loses its continuity and the charge carriers transport process is blocked by the inert filler grains. Thus, a more general model is needed to predict the conductivity of the composite sample. For this purpose various semiempirical models can be applied originating from the knowledge of the sample phase composition, geometry of phase distribution, and the properties of particular phases present in the sample. This data can be achieved either from the experiments or from more detailed calculations based on the molecular scale models described later. The semiempirical models work in the space resolution of the micrometer range and thus are called mesoscale models (MSM) as ones being between the molecular ones working with nano- or even picometer resolution and the thermodynamic ones for which the calculations are held for a sample treated as a single entity. The MSM family can also be divided into two subgroups. The first of them contains percolation process-based models such as the ones based on the effective medium theory (EMT). The other is based on dividing the whole sample into single units, which are later treated as homogenous elements of defined properties. The average material parameters are calculated on the basis of properties of particular components, their fraction in the material, and the geometrical arrangement. Here, depending on the method of sample division and the algorithm used leading to the final values, one can distinguish between the finite element, finite gradient, and random resistor network type of simulations. All of the above-mentioned methods were widely applied to the simulation of the electrical properties of the composites. In contrast, the first two have not been applied yet to the composite polymeric electrolytes but can be. Thus, only a short description of these algorithms will be presented.

8.1 Finite Gradient and Finite Element Approach

The two methods shortly described below have not been yet applied to conductivity prediction of composite polymeric electrolytes. The authors have decided to describe them here to, first, give a wider background for the later described and belonging to the same family of approaches, random resistor network (RRN) models and, second, to present their features as for some composite geometries (oriented systems, layered structures) their computational potential is much stronger in comparison to the RRN calculations.

A finite element method (FEM) is commercially used for mechanical construction analysis. It was also developed as a tool for the prediction of electric properties of the ionically conductive inorganic composites. The simplest description of the system geometry introduces a brick-layer approach. Uniform elements are then placed in a cubic network forming units of constant size. Those assumptions are not realistic for systems of the practical meaning. Additionally, stress must be put on the influence of geometrical and electrical properties of grain boundaries present within one sample on its properties. To overcome these limitations some more realistic images of the interface can be introduced into the model. As a starting point in [119], a two-dimensional model of the composite sample is generated. In this case an oversimplifying assumption that normal current density is always perpendicular to the phase boundary present in the simulated sample is introduced. A potential drop is observed on the low-conducting grain interfaces. Thus, the calculations solving an appropriate Laplace equation yield to frequency-dependent voltage distribution in the sample. A commercial software package FLUX-EXPERT was used for the calculations. More detailed description of the simulation engine can be found in [120, 121].

The method was extended to three-dimensional systems in [122]. A simulated polycrystal under consideration is built of a three-dimensional network of the cubic grains. Because of its symmetry regime, the complete information about the sample impedance can be yielded from the analysis of one grain including its boundaries. The main advantage of the finite element method is here related to the existence of large local variations of the gradient. In this case the varying sizes of finite elements can be easily implemented. The resulting density of the test mesh is adapted dynamically to local field gradient values being increased on the boundaries. As a result a single contact spot is typically described with a few hundred elements. Some additional extensions to the proposed model are put forward to adapt the model to the studies of a real sample of the genuine composite systems: varying thickness of the insulating phase, non-zero transfer resistance at the grain-to-grain contact and a non-zero conductivity of the insulating layer.

Alternatively, the finite gradient method can be applied to similar systems. Coverdale et al. [123] develop a computational method for simulating frequency dependence of the impedance response in materials being a multiphase composite. The approach utilizes the digital image of a studied microstructure together with the experimentally gathered frequency-dependent electrical properties of the individual phases as the input data. The initial microstructure of a virtual sample is obtained on the basis of real 2D-microphotographs and X-ray tomographs of the real sample under consideration. The previously applied algorithm [124, 125] is here developed by the introduction of an extension of the computational scheme to three-dimensional samples. Finally, a conjugated gradient-oriented finite difference approach is used.

The main goal of the introduced alterations is to make the computational scheme quicker and more effective in comparison with the typical Fogelholm [126] algorithm. The main difference in calculation times and efficiencies can be observed for systems in which the highly conductive phase is far from the percolation point where the original approach becomes extremely slow. The authors limit their task to studies of the binary system claiming that the proposed model is also valid for more complicated sample geometries. The simulation results obtained for samples of "experimental" geometry are later compared with those reached for some simple test geometric patterns. These show cases include a single phase system, two phase systems consisting of two halves with serial, and parallel arrangement of the phases and a cubic sample filled with phase B with a single sphere of phase A located in the center of the cube. As the most complicated test pattern a two-phase interpenetrating network is studied. In this case both phases percolate in three dimensions in a quasi-random way.

8.2 *Effective Medium Theory*

Coming back to the polymer composite issues, two different classes of mesoscale models must be taken into consideration. One is the RRN, which belongs to the same family as approaches presented above. The other one is the EMT approach, which allows calculating the average material parameters for multiphase solids. It has been successfully applied to describe electrical properties of various heterogeneous systems. EMT is a very useful tool not only for modeling the experimental data but also for predicting the properties of the material studied from a limited number of experiments. Although models based on several fitted parameters are often in a better quantitative agreement with experiments, the approaches presented here will be based on a limited number of parameters, all of which can be obtained experimentally. The EMT model of the composite polymeric electrolyte assumes that there are three phases present in the sample. Each of them is of different electrical properties. These are (see Fig. 12a)

1. Highly conductive interface layers coating the surfaces of grains
2. Dispersed insulating grains
3. Matrix polymer ionic conductor

In step one, a dispersed grain and the layer covering it are considered as a unit called later the composite grain and calculates the conductivity of this unit (σ_c) according to the Maxwell-Garnett mixture rule [127] (see (7)). The input data used here are σ_1 and σ_2 being, respectively, the conductivity of the interface layer and that of the dispersed grain. $Y = (1 + (t/R))^{1/3}$ is a volume fraction of a dispersed phase in a composite unit.

$$\sigma_c = \sigma_1\left(2\sigma_1 + \sigma_2 + 2Y\left(\sigma_2 - \sigma_1\right)\right)/\left(2\sigma_1 + \sigma_2 - Y\left(\sigma_2 - \sigma_1\right)\right). \tag{7}$$

According to the assumption above, the composite electrolyte can be treated as a quasi two-phase mixture consisting of an ionically conductive pristine polymer matrix with dispersed composite units. Such a quasi two-phase mixture can simply be described by the Brugeman equation [128] or the simple effective medium equation introduced by Landauer [129]. These simple approaches ignore local field effects and are suitable only for the description of quasi two-phase mixtures for volume fractions of composite units lower than 0.1. In our system as the volume fraction of composite units increases, the composite grains join each other and form complicated clusters for which local field effects have to be considered. Therefore, the simple Landauer approach was improved according to the method proposed by Nan [130, 131] and Nakamura [132]. Two limiting situations are considered:

1. The matrix phase is a polymeric electrolyte and contains dispersed composite units.
2. The matrix phase is a composite unit and contains dispersed nonconductive grains.

Only for these two boundary situations used for calculations of the corrected value of the σ_c and σ_3, the local field effect is a geometrical effect and has no relation to the conductivity of the two phases. "The matrix phase" screens in these two cases the effect of the "dispersed phase" in the limiting composite unit [131]. Thus, the conductivity limits of the two limiting cases mentioned above can be calculated according to a simplified form of Nakamura's equations (8 and 9). Here, the corrected values of conductivity are calculated on the basis of the additional sample parameters:

$$\sigma_c^a = \sigma_c(d-1)V_c/(d-V_c) = \sigma_c 2V_c/(3-V_c), \tag{8}$$

$$\sigma_3^a = \sigma_3((d-1)-(d-1)V_c)/(d-1+V_c) = \sigma_3(2-2V_c)/(2+V_c), \tag{9}$$

where $V_c = V_2/Y$ is here the volume fraction of composite units, V_2 is the volume fraction of the dispersed grains, σ_3 is the conductivity of the matrix polymeric electrolyte, and d is the dimensionality of the system that is equal to 3 for spherical grains. Therefore, after introducing the improved conductivity parameters in the self consistent EMT equation one obtains two equivalent equations (10) or (11) allowing to find σ_m (the conductivity of the composite polymeric electrolyte).

$$V_2/Y\left(\sigma_c^a-\sigma_m\right)/\left(\sigma_c^a+(1/p_c-1)\sigma_m\right) + \left(1-V_2/Y\right)\left(\sigma_3^a-\sigma_m\right)/\left(\sigma_3^a+\left(1/p_c-1\right)\sigma_m\right)=0 \tag{10}$$

or

$$V_2/Y\left(\sigma_c^a-\sigma_m\right)/\left(\sigma_m+p_c\left(\sigma_c^a-\sigma_m\right)\right)$$
$$+\left(1-V_2/Y\right)\left(\sigma_3^a-\sigma_m\right)/\left(\sigma_m+p_c\left(\sigma_3^a-\sigma_m\right)\right)=0. \tag{11}$$

Here, a continuous percolation threshold for the composite grains (p_c) must be defined. Since the composite grains are allowed to overlap in the region of the interface layer, p_c can be taken to be equal to 0.28. The highest enhancement of ionic conductivity is reached for the volume fraction of the dispersed grains equal to

$$V_2^* = (1+t/R)^3. \tag{12}$$

For concentrations of grains exceeding V_2^* the quasi two-phase system consists of a mixture of composite grains and dispersed bare insulating grains. Thus, conductivity is calculated from (13) for which the corrected input parameters are calculated according to (14) and (15).

$$\left(1-V_2\right)\left(\sigma_c^a-\sigma_m\right)/\left(\left(\sigma_m+P_c\right)\left(\sigma_c^a-\sigma_m\right)\right)$$
$$+\left(V_2-V_2^*\right)\left(\sigma_2^a-\sigma_m\right)\left(\sigma_m+P_c\right)\left(\sigma_2^a-\sigma_m\right)=0 \tag{13}$$

$$\sigma_c^a=\sigma_c\left(2-2V_2+2V_2^*\right)/\left(2+2V_2-V_2^*\right) \tag{14}$$

$$\sigma_2^a=\sigma_2\left(2V_2-2V_2^*\right)/\left(3-V_2+V_2^*\right) \tag{15}$$

An additional parameter must be introduced here, i. e., the percolation threshold of the dispersed bare insulating grains that are not allowed to overlap (P_c). Therefore P_c is different from the previously defined p_c and was taken as the percolation threshold for a general random mixture equal to 0.15. Three characteristic volume fractions of the dispersed grains can be observed:

$V_2' = P_c/(1 + t/R)^3$ is the critical volume fraction at which composite grains join to form a composite cluster and thus a continuous percolation threshold is reached.

$V_2'' = 1/(1 + t/R)^3$ is the volume fraction of the filler at which the continuous composite grain cluster appears to fill the total electrolyte volume at which point the conductivity of a composite electrolyte reaches the maximum.

$V_2''' = 1 - P_c + P_cV_2$ is the volume fraction of the filler at which the conductor-to-insulator transition occurs.

In the real systems the drop of conductivity for grain concentrations higher than V_2" is much quicker than the one predicted by the Nan and Smith model. To improve the goodness of the fit, it was assumed in the model proposed by our group [97] that the conductivity of the amorphous shell changes with the change of grain concentration. From DSC data, it is known that the glass transition temperature (T_g) of composite electrolyte increases with the increase of the ceramic grain concentration. Assuming that the thermal dependence of the amorphous shell conductivity would follow the VTF equation (16)

$$\sigma = \sigma_0 \exp(-B/(T - T_0)) \tag{16}$$

and that the pseudoactivation energy B and the preexponential factor σ_0 were independent of filler concentration. The last statement can be confirmed by our previous results based on the application of the Almond–West Formalism to the composite polymeric electrolytes. Therefore, the interface layer conductivity is only dependent on T_g and decreases with an increase in T_g. Values of T_g taken from DSC experiments were used for the calculation of σ_1 for electrolytes with various concentrations of the filler. Using the variable σ_1 and t/R parameters the corrected model was constructed.

To generalize the model applied, the variation of T_g as a function of grain sizes and concentration should be considered. The effect of the dissolved salt on T_g should also be taken into account. The following empirical equation (17) for the T_g of immiscible blends and mixed phases systems is assumed as the basis for the model:

$$T_g = K_0 + K_1\left[f(V_2/S_2)\right] + K_2\left[f\left(c_s, V_2\right)\right], \tag{17}$$

where, V_2 and S_2 are respectively volume fraction and surface area of the dispersed phase, c_s is the molar concentration of the dopant salt added, and K_0, K_1, and K_2 are adjustable constants.

In the next step the temperature dependence of conductivity of mixed phase composite electrolytes is calculated on the basis of EMT schemes. It was assumed [133] that the temperature dependence of conductivity of the polymer matrix follows the Arrhenius equation. The activation energy and the preexponential factor are calculated from the temperature dependence of conductivity measured for the respective systems studied. The conductivity of the dispersed phase was assumed to be temperature independent. Both the VTF and Arrhenius equations were used to describe the temperature dependence of the conductivity of the surface layer. In the former case the VTF parameters calculated from the conductivity data measured for amorphous EO copolymer-based electrolytes were used for the calculations. The activation energy and the preexponential factor for the Arrhenius dependence were calculated from moderate temperature conductivity data experimentally obtained for the respective electrolytes.

The a.c. behavior of conductivity of composite electrolytes can also be predicted [134, 135]. For a.c. conduction behavior, the conductivity parameters σ_{i} in the corresponding equations must be replaced by frequency-dependent complex conductance parameters expressed according to the equation:

$$\sigma_{\mathrm{i}}(\omega) = \sigma_{\mathrm{idc}} - i\omega\varepsilon_{\mathrm{i}}, \tag{18}$$

here $\sigma_i(\omega)$ represents the complex conductance of a particular phase, i = 1, 2, 3, c, etc; σ_{idc} is the direct current conductivity for this phase; ε_i is the dielectric constant of the *i*th phase; and ω is the frequency applied.

The typical set of dependencies obtained from DC calculations is presented in Fig. 19. Three characteristic critical volume fractions (defined previously in the theoretical section) of the filler can be easily observed. V_2' observed at the lowest grain concentration represents the positive percolation threshold of the composite grains. The value of this parameter is reversely proportional to the *T*/*r* parameter. The V_2'' value represents the maximum of conductivity and thus the grain concentration at which the whole volume of the sample is filled by the composite grains. The V_2' value is also strongly dependent on the *T*/*r* value and is increasing with its decrease. V_2'' value represents the negative percolation threshold and thus the grain concentration at which the conductor to insulator transition occurs. It is also clearly seen that the value of the maximal conductivity of the sample is strongly dependent on the *T*/*r* value. The higher the *T*/*r* the higher is also the maximal conductivity. For the purpose of quantitative calculations the *T*/*r* ratio can be easily found from experimental data (DSC or XRD experiments). The data for the different systems studied are collected in Table 7. The PEO–salt–Al_2O_3 system does not obey the rule of the constant *t*/*R* parameter for various filler sizes. The experimentally calculated *t*/*R* parameter varies from 1.16 for the smallest grains to 0.40 for the biggest ones. A set of plots for this system is presented in Fig. 20. The *T*/*r* values for which the calculations were performed were chosen on the basis of the experimental data for the system simulated. Similar observations can be made for PEO–NaI–SiO_2 systems. As can be seen, contrastively to the inorganic dopants where *T*/*r* is a strong function of the grain concentration. In contrast in the case of PAAM it is almost constant in the dopant concentration range $V_2 = 0.05–0.40$.

The AC data are shown in Fig. 21. The frequency dependence of conductivity was calculated for the $(PEO)_{10}$NaI–20% Θ-Al_2O_3 system for the fixed contents of the dispersed phase. The electrical properties predicted here are insensitive to changes of frequency in the low frequency range and strongly dependent on the change of ω at high frequencies. Above the limiting frequency conductivity starts to increase, whereas the dielectric constant decreases considerably. The higher the conductivity the higher is the frequency at which the conductivity increase is noticed. The observed behavior is similar to that observed for numerous composite polymeric electrolytes with similar values of the obtained frequency exponent. Thus, an assumption can be derived that according to this model polymeric electrolytes behave as ideal dielectrics.

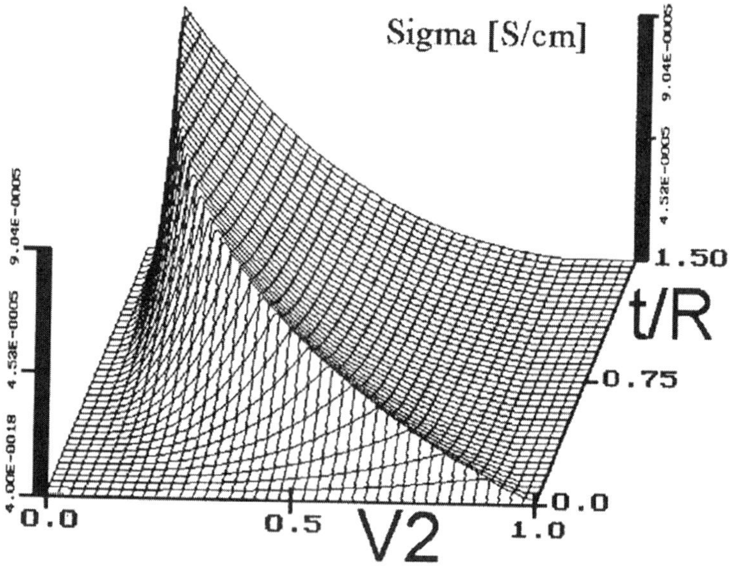

Fig. 19 Ionic conductivities calculated on the basis of the EMT model for PEO–PAAM–$LiClO_4$ composite electrolyte versus PAAM concentration in volume fraction (V_2) and (t/R) parameter

Table 7 Typical parameters of composite polymeric electrolytes used as input for simulations [97, 133]

System	Filler concentration (wt%)	Range of grain sizes (μm)	Average grain size (μm)	*t*/*R* ratio	Average thickness of the shell
$(PEO)_{10}NaI$–Θ-Al_2O_3	10	~2	~2	1.16	2.32
$(PEO)_{10}NaI$–Θ-Al_2O_3	10	2–4	~3	0.88	2.64
$(PEO)_{10}NaI$–Θ-Al_2O_3	10	4–10	~7	0.38	2.66
$(PEO)_{10}NaI$–Θ-SiO_2	5	~5	~5	1.28	6.40
$(PEO)_{10}NaI$–Θ-SiO_2	10	~5	~5	0.92	4.60
$(PEO)_{10}NaI$–Θ-SiO_2	20	~5	~5	0.66	3.30
$(PEO)_{10}LiClO_4$–PAAM	5	<3	–	0.55	–
$(PEO)_{10}LiClO_4$–PAAM	40	<3	–	0.65	–

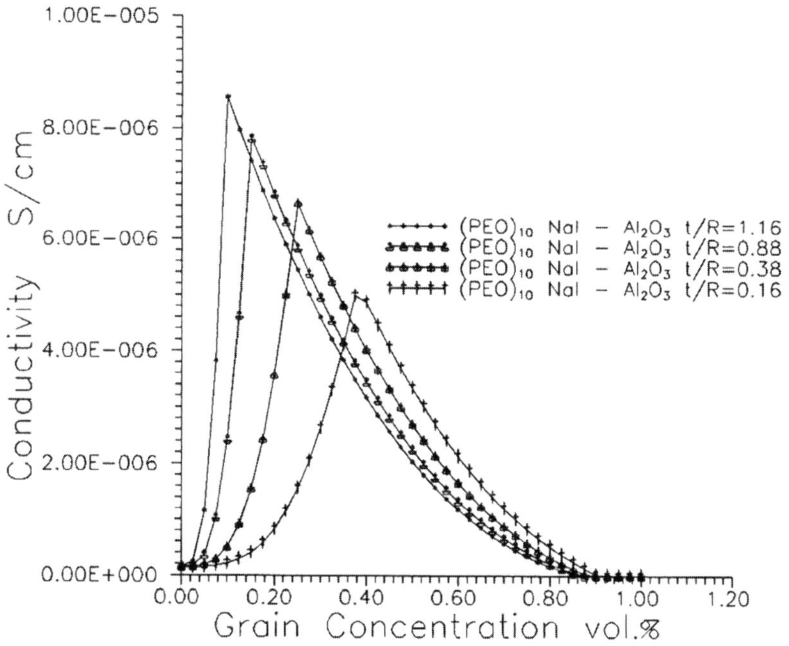

Fig. 20 Ionic conductivities calculated on the basis of the EMT model for composite electrolyte $(PEO)_{10}NaI–Al_2O_3$ versus filler concentration in volume fraction (V_2) and (t/R) parameter

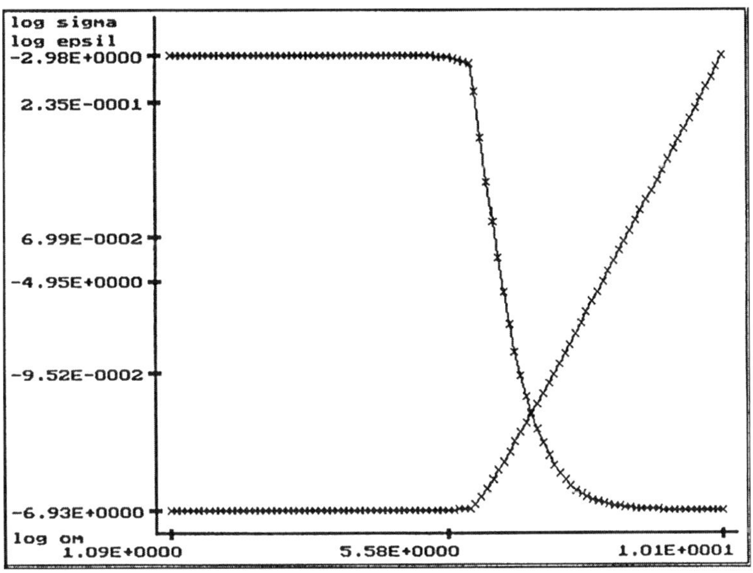

Fig. 21 Frequency dependence of ionic conductivity and dielectric constant calculated for the $(PEO)_{10}NaI–Al_2O_3$ system by mean of the EMT approach

8.3 Random Resistor Networks

As the second class of models of conductivity in polymeric composites a random resistor network approach (RRN) was successfully introduced being on one hand much more versatile in comparison with EMT, but on the other, much more computational power demanding. Previously to the application described here, RRN modeling was a widely recognized approach applied to studies of different properties of various heterogeneous materials consisting of regions of nonidentical properties. Apart from electrical phenomena, magnetic properties of materials [136] or even studies of sample destruction under critical mechanical stress [137] or geophysical studies of earthquakes [138] can be successfully performed by means of this computational tool. When applied to electrical properties of heterogeneous systems, the method relates the value and type of a particular element to local properties of the sample. Resistivities, capacitances, and parallel R-CPE connections can be used being adequate to conductors, insulators, and other types of materials. Most of already published works (e. g. [139–141]) deal with very simple so-called binary systems being the image of a two-compound system. Usually, one of the components of the mixture is an insulator while the other is a conductor. Electrical parameters of the branches (network elements) have only two valid values correlated to either a path (resistor) or an insulator (break). For such models an abrupt percolation threshold can be observed for a given composition of the sample represented by the particular number of either conducting (filling conductors into isolative sample) or insulating (breaking the already conducting system) branches in the complete network.

Different more complicated systems were also studied by the means of the RRN approach [142–147]. In [147] Wang et al. predict conductivity in composites consisting of the particles of metal and solid electrolyte. The system simulated consists of units of two different types to which either ionic or electronic conductivity can be attributed. To represent different types of materials present in the sample, the authors use three resistive elements (metal–metal, metal–electrolyte, and electrolyte–electrolyte connections) and two capacitors (geometrical capacity of the electrolyte and the double layer capacity of the electrolyte–metal junction). The phenomenon of double layer creation together with that of respective capacity presence is revealed at grain boundaries of different materials. An increase of average sample conductivity can be observed for compositions in which the percolation of metallic grains is observed.

A model of a system with similar geometry such as composite polymeric electrolytes with nonconductive filler is presented in [139]. Inorganic composites based on LiI/Al_2O_3 and CuCl/Al_2O_3 are studied here. The authors assume that the highly conductive interfacial layer is built on the grain–matrix border. The main difference between the studied system and the polymeric systems lies in the fact that the isolative grains of the inorganic composite are not fully randomly located in the volume of the real sample. In the presented model randomly located (thus, the model is not in agreement with the real geometry) nodes (grains) are insulators. Sites not filled with the grain and, thus, belonging to the matrix exhibit ionic conductivity. The resistivity value for a branch is calculated on the basis of the status of its end nodes. If both belong to the grains, the branch is broken (infinite resistivity) and, if both belong to the matrix, the branch is the pristine conductor. In the case of a branch with ends placed in the nodes attributed to two different materials, a high conductivity value is attributed. Finally, we must conclude that models presented in the literature from the point of view of the application in composite polymeric electrolytes are strongly oversimplified and thus cannot be applied without further elaboration. To tailor the model to the polymeric systems [148, 149], an assumption is introduced that the thickness of the shell formed on the grain boundary is finite and higher than one network unit together with shell parameters dependency on the grain filler type, size and sample composition. Additionally, the consideration that the highly conductive shell is nonhomogenous can be used as an extension to the previously presented EMT approach. Electrical parameters of the material in the amorphous region may vary within the shell depending on the distance from the grain–matrix boundary. Also, as declared previously changes of sample composition can affect shell conductivity. To make the model more realistic, a distribution of filler grain sizes and t/R ratio in a given range can be introduced.

Here it must be clearly stressed that realization of so complicated calculations was possible only because of significant lowering of prices for high speed processors and large capacity memories in the last few years. Finally, it is possible to deal with a 300^3 matrix filled with impedance data for the virtual sample represented as complex numbers. The initial part of the algorithm is related to the procedure of creation of a three-dimensional resistor network being a representation of morphology and phase structure of a sample of the simulated composition. At the beginning a virtual sample is built of empty cubes. Then the location of spherical grains within the matrix is started (Fig. 22a). The procedure is based on random attributing a set of coordinates to the grain center. If a particular simulation assumes that grains are not identical, a radius of a particular grain is also randomly chosen from a given range. Next, all cubic clusters (Fig. 22b) belonging to the grain are marked using the following rule: the cluster belongs to the grain if the center of the cube lies inside it. This procedure is repeated until the assumed volume fraction of the filler is achieved. As the ceramic material is not elastic and grains cannot overlay, each spatial unit can be attributed to only one grain. If a conflict of spatial restrictions is observed, during the next grain location an additional procedure changes the grain center coordinates and the check is performed once again. The algorithm also takes into consideration the insertion of grain fragments in the border area of the sample to observe the condition that the fragment analyzed is a valid representation of a larger sample. Grain insertion is restricted here so that the coordinates of the grain center must be located within the sample. The residual part of the virtual sample (not belonging to grains) is then attributed to the shells (respecting the assumed shell thickness) and polymer

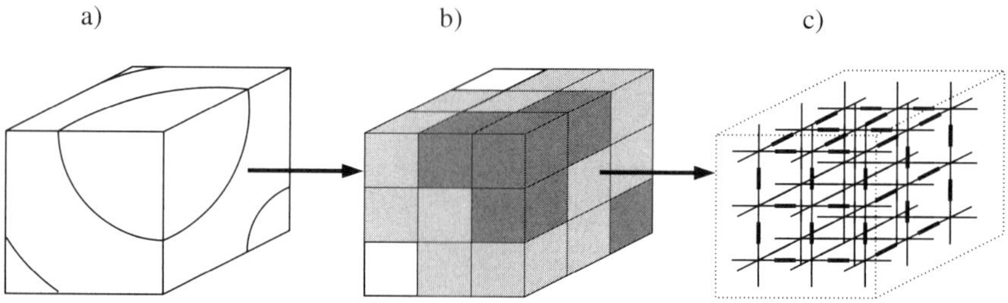

Fig. 22 The procedure of the resistor network generation: (**a**) continuous model, (**b**) discrete model, and (**c**) resistor network

electrolyte matrix (all cells belonging to neither the grain nor the shell). Contrary to the grain–grain situation, shells can overlay with other shells (one unit can belong to the shells of two or more grains).

The values of characteristic impedances [150, 151] are attributed to all unitary cells on the basis of the assumed material parameters of all phases present in the system. The equivalent circuit used for the representation of each cell was a parallel combination of a resistor and a capacitor. At the end of this stage the virtual material is described as a three-dimensional network of impedances Z_{xyz} (Fig. 22c). Each cubic unit is electrically connected to its six neighbors (only the connections by the common walls are taken into consideration). The cell-oriented structure is then converted into a branch-oriented one (Fig. 23a) with consequent location of the network nodes in the centers of the units (compare Figs. 23b and c). Each branch is, thus, characterized by an effective impedance value calculated as a serial connection of fractions originating from both neighboring cells (Figs 23b and 24).

A branch-oriented three-dimensional network obtained in this way can be solved in various ways, including both analytical and numerical approaches. For a matrix of the size used, only numerical iteration procedures can be applied. The rules of energy and charge preservation for the previously generated three-dimensional network leads to two basic equations (19) and (20) describing the current flow within the sample.

$$j(r) = \sigma(r)\text{grad}V(r), \tag{19}$$

$$\text{grad}j(r) = 0, \tag{20}$$

where $j(r)$ is the local current density, $V(r)$ is electrical potential, and $\sigma(r)$ is electrical conductivity or admittance characteristic for the material filling unit number r. The applied algorithm uses a modification of the method introduced by Kirkpatrick [141] to find mean parameters of the heterogeneous network. To find the impedance

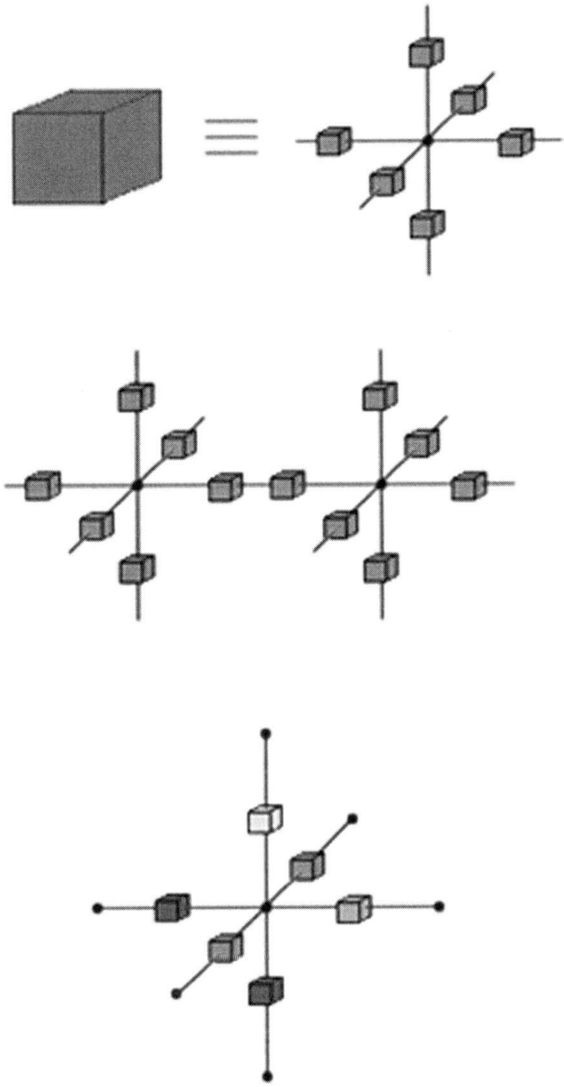

Fig. 23 Conversion of cubic units into resistivities

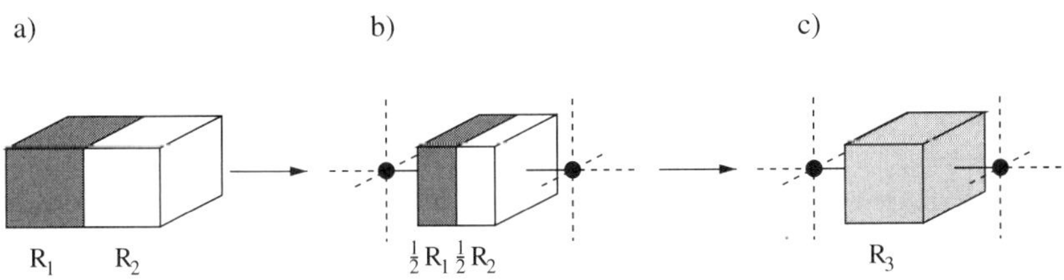

Fig. 24 Addition of the branch resistivities

of the sample, the overall current must be calculated for a given test potential difference applied to the network. The current value is equal to the one found for any of the perpendicular layers of nodes in the sample. This value is found as a sum of currents in all branches located between nodes belonging to two neighboring layers. This observation is valid only in the stationary condition when the first Kirchhoff's law is obeyed for each node (21).

$$\sum_i I_i = 0 \tag{21}$$

The network balancing is obtained by the attribution of proper potential values to all nodes in the network. Initial potential values must be found as a starting point for the calculations. All the nodes located in the first and the last layer and, thus, attributed to the electrodes, are fixed to potential values equal 0 and U respectively. The values of the intralayer branch impedances in these layers are equal to zero and the potentials are time independent. Between them monotonic change of the potential is maintained by creating a set of equipotential layers of nodes. The value of the potential change between two neighboring layers is at this step constant. The initial potential distribution obtained in this way is later modified by the iteration procedure. In each of the iteration steps, the value of each node potential is determined. A sum of currents for a node is calculated on the basis of the actual potential of the node, potentials of the neighboring nodes, and the impedances of the braches located between them. Consequently, the potential of the node is changed in a manner allowing the fulfillment of (21). If it is not possible (as the value of the new node potential must lie in the range of the neighboring nodes voltages) the modified value is chosen to balance the node in the best possible manner. After performing the potential shifts for all nodes the whole procedure is repeated until the voltage distribution is stabilized. Consequently, the stationary state of the branch currents for all of the nodes should be reached in this way. The main disadvantage of the procedure described above lies in the fact that a quick stabilization of the voltage distribution is reached, unfortunately, in a way which prevents global current stabilization. Therefore, an additional subprocedure was created. It takes into consideration the interlayer currents forcing an additional modification of the node potentials in the beginning of the iteration process.

The total equalization of the currents in a 100^3 network is obtained in 10^3–10^4 iterations [152] (see Table 8). This value is in the upper limit range, taking into consideration the acceptable calculation time. Thus, for the used size of the matrix equal to 300^3 there was a need to change the stop criteria. The analysis of current changes for the following iteration steps in connection with an extrapolation procedure to obtain the final current value of the balanced network leads to at least one order of magnitude decrease in the needed calculation time allowing for the calculations of large samples (see Fig. 25).

Table 8 Progress of the iteration process for a 100^3 RRN matrix

Step no.	I_{max}	I_{min}	I_{av}	ΔI	Nodes (%)
2	154.174	102.530	126.553	40.81	1.21
10	150.938	105.884	125.227	35.98	1.49
20	148.090	106.999	124.612	32.98	1.51
50	141.927	109.488	123.712	26.22	1.76
100	135.169	113.027	122.895	18.02	1.71
200	127.841	117.293	121.094	8.64	1.53
300	124.483	119.354	121.765	4.21	1.40
400	122.912	120.415	121.623	2.05	1.47
500	122.173	120.942	121.560	1.01	3.86
600	121.825	121.203	121.530	0.51	5.67
700	121.660	121.332	121.517	0.27	12.53
800	121.584	121.398	121.511	0.15	24.59
900	121.550	121.433	121.508	0.10	56.39
1000	121.538	121.452	121.507	0.07	73.52
1500	121.526	121.485	121.506	0.03	94.55
2000	121.516	121.496	121.506	0.02	97.84
2500	121.511	121.501	121.506	0.01	99.12
2820	121.509	121.503	121.506	0.00	99.36

8.4 *Comparative Study of Results*

Despite the similarities of the two presented approaches, one of the main differences between EMT and RRN lies in the fact that assuming identical electrical parameters of the present phases and identical fraction of the dispersoid, the first model uses only a single parameter (t/R) to define spatial relations between the grains and the shells, while the other can independently define the grain diameter ($d = 2R$) and the thickness of the amorphous shell t. Thus, a sample single from the point of view of EMT can be in RRN described as a whole family of materials holding the same t/R value. The EMT model allows for conductivity prediction for samples of very high filler concentration predicting the negative percolation phenomenon. In contrast, the RRN model produces results only up to about 40 vol% of the filler. For higher ones the problem of the next grain location occurs. Figure 26 [153] shows data concerning the number of trials needed to locate the next grain in a matrix for increasing amount of the filler. One can easily observe that above about 40% of this value increases rapidly to more than a hundred making further filling of the matrix practically impossible.

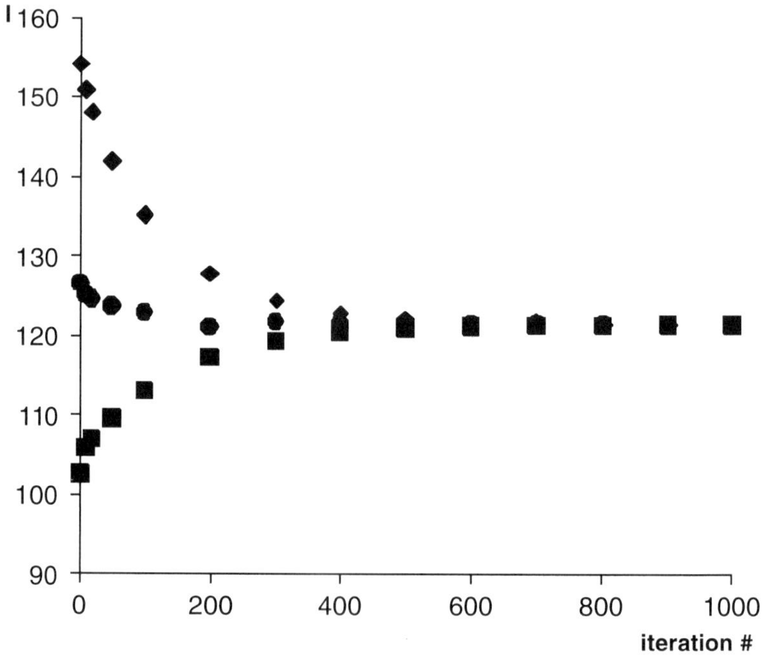

Fig. 25 Progress of the iteration process for a 100^3 RRN matrix in terms of the average node current stabilization: (*filled diamond*) maximal current values, (*filled circle*) average current values, and (*filled square*) minimal current values

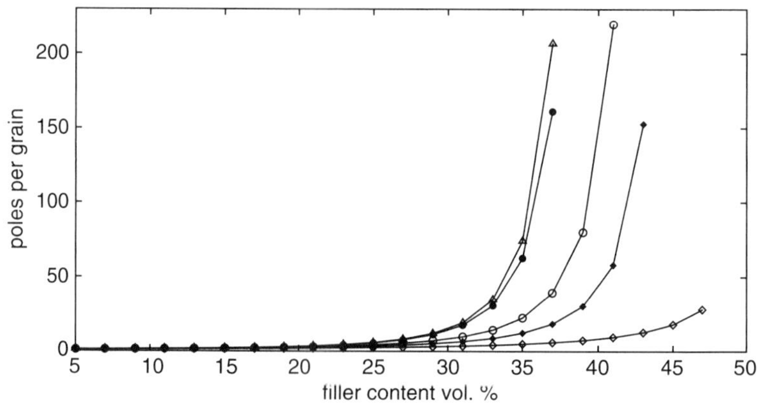

Fig. 26 The dependency of the number of poles needed to put one grain into the matrix as a function of the grain volume fraction for different grain sizes (matrix size equal to 900^3 units): (*open diamond*) $d = 0.75\,\mu m$, (*filled diamond*) $d = 1.25\,\mu m$, (*open circle*) $d = 1.75\,\mu m$, (*filled circle*) $d = 2.25\,\mu m$, and (*open triangle*) $d = 2.75\,\mu m$

For a typical set of the RRN simulations an assumption of the constant t value (independent of the grain size) was made. The two-parameter model allows to introduce this much more realistic condition contrary to the constant t/R assumption typical for EMT calculations. The basis of this computational scheme can be easily explained by the fact that the strength of the grain–matrix interaction and, thus, the depth of the polymer microstructure change measured from the surface of the grain are only a function (for the same matrix) of the grain material and the grain surface condition (surface area, hydrophobic/hydrophilic, surface groups). Therefore, in this type of calculations the t/R value is not constant and decreases with the increase of the grain size.

The dependencies of the computed values of conductivity on the additive amount are shown in Fig. 27. The curves depicted were plotted for a set of grain sizes varying in 3–11 μm range. Contrary to the previously presented results, a constant value of shell thickness equal to 5 μm was assumed instead of the constant t/R parameter value. The size of the virtual sample was equal to 300^3 units. Observing the plots, one can easily find that for almost the whole range of the filler grain size, the computed conductivity values increase with the filler volume ratio to reach a maximum and later drop again. A strong dependency between the grain size, maximal conductivity, and the curve shape can be observed. The smaller is the grain diameter, the higher is maximal conductivity. Both the increase of conductivity and the later decrease become more abrupt when the grain diameter becomes smaller. Additionally, the filler concentration to which maximal conductivity is

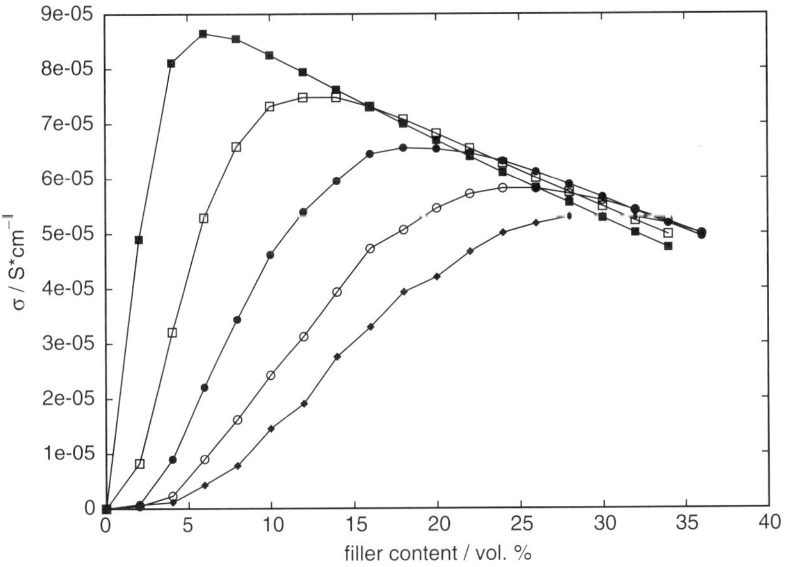

Fig. 27 The dependency of the calculated conductivity of the sample ($(PEO)_{10}$ $LiClO_4$–PAAM) as a function of the filler volume fraction for the constant t value equal to 5 μm and different grain sizes (isotropic shell conductivity distribution): (*filled square*) d = 3.0 μm, (*open square*) d = 5.0 μm, (*filled circle*) d = 7.0 μm, (*open circle*) d = 9.0 μm, and (*filled diamond*) d = 11.0 μm

attributed shifts to higher filler contents with the increase of the grain size. For the largest analyzed grain, the maximum was out of the studied range of filler contents. Comparing the EMT and RRN results, one can say that both methods produce the results being in semiqualitative agreement when taking into consideration samples of similar *t*/*R* value.

Figure 28 reveals some limitations of the RRN approach related to the ambiguous meaning of the *t*/*R* parameter. The depicted conductivity dependencies are gathered for the constant filler grain diameter $d = 5\,\mu m$ and for different shell thicknesses (t). Similar to the previous case, an increase of *t*/*R* this time reached by a change of shell thickness leads to an improvement of the maximum conductivity value with simultaneous shifting the filler contents for which the maximum is reached to lower values. Also the curve shape changes and the slope of the increasing branch of the dependency are higher. The main discrepancy can be observed for the decreasing branch of the dependency. The set of the values predicted for the samples containing more filler than is needed for reaching the conductivity maximum has a slope independent of the assumed t value. Additionally, in this region the curves for all *t* values overlay one another. This can be attributed to the fact that above the conductivity maximum all polymeric material present in the sample should belong to grain shells. Thus, the number of the present phases is reduced and the sample is filled only with grains and shells. The only occurring change of phase composition is related to polymer dilution by the addition of excessive filler grains. Thus, the amount of conductive polymeric material in the sample is independent of

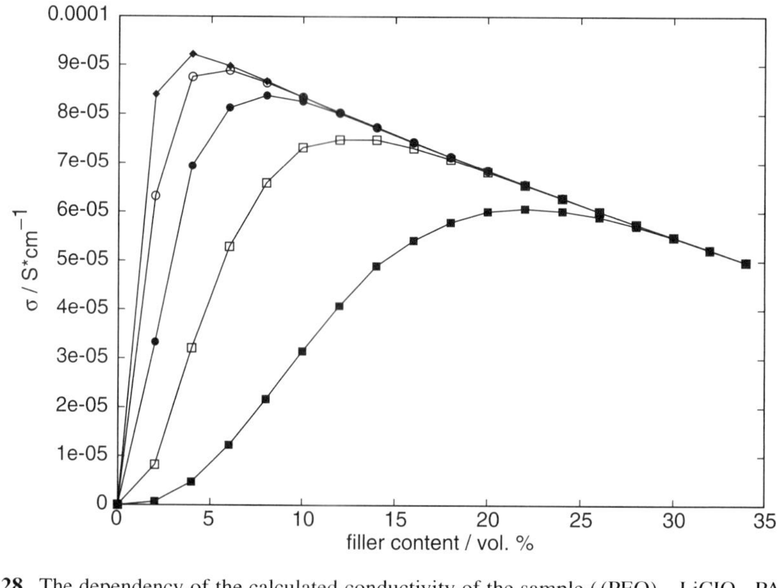

Fig. 28 The dependency of the calculated conductivity of the sample ($(PEO)_{10}$ $LiClO_4$–PAAM) as a function of the filler volume fraction for the constant *d* value equal to $5\,\mu m$ and different shell thicknesses (isotropic shell conductivity): (*filled square*) $t = 3.0\,\mu m$, (*open square*) $t = 5.0\,\mu m$, (*closed circle*) $t = 7.0\,\mu m$, (*open circle*) $t = 9.0\,\mu m$, and (*closed triangle*) $t = 11.0\,\mu m$

the *t* parameter and related only to the amount of the filler added. This observation also confirms the assumption that the model utilizing constant thickness of shell type of dependency is closer to reality.

The comparison of results obtained by means of the EMT and RRN approach for $(PEO)_{10}LiClO_4$–PAAM system was done upon the assumption of the constant *t*/*R* parameter for the whole filler concentration range. The average value of the *t*/*R* parameter was equal 0.6 for both EMT and RRN calculations. The above described EMT model was used in this case yielding a good fit to the experimental data. A simulated curve (see Fig. 29) was calculated assuming the following parameters: $\sigma_1 = 10^{-4}$ S cm^{-1} (conductivity of the shell assumed on the basis of ambient temperature conductivity of the fully amorphous analog of PEO (random EO–PO copolymer)), $\sigma_2 = 10^{-12}$ S cm^{-1} (conductivity of the filler), and $\sigma_0 = 6 \times 10^{-7}$ S cm^{-1} (experimentally found value for the PEO–$LiClO_4$ system). The only difference between the experimental and the EMT fitted data is observed for the electrolyte containing 50 vol% of the filler. This effect can be explained by changes of the *t*/*R* parameter with an increase of the amount of the filler added. The *t*/*R* parameter can be assumed to be almost constant up to a volumetric concentration of the filler equal to V_2'' (i. e. to the volume fraction at which the composite units filled the total volume of the electrolyte). For volume fractions of the filler exceeding V_2'',

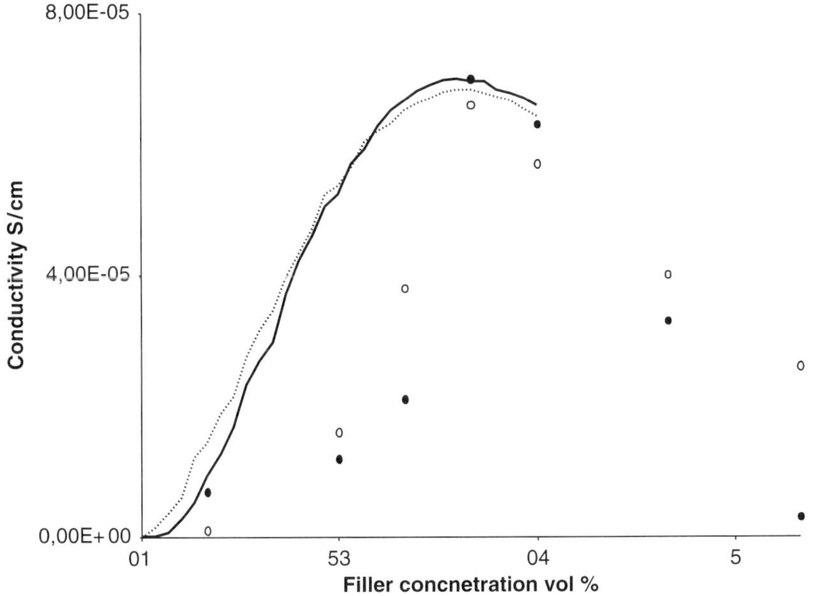

Fig. 29 Comparison of measured values (*closed circle*) of ionic conductivity of $(PEO)_{10}$ $LiClO_4$–PAAM system measured for various PAAM volume concentration and theoretical values obtained from calculations for the same system by means of the EMT (O) (*t*/*R* ratio is equal to 0.6), and RRN (t = 3 µm and d = 5 µm) (*solid line*) and RRN calculations (t = 5 µm and d =9 µm) (*dashed line*)

the t/R ratio starts to decrease due to an increase in the concentration of the insulating grains. Therefore, the experimentally measured conductivities fall faster than the ones calculated by the model that assumes a constant value of t/R. Since in the case presented t/R varies only slightly the effect is not very pronounced. Another explanation of the conductivity drop is a decrease in the surface layer conductivity with an increase in the grain concentration due to a disturbance of the segmental chain motions by the stiff dispersoid. The RRN curve (see also Fig. 27) obtained for similar input parameters (t = 3 μm and d = 5 μm or t = 5 μm and d = 9 μm) is limited to the filler concentration range 0–30 vol% due to the previously mentioned limitations of the method. As one can see, the model is in a good fit as it comes to the maximal conductivity value and the maximum location yielding high simulated conductivities for the lower concentration range.

For the EMT calculations the t/R parameter has a fixed value for all grains in the sample, while the additional feature of the RRN model lies in the fact that a more realistic image of the sample can be built by the introduction of the grain diameter dispersion. Figure 30 shows the values of conductivity simulated for three average grain diameters with over-imposed Gaussian distribution. In general, the shape of the dependence is similar to the one obtained for simulations performed in the constant grain diameter regime. When comparing curves depicted in Figs. 27 and 30 one can easily see that the introduction of grain size distribution leads to a significant shift of the conductivity maximum to higher filler concentrations (about 10 vol%). Additionally, the observed maximum is flatter with the decrease of the

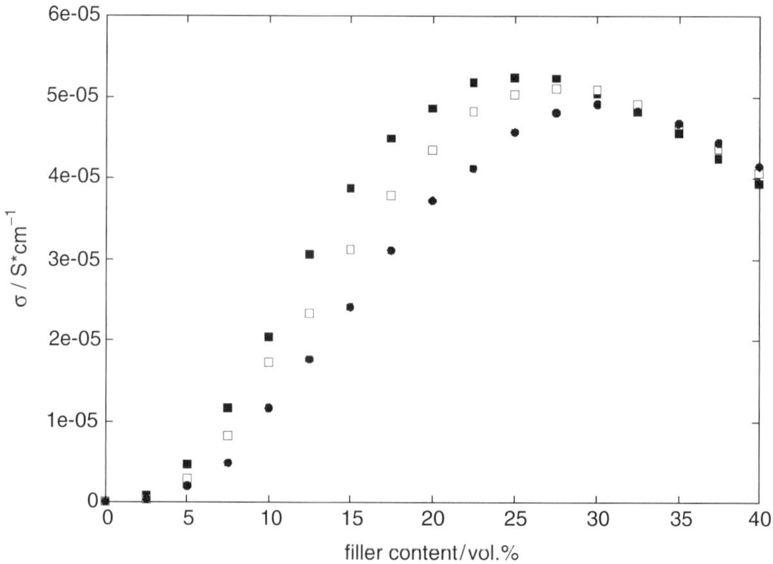

Fig. 30 A set of RRN curves for different average diameters of the filler grain for $(PEO)_{10}$ $LiClO_4$–PAAM system (t constant and equal to 5 μm): (*filled square*) d = 4 μm (d_{max} = 6 μm, d_{min} = 2 μm), (*open square*) d = 5 μm (d_{max} = 7 μm, d_{min} = 3 μm), and (*filled circle*) d = 6 μm (d_{max} = 8 μm, d_{min} = 4 μm)

highest conductivity value by factor lying in range 1.5–2. This observation shows that the introduction of a fraction of larger grains and, thus, a partial decrease of the average *t*/*R* ratio cannot be balanced by the presence of small grains.

Ambient temperature conductivities of a PEO–salt–Al_2O_3 system can be fitted according to various models and compared with experimentally measured conductivities as shown in Fig. 31. The initial estimates for adjustable parameters in these calculations were selected as $\sigma_1 = 1\times10^{-5}$ S cm^{-1} obtained by the extrapolation of moderate temperature conductivities measured for the amorphous melted PEO–NaI electrolyte to the ambient temperature region, $\sigma_0 = 1.5\times10^{-7}$ S cm^{-1} measured experimentally for the $(PEO)_{10}NaI$ electrolyte, and $\sigma_2 = 1\times10^{-20}$ S cm^{-1} which is the estimated conductivity of nonconductive grains. The value of *t*/*R* is chosen as equal to 1.16, which is the value calculated for the V_2'' concentration of the filler added on the basis of X-ray and DSC investigations. A Nan and Smith model alteration, in which the *t*/*R* ratio was calculated independently for each of the volume fractions V_2 studied, was also tested. Both EMT models based on the Nan and Smith approach agreed with the experimental data for volume fraction of the filler lower than V_2'' (the filler concentration at which the maximum of conductivity is obtained). For higher concentrations of inorganic grains, a plateau is observed in the calculated theoretical curves, whereas the experimentally measured values of

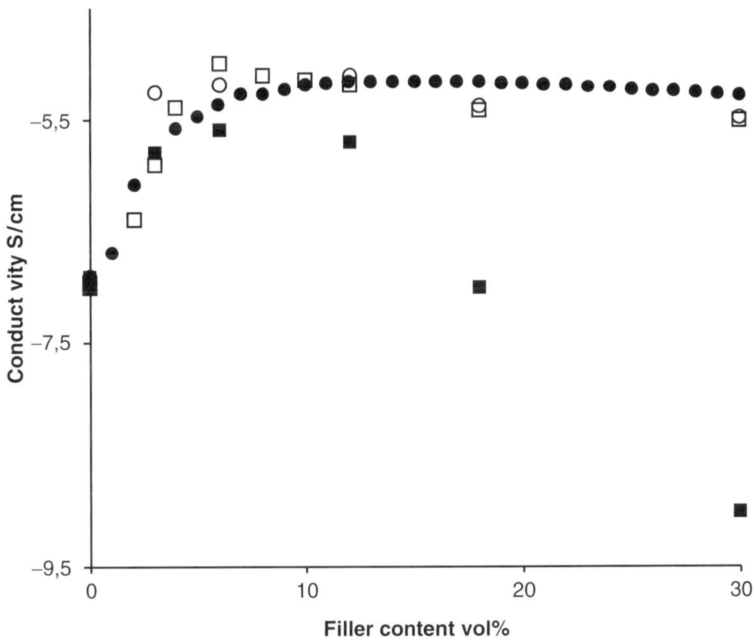

Fig. 31 A comparison of experimental data (logarithmic scale) for the $(PEO)_{10}$ NaI–SiO_2 system (with simulations obtained by the EMT and RRN models. (*Filled square*) experimental data, (*open square*) basic EMT model, (*open circle*) EMT model with various *t*/*R*, and (*filled circle*) basic RRN model

conductivity decreased in this filler concentration range. Even considering the drop in the t/R ratio, the calculated theoretical values are still much higher than the experimental ones. A basic RRN model starting from similar input parameters gives an even flatter conductivity curve. The observed maximum is even wider than for the EMT data and the negative slope observed above the conductivity maximum is a bit smaller. The data obtained are also in a semiquantitative agreement with the experimental values only for low filler concentrations.

Contrastively, Fig. 32 demonstrates results of simulation for the same system obtained by applying the corrected EMT and RRN models. The correction of the EMT approach is based on both the correction of the amorphous phase conductivity and change of the t/R value for samples of changing composition. For several systems fits of experimentally measured T_g values to (17) have been found. For constant concentration of the dopant salt, it was found that T_g is well approximated by the simplified equation having the following form:

$$T_g = K_0 + K_1V_2 + K_2V_2^2 \tag{22}$$

or namely for $(PEO)_{10}NaI–SiO_2$

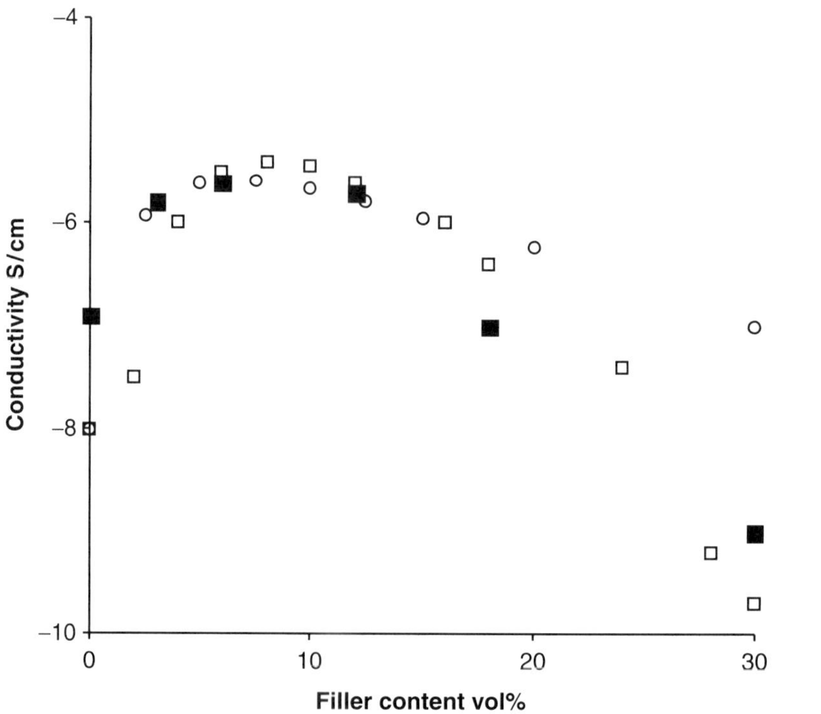

Fig. 32 A comparison of experimental data (logarithmic scale) for the $(PEO)_{10}$ $NaI–SiO_2$ system with simulations obtained by the EMT and RRN models. (*filled square*) experimental data, (*open square*) corrected EMT model, and (*open circle*) corrected RRN model

$$T_g = K_0 + K_1V_2 + K_2V_2^2 \tag{23}$$

K_0 was found equal to the glass transition temperature of the pristine polymeric electrolyte without the dispersed inorganic phase, K_1 describes the influence of the dispersoid added to the T_g of the composite system and K_2 is connected with the polymer-dispersed grain–salt interaction. The results show the narrow plateau region with a conductivity maximum followed by an abrupt decrease at higher concentrations of the filler. Therefore, the model proposed is better than the original one based on the basic Nan and Smith approach. The strength of the improved one lies mainly in the consideration of the effects of the flexibility of the amorphous phase as well as the variation of the *t*/*R* ratio. Both introduced corrections find confirmation in the experimental data and are based on the empirical parameters found for a set of samples with various compositions.

Similar corrections can be introduced to the RRN model. A distribution of conductivity within the highly conductive shell can be incorporated into the developed model. The introduction of this feature allows taking into consideration both the stiffening effect of the filler on the amorphous phase of the polymer and the decrease of the strength of interaction between the matrix and the filler at a large distance from the grain surface. The suggested profile of conductivity is built on the assumption that the stiffening effect is caused by the interaction of the polymer chains with the grain surface and can be observed only in its close vicinity while the effect of the crystallization hindering reaches farther. In this situation, starting from the surface of the grain the local conductivity value increases as the stiffening effect vanishes (see Fig. 33). At a given distance (being equal to half of the shell

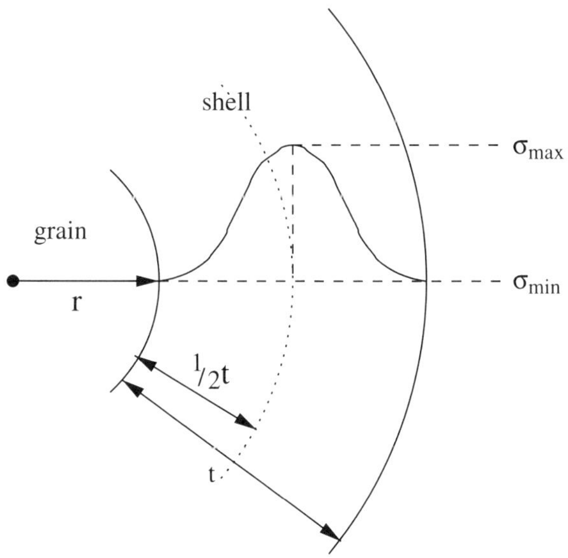

Fig. 33 The visualization of conductivity distribution within the shell

thickness as a symmetrical Gaussian function was used to describe the changes) a maximum is reached. At this point the conductivity value should be similar to one observed for the pure amorphous phase not being in contact with the hard filler surface. For larger distances a consequent decrease in conductivity should be assumed as the effect of the polymer phase disordering also vanishes. Finally, because of the continuity of the material on the outer border of the shell, the local properties of the electrolyte should be identical with the ones of the pristine matrix. For these particular simulations, the minimum and maximum conductivity within the shell were equal to $\sigma_{min} = 10^{-7}$ S cm^{-1} and $\sigma_{max} = 10^{-5}$ S cm^{-1}, respectively. The application of the correction (without taking into consideration the change of the average t/R with the increasing filler content) allows to predict conductivity data in quantitative agreement with the experimental point in the filler content range 0–20 vol%. The last experimental point $V_2 = 0.3$ yields a conductivity value lower than that predicted. This discrepancy can be explained first with a strong change of the t/R for this composition of the sample, second by the fact that the RRN model works here close to its limits and, finally, by the fact that here a slightly different conductivity profile should be used locating the conductivity maximum closer to the grain surface as an average shell belonging unit can be stiffened by more than one grain. The model assumes that total stiffening is as strong as the strongest of all involved grains, while in this situation an additive function cumulating interactions of all grains in the neighborhood should be used instead.

Finally, some interest should be focused on the comparison of a.c. conductivity data produced by EMT and RRN models. By substituting the d.c. conductivity value by the complex, frequency-dependent a.c. permitivity the models can be extended to the a.c. range. The calculations were performed using the previously constructed models. The frequency dependence of conductivity calculated for PEO–NaI–Θ-Al_2O_3 system (for the fixed volume fraction of the dispersed phase) can be always depicted for the EMT simulation as a single time constant process (see Fig. 21). The electrical properties are insensitive to changes in frequency at low frequencies and strongly depend on the change in ω at high frequencies. According to the EMT model, polymeric electrolytes behave as ideal dielectrics at sufficiently high frequencies independently of their phase compositions.

The RRN calculations are only in the partial agreement with the observations above. For this model not all the samples behave as ideal dielectrics and depressed and merged semicircles are observed. A second time constant can be found in some impedance spectra of the slightly modified samples for filler amount values around the percolation threshold (see Fig. 34). The conductivity maximum can be attributed to the total conversion of the sample volume (disregarding the units belonging to the grains) into highly conductive amorphous shells. No more spatial units can be attributed to the pristine polymeric matrix. In the a.c. image, this range can be described as a well-defined single semiarc-type spectrum (see Fig. 34). The following conductivity drop is related to a decrease in the number of clusters containing the polymeric material due to the dilution effect of excessive filler grains. In the a.c. spectra, this phenomenon can be attributed to the depression of the previously observed single semiarc.

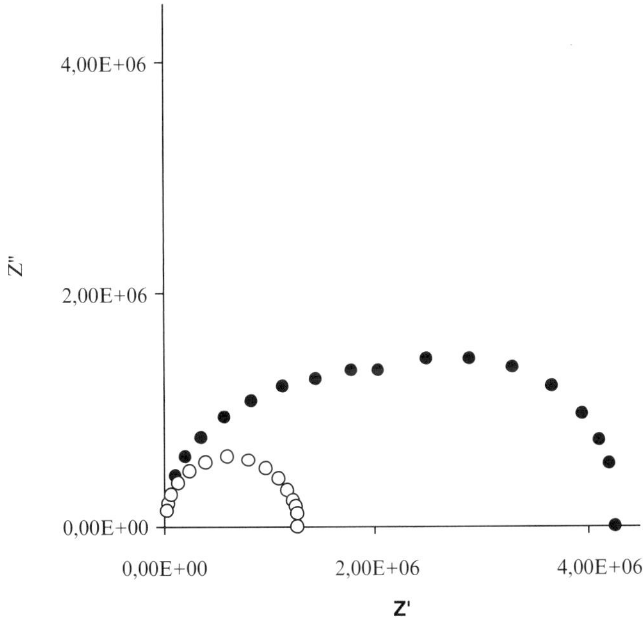

Fig. 34 A comparison of simulated high frequency dielectric response of the 10 vol% composite sample (nonstiffening filler): (*filled circle*) grain diameter $d = 5\,\mu m$, shell thickness $t = 3\,\mu m$ (sample close to the percolation threshold), (*open circle*) grain diameter $d = 5\,\mu m$, shell thickness $t = 7\,\mu m$ (the sample far away from the percolation threshold)

9 Molecular Scale Models

At the beginning, one must notice that molecular scale models cannot deal with the composite electrolyte as a whole complete system due to the fact that its complexity is far beyond the approach limitations. Quantum mechanics is even more rigorous as it comes to limiting the number of atoms in the studied system to about few hundred while molecular dynamics can play with sets of thousands. Nevertheless, both methods can deliver valuable data on ion dynamics, ionic pair formation, salt solvatation processes, vibrational spectra prediction, etc. The behavior of polymer electrolytes in lithium batteries is reviewed in the context of molecular scale models as well as on the system scale. It is shown [154] how the molecular structure of the electrolyte strongly influences ion transport through the polymer as well as across the interfaces and determines the values of a number of parameters needed for system models that can predict the performance of a battery. The molecular level models may be combined with the rheological models to provide workable models of interfaces and bulk electrolyte dynamics that can, in turn, be used to provide a more accurate level of performance prediction than the system models. This connects molecular structure with battery performance and guides the design and synthesis of new and better materials.

9.1 Quantum Mechanics Applied for Polymer–Salt–Filler Complexes

A small review of the use of computer simulations and especially the use of standard quantum-mechanical ab initio electronic structure calculations to rationally design and investigate different choices of chemicals/systems for lithium battery electrolytes is presented in [155].The presented case studies show that the calculations can reveal the barriers of conformational changes for polymers used to make polymer electrolytes. Finally, they claim that the use of modeling techniques can now follow almost any new experimentally launched concept and contribute with accurate and cost-effective data to assist interpretation. They show application of the QM methods in vibrational spectra prediction, preparation of the field potentials for molecular dynamics calculations, design of new salts, ionic pair formation studies, and analysis of anion filler interaction exemplified by the BF_4^- anion and TiO_2 particle surface. Johansson in [156] reports a first attempt of investigating local surface interactions affecting macroscopically observed properties using a rutile TiO_2-cluster via ab initio Hartree–Fock (HF) and density functional theory (DFT) (B3LYP) type calculations. As an exploratory work, a small $Ti_{11}O_{22}$ atomistic cluster was used being embedded in point charges (PCs) and the (110) surface for the adsorption process. Out of lithium cations (Li^+), tetrafluoroborate anions (BF_4^-), and dimethyl ethers ($O(CH_3)_2$), BF_4^- is found to be the preferred adsorbate (see Fig. 35). The ether has a comparable energy of adsorption when adjusting for bidentate coordination. The present results support the possibility of creation of new diffusion pathways for the lithium ion at the nanoparticle surfaces and emphasize the possible role of anion adsorption. The vibrational spectra change upon addition of TiO_2 nanoparticles (25 nm) to lithium–diglyme were studied in [157]. This process causes only minor changes to molecular vibrations of the diglyme species in the spectral region that is diagnostic of the diglyme conformation. In the low-frequency region, however, the spectrum of lithium diglyme without nanoparticles shows evidence of considerable dispersion, which is removed upon the addition of nanoparticles. QM calculations show here that nanoparticles improve the conductivity of polymer electrolytes via an interaction with the ions rather than with the polymer.

Later Johanson et al. show that interaction of anions and neutral species can be simulated as a starting point for the search of new supramolecular receptors [158]. The complexation of simple anions ($F^- \leftarrow$ and Cl^-) to different neutral species being anion-coordinating agents has been here studied using electronic structure calculations. The obtained changes in equilibrium constants for salt dissolution reactions in different typical electrolyte systems are also reported. In addition the lithium ion affinities of the obtained anionic complexes have been calculated. Using the present results, the authors discuss strategies for future usage of anion complexing agents and make recommendations of salt and agent combinations for better lithium battery electrolyte performance. Another type of cryptand compounds together with the implications of using these cryptands in polymer electrolytes is studied in [159], where complexation of different simple anions (F^-, Cl^-,

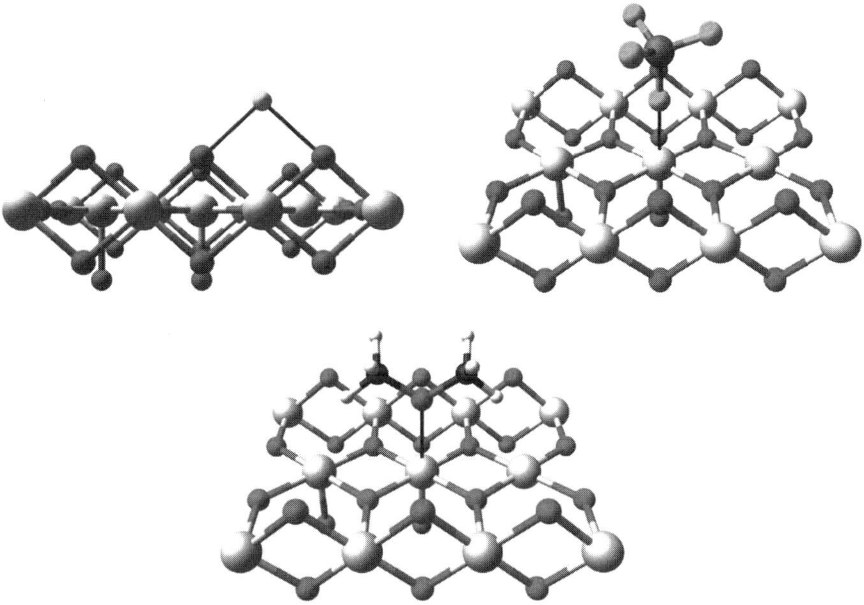

Fig. 35 A calculated structure of the TiO_2 absorbate with (**a**) Li^+ cation, (**b**) BF_4^- anion, and (**c**) $O(CH_3)_2$ being a model molecule for PEO

BF_4^-, and ClO_4^-) with a neutral cryptand, with neutral B–N dipoles as functional units and modifications thereof has been studied using ab initio Hartree-Fock calculations. F^- is the most preferred anion for all cryptand versions, except for the largest cavity cryptand for which Cl^- is preferred. The formed complexes show very low lithium ion affinities in general, much lower than those for anions such as PF_6^- and TFSI (by up to 50%). The lowest lithium affinity is obtained for the combination of the smallest cryptand and the F^- anion.

Additionally, a quantum chemical study of the binding of Li^+ cation to polyalkyloxides has been carried out [160]. The lithium cation interaction with three polyalkyloxides (polyethylene oxide (PEO), polytrimethylene oxide (PTMO), and polypropylene oxide (PPO)) has been investigated using the ab initio molecular orbital theory at the HF/6–31G level with molecular models for the polymers. Coordination by one to six oxygens was considered. In addition, higher level calculations were carried out using G3(MP2) theory for coordination of Li^+ by one oxygen. For coordination of lithium by one oxygen, the binding energy ordering is PTMO > PPO > PEO, with PTMO having the largest lithium cation affinity. The same ordering is found for larger coordination numbers with the exception of coordination by six oxygens, where the ordering changes due to steric interactions.

The electronic structure and vibrational spectra of the metal ion–PEO oligomer represented by $M–CH_3O(CH_2CH_2O)_nCH_3$ (n = 2–7) (M = Na^+, K^+, Mg^{2+}, and Ca^{2+}) complexes can be obtained by employing the ab initio and hybrid density functional methods [161]. Structural and spectral impacts of the chain length of the PEO oligomers on the alkali- and alkaline earth-metal ion coordination in different chemical environments have been presented. In metal ion–PEO oligomer ion-pairs, the Mg^{2+} ion binds strongly to ether oxygens of polyether complexes. The density functional calculations predict that the heptaglyme ($n = 7$ in the above series) binds strongly to the metal ion. The frequency shifts of characteristics vibrations induced by the metal ion-coordination in $CH_3O(CH_2CH_2O)_nCH_3$–M ($n = 2$–7) complexes were also discussed.

Also the crystalline solid polymeric electrolytes are studied by the QM methods. In [162] the lithium ion transport in crystalline $LiPF_6PEO_6$ has been modeled by ab initio electronic structure calculations. It has been proved that the activation energy values obtained using a molecular model corresponds excellently to experimental values from conductivity measurements. On the basis of the present results, structural modifications in order to further enhance the single ion conductivity are suggested.

Finally, the interaction between quantum mechanics and molecular dynamics can be presented [163] on the basis of the forcefield development for a novel salt applied in crystalline PEO-based complexes. Both ab initio calculations and available empirical data have been used. The forcefield has been verified in simulations of the crystal structure of Li_2SiF_6 in two different space groups: P321 and P3m1. The use of MD simulation to assess the correct space group for Li_2SiF_6 shows that it is probably P321.

9.2 *Molecular Dynamics of Polymeric Systems*

Contrastively to previously mentioned quantum mechanics simulations, molecular mechanics do not utilize the Schrödinger equation but treats a system as being governed by Newton's dynamics rules. In general all atoms and bonds are represented as weights and springs. Interactions between particles are approximated by a classical potential landscape and all atoms interact by a set of forces. The functional form of the forces and parameters is called a forcefield. In molecular dynamics, which is an extension of the presented approach, the laws of classical mechanics are used to compute the motion of the particles. The forcefield is defined here as a function of energy to distance (24) (potential energy E_{total}) with main constituents coming from E_{valence} (25), $E_{\text{crossterm}}$, and $E_{\text{non-bond}}$ (26):

$$E_{\text{total}} = E_{\text{valence}} + E_{\text{crossterm}} + E_{\text{nonbond}} \tag{24}$$

The valence term is defined as

$$E_{\text{valence}} = E_{\text{bond}} + E_{\text{angle}} + E_{\text{torsion}} + E_{\text{oop}} + E_{\text{UB}}, \tag{25}$$

and the nonbond interactions are the sum of

$$E_{\text{nonbond}} = E_{\text{vdW}} + E_{\text{Coulomb}} + E_{\text{Hbond}}, \tag{26}$$

where $E_{\text{crossterm}}$ is related to crossterm interactions, E_{oop} to the out-of-plane interactions, E_{vdW} represents Van der Waals forces, E_{UB} the Urey–Bradley term for interactions between atom pairs involved in 1–3 configurations, and E_{Hbond} represent the hydrogen bonding.

In general this approach is less accurate but less computer time consuming in comparison with QM and so it can be used for studies of larger systems. Some works [164, 165] were done in comparison with results obtained by both ab initio quantum mechanics (QM) and classical molecular dynamics (MD) simulations. They can be employed to model an electrolyte composed of polyphosphazene (PP), lithium triflate ($LiCF_3SO_3$), and water. Structures and energetics are systematically studied by QM for binary complexes of Li^+, $CF_3SO_3^-$, and Li^+ and $CF_3SO_3^-$ with water or PP fragments, and for their ternary combinations. Classical MD simulations qualitatively reproduce the results of QM calculations and provide details about the Li^+ distribution in a larger system. The results of the QM and classical MD calculations suggest a model for the microstructure of the polyelectrolyte in which Li^+ interacts most strongly with the backbone nitrogen of PP, somewhat weaker (and comparably) with ether oxygens on PP side chains and with water oxygens. This indicates that Li^+–N interactions should significantly affect migration of Li^+ in PP polymer electrolytes. The calculated coordination patterns of Li^+ with the poly(ethylene oxide) model (ethylene oxide)$_6$ [(EO)$_6$] agree with experimental results in which Li^+ is strongly coordinated with five oxygens in PEO.

The changes of local cation dynamics upon the addition of the TiO_2 nanoparticles into the polymeric electrolyte were studied by van Eijck et al. [166]. They use a model system that consists of NH_4I dissolved in a deuterated oligomer of PEO so that they can use inelastic and quasi-elastic neutron scattering to study rotational dynamics of the NH_4^+ ion without the complicated signal from the oligomer. The analysis of the local rotational potential of the NH_4^+ ion by comparing the experimental data with the H atom trajectories from a molecular dynamics simulation reveals an overall reduction of the interaction between the NH_4^+ ion and the ether-oxygens of the diglyme. Although the extension of the results obtained from the model cation diffusion in PEO cannot be made directly, the authors claim that the results are entirely consistent with the observed increase in conductivity upon the addition of nanoparticles. The experimental neutron-scattering data and the MD simulations are all consistent with the presence of groups of two basic environments for NH_4^+ ions, at least in the diglyme system. On a timescale of about 10 ps these environments are visited by all NH_4^+ ions and the addition of nanoparticles softens the more hindered environments making the two groups more similar to each other. The results obtained strongly suggest that the observed increase in conductivity of polymer electrolytes on the addition of nanoparticles is due to a reduction of the interaction between ether-oxygens and the cations, rather than an increase in the polymer dynamics. Similar problems were studied in [167, 168] where the amorphous $LiBF_4(PEO)_{20}$ system was simulated alone and containing a ca. 1.4 nm diameter Al_2O_3 nanoparticle and in just a position with a ca. 6.5-nm-thick α-Al_2O_3 slab at a nominal temperature of 293 K by molecular dynamics (MD) methods. The Li-ion mobility in the poly(ethylene oxide) (PEO) host was found to increase on the addition of the nanoparticle; the effect is also noticeable for the alumina slab. This can be seen as the theoretical confirmation of the positive influence of nanoparticles on ion mobility in a PEO–salt system as observed earlier experimentally. The other effects observed are related to this Li-ion mobility enhancement: the PEO forms an immobilized coordination sphere around the particle and an immobilized layer at the surface of the α-alumina slab. This observation concerning the behavior of the polymer host stays in good agreement with the stiffening approach proposed for the mesoscale models previously in this chapter. No Li^+ ions are found near the particle or at the slab surface. Instead, two to three unpaired BF_4^- anions are found attached to the particle within the region of immobilized PEO and at least one is found immobilized on the slab surface, leaving free Li ions in the regions away from the particle and slab surfaces. No more than 60% of the Li^+ ions form ion pairs and ion clusters in the regions away from the particle surface and up to 87% of Li^+ ions form ion pairs and ion clusters in the regions away from the slab surface. In [168] a $(PEO)_{10}LiBr$ system was studied. As a result a geometrical arrangement of the PEO chain around a 1.4 nm particle of Al_2O_3 in $LiBr(PEO)_{10}$ was obtained (see Fig. 36), showing the detailed structure around the Li^+ ion bound to the particle and an example of an ion cluster.

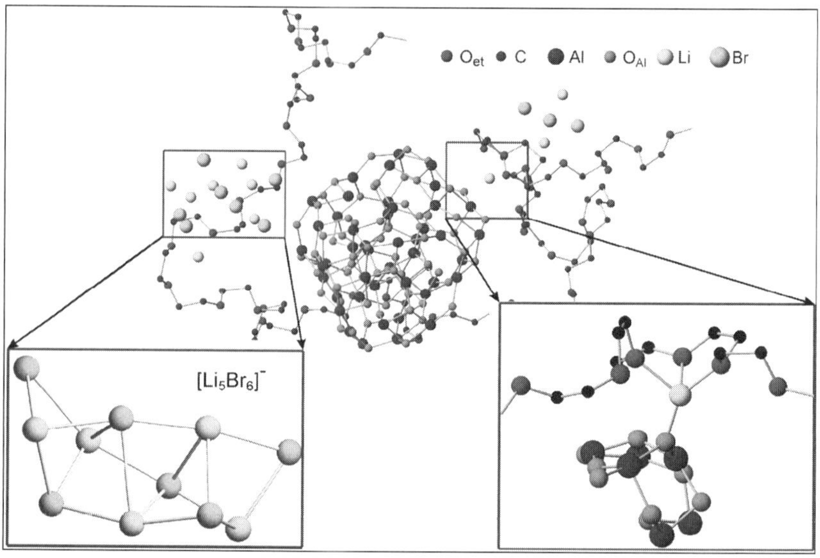

Fig. 36 A geometrical arrangement of the PEO chain around the 14 Å particle of Al_2O_3 in $LiBr(PEO)_{10}$

Another model system for the nanocomposite electrolyte was studied in [169]. Here the MD simulations were performed for polyethylene oxides (PEO) sandwiched between two TiO_2 surfaces revealing that the PEO density was significantly perturbed by the TiO_2 surfaces. It was found that the addition of a nanoparticle with soft-repulsion interactions with PEO resulted in the formation of a PEO interfacial layer with reduced PEO density but increased ion concentration. The results show that PEO and ion mobility in the interfacial layer were higher than those of bulk SPE by 20–50%, whereas conductivity of SPE's increased only by 10%. On the basis of the presented examples generally it must be concluded that with the growing power of the computational tools, molecular models of composites can deliver important information leading to better understanding of the nature of experimentally observed phenomena.

10 Summary

Proper design of inorganic or organic filler, the properties affecting ion transport phenomena, can be modified in a variety of polyether-based electrolytes. The addition of fillers results in an increase in ionic conductivity for amorphous and crystalline polymeric matrices, enhancement in lithium transference number, and, on top of this, with stabilization of lithium electrode–polymer electrolyte interfacial resistivity. All the models used so far are generally equivalent for the same set of systems studied. In the present paper, we have tried to relate the increase in conductivity and lithium transference number to the changes in ion–ion and ion–polymer interactions caused by the filler. The first class of reactions seems to be particularly important in the explanation of phenomena occurring in polyether-based electrolytes. Moreover, the experimentally obtained data can be confirmed by various theoretical models based on both mesoscale and molecular scale approaches. The calculated data not only allow for better understanding of the studied material properties but also enable a virtual material design process to be realized in a computer memory. The feedback between the experimental observations and theoretical predictions gives hope for both deeper knowledge and system property tailoring in the near future.

References

1. P.G. Bruce (Ed.), in: Solid State Electrochemistry, Cambridge University Press: Cambridge, UK, 1995.
2. B. Scrosati (Ed.), in: Applications of Electroactive Polymers, Chapman and Hall: London, 1993.
3. J.A. Connor, in: F.M. Gray (Ed.), Polymer Electrolytes, The Royal Society of Chemistry: UK, 1997.

4. F. Croce, R. Curini, A. Martinello, L. Persi, F. Ronci, B. Scrosati, R. Caminiti, J. Phys. Chem. 103 (1999) 10632.
5. F. Croce, G.B. Appetecchi, L. Perci, B. Scrosati, Nature 394 (1998) 456.
6. W. Wieczorek, K. Such, H. Wycislik, J. Plocharski, Solid State Ionics 36 (1989) 255.
7. W. Wieczorek, Z. Florjańczyk, J.R. Stevens, Electrochim. Acta 40 (1995) 2251.
8. A.J. Bhattacharyya, J. Maier, Adv. Mater. 16 (2004) 811.
9. M. Marcinek, A. Bac, P. Lipka, A. Zalewska, G. Zukowska, R. Borkowska, W. Wieczorek, J. Phys. Chem. B 1004 (2000) 11088.
10. M. Kalita, M. Bukat, M. Ciosek, M. Siekierski, S.H. Chung, T. Rodríguez, S.G. Greenbaum, R. Kovarsky, D. Golodnitsky, E. Peled, D. Zane, B. Scrosati, W. Wieczorek, Electrochim. Acta 50 (2005) 3942.
11. L. Sannier, A. Zalewska, M. Marczewski, H. Marczewska, W.Wieczorek, Patent Application B0783FR2.
12. J. Plocharski and W. Wieczorek, Solid State Ionics 28–30 (1988) 979.
13. W. Wieczorek, K. Such., J. Przyluski, Z. Florjanczyk, Synthetic Met. 45 (1991) 373.
14. W. Wieczorek, Mater. Sci. Eng. B 15 (1992) 108.
15. F. Croce, F. Bonino, S. Panero, B. Scrosati, Philos. Mag. B 59 (1989) 161.
16. B. Scrosati, J. Electrochem. Soc. 136 (1989) 2774.
17. F. Croce, S. Passerini, A. Selvaggi, B. Scrosati, Solid State Ionics 40–41 (1990) 375.
18. F. Croce, F. Capuano, A. Selvaggi, B. Scrosati, G. Scibona, J. Power Sources 32 (1990) 381.
19. S. Skaarup, K. West, B. Zachau-Christiansen, Solid State Ionics 28–30 (1988) 975.
20. S. Skaarup, K.West, P.M. Julian, D.M. Thomas, Solid State Ionics 40–41 (1990) 1021.
21. J.R. Stevens and B.E. Mellander, Solid State Ionics 21 (1986) 203.
22. C. Berthier, W. Gorecki, M. Minier, M.B. Armand, J.M. Chabagno, P. Rigaud, Solid State Ionics 11 (1983) 91.
23. C.C. Liang, J. Electrochem. Soc. 120 (1973) 1289.
24. J. Weston, B.C.H. Steele, Solid State Ionics 7 (1982) 75.
25. H. Yuankang, C. Zhushang, Z. Zhigi, W. Chaoying, C. Liquan, Chem. J. Chinese Univ. (English Edition) 2(1) (1986) 97.
26. C. Linquan, in: B.V.R. Chowdari and S. Radhakrsishna (Eds.), Materials for Solid State Batteries, World Scientific Publication: Singapore, 1986, p. 69.
27. J. Plocharski, W. Wieczorek, J. Przyluski, K. Such, Appl. Phys. A 49 (1989) 55.
28. F. Capuano, F. Croce, B. Scrostai, J. Electrochem. Soc. 138 (1991) 1918.
29. W. Gang, J. Ross, D. Brinkmann, F. Capuano, F. Croce, B. Scrosati, Solid State Ionics 53–56 (1992) 1102.
30. F. Croce, F. Capuano, A. Selvaggi, B. Scrosati, J. Power Sources 32 (1990) 389.
31. S. Panero, B. Scrosati, S.G. Greenbaum, Electrochim. Acta 37 (1992) 1533.
32. J. Przyluski, W. Wieczorek, in: B.V.R. Chowdari and S. Radakrishna (Eds.), Proc. II Int. Symposium on Solid State Ionic Devices, World Scientific Publication: Singapore, 1988, pp. 475–480.
33. J. Przyluski, Z. Florjańczyk, K. Such, H. Wycislik, W. Wieczorek, Synthetic Met. 35(1–2) (1989) 241.
34. W. Wieczorek, K. Such, J. Plocharski, J. Przyluski, in: B. Scrosati (Ed.), Proc. II Int. Symposium on Polymeric Electrolytes, Siena, 1989, Elsevier Applied Science: London, New York, 1990, pp. 339.
35. F. Croce, B. Scrosati, G. Mariotto, Chem. Mater. 4 (1992) 1134.
36. G. Sandi, R. Kizilel, K.A. Carrado, R.F. Saavedra, N. Castagnola, Electrochim. Acta 50 (2005) 2891.
37. H. Zhou, P.S. Fedkiw, Solid State Ionics 166 (2004) 275.
38. D. świerczyński, A. Zalewska, W.Wieczorek, Chem. Mater. 13 (2001) 1560.
39. P. Aranda, E. Ruiz-Hitzky, Chem. Mater. 5 (1993) 1395.
40. R.A. Vaia, K.D. Jandt, E.J. Kramer, E.P. Giannelis, Macromolecules 28 (1995) 8080.
41. M.W. Riley, P.S. Fedkiw, S.A. Khan, J. Electrochem. Soc. 149 (2002) 667.

42. Y. Uchimoto, Z. Ogumi, Z. Takehara, Solid State Ionics 40/41 (1990) 624.
43. M.R. Allock, P. Austin, J.X. Neenan, J.T. Sisko, P.M. Blonsky, D.F. Shriver, Macromolecules 19 (1986) 2508.
44. O. Olabisi, L.M. Roleson, M.T. Shaw, Polymer-Polymer Miscibility, Academic Press: New York, 1979.
45. D.J. Bannister, G.P. Davies, I.M. Wars, J.E. Mc Intyre, Polymer 25 (1984) 1291.
46. F.M.Gray, J.R. Mac Callum, C.A. Vincent, Solid State Ionics 18/19 (1986) 252.
47. A. Le Mechaute, G. Crepy, G. Marcellin, T. Hameide, A. Guyid, Polymer Bull. 14 (1985) 233.
48. T. Yamamoto, M. Inai, T. Kanbara, in: S. Hoshino and T. Takahashi (Eds.), Proc. VII Int. Conference on Solid State Ionics Hakone, November 1989, p.185.
49. K. Such, Ph.D. Thesis, Department of Chemistry, Warsaw Technical University, 1990.
50. K. Such, Z. Florjanczyk, W. Wieczorek, J. Przyluski, in: B. Scrosati (Ed.), Proc. II Int. Symposium on Polymeric Electrolytes, Siema 1989, Elsevier Applied Science: London, New York, 1990, p.119.
51. K. Such, Z. Florjanczyk, W. Wieczorek, J. Przyluski, in: S. Hoshino and T.T. akahashi (Eds.), Proc. VII Int. Conference on Solid State Ionics, Hakone November 1989, p.187.
52. Z. Florjanczyk, K. Such, W. Wieczorek, M. Wasiucionek, Polymer 32 (1991) 3422.
53. Z. Florjanczyk, K. Such, W. Wieczorek, J. Macromol. Sci. Pure Appl. Chem. A 29 (1992) 10.
54. M.L. Kaplan, E.A. Rietman, R.J. Cava, L.K. Holt, E.A. Chandross, Solid State Ionics 25 (1987) 37.
55. G. Haichun, F. Changming, Z. Yi, in: B.V.R. Chowdari, S. Chandra, S. Singh and P.C. Srivastava (Eds.), Solid State Ionics – Materials and Applications, World Scientific Publication: Singapore, 1992, p. 539.
56. J.R. Stevens, Lecture presented at the Int. Meeting on Polymer Ionics, Gothenburg, Sweden, August 1992; T. Mani, R. Mani, J.R. Stevens, Solid State Ionics 60 (1993) 113.
57. W. Wixwat, J.R. Stevens, A.M. Andersson, C.-G. Granqvist, B. Scrosati (Ed.), Proc. II Int. Symposium on Polymeric Electrolytes, Siema, 1989, Elsevier Applied Science: London, New York, 1990, p. 461.
58. K.Such, J.R. Stevens, W. Wieczorek, M. Siekierski, Z. Florjanczyk, J. Polym. Sci. B 32 (1994) 2221.
59. W. Wieczorek, K. Such, Z. Florjańczyk, J.R. Stevens, J. Phys. Chem. 98 (1994) 6840.
60. W. Wieczorek, K. Such, S.H. Chung, J.R. Stevens, J. Phys. Chem. 98 (1994) 9047.
61. W. Wieczorek, Z. Florjańczyk, J.R. Stevens, Electrochim. Acta 40 (1995) 2251.
62. W. Wieczorek, K. Such, Z. Florjańczyk, J.R. Stevens, Electrochim. Acta 40 (1995) 2417.
63. V. Gutmann, in: Coordination Chemistry in Nonaqueous Solutions, Springer: Verlag, Berlin, 1968, Chap. 2.
64. W.B. Jensen, The Lewis Acid–Base Concepts, Wiley: New York, 1980, Chap. 7.
65. W.Wieczorek, A. Zalewska, D. Raducha, Z. Florjańczyk, J.R. Stevens, A. Ferry, P. Jacobsson, Macromolecules 29 (1996) 143.
66. A. Blazejczyk, W. Wieczorek, R. Kovarsky, F. Golodnitsky, E. Peled, L.G. Scanlon, G.B. Appetecchi, B. Scrosati, J. Electrochem. Soc. 151 (2004) A1762.
67. A. Blazejczyk, M. Szczupak, P.Cmoch, W.Wieczorek, R. Kovarsky, F. Golodnitsky, E. Peled, L.G. Scanlon, G.B. Appetecchi, B. Scrosati, Chem. Mater. 17 (2005) 1535.
68. J. Xi, X. Tang, Chem. Phys. Lett. 393 (2004) 271.
69. M. Moskwiak, I. Giska, R. Borkowska, A. Zalewska, M. Marczewski, H. Marczewska, W. Wieczorek, J. Power Sources 159 (2006) 443.
70. F. Croce, S. Sachetti, B. Scrosati, J. Power Sources 162 (2006) 685.
71. F. Croce, L. Settimi, B. Scrosati, Electrochem. Commun. 8 (2006) 364.
72. M. Marczewski, A. Jakubiak, H. Marczewska, A. Frydrych, M. Gontarz, A. Sniegula, Phys. Chem. Chem. Phys. 6 (2004) 2513.
73. A.F. Danil de Namor, R.G. Hutcherson, F.J.S. Verlade, M.L.Z. Ormachera, L.E.P Salazar, I.A. Jammaz, N.A. Rawi, Pure Appl. Chem. 70 (1998) 769.

74. H.S. Lee, X.Q. Yang, C.L. Xiang, J. Mc Breen, L.S. Choi, J. Electrochem. Soc. 145(8) (1998) 2813.
75. X. Sun, H.S. Lee, S. Lee, X.Q.Yang, J. Mc Breen, J. Electrochem. Solid State Lett. 1(6) (1998) 239.
76. F. Zhou, D.R. Mac Farlane, M. Forsyth, Electrochim. Acta 48(12) (2003) 1749.
77. J. Mc Breen, H.S. Lee, X.Q. Yang, X. Sun, J. Power Sources 89 (2000) 163.
78. H.S. Lee, X.Q. Yang, X. Sun, J. Mc Breen, J. Power Sources 97–98 (2001) 566.
79. H.S. Lee, X.Q. Yang, J. Mc Breen, L.S. Choi, Y. Okamoto, Electrochim. Acta 40(13–14) (1995) 2353.
80. X. Sun, H.S. Lee, X.Q. Yang, J. Mc Breen, J. Electrochem. Soc. 149 (2002) A355.
81. H.S. Lee, X. Sun, X.Q. Yang, J. Mc Breen, J. Electrochem. Soc. 149 (2002) A1460.
82. H.S. Lee, Z.F. Ma, X.Q. Yang, X. Sun, J. Mc Breen, J. Electrochem. Soc. 151 (2004) A1429.
83. T.N. Gu, Y. Iriyama, Z. Ogumi, J. Fluorine Chem. 123 (2003) 279.
84. G.J.P. Britovsek, J. Ugolotti, A.J.P. White, Organometallics 24 (2005) 1685.
85. M. Marcinek, G.Z. Zukowska, W. Wieczorek, Electrochim. Acta 50 (2005) 3934.
86. F. Zhou, D.R. Mac Farlane, M. Forsyth, Electrochim. Acta 48 (2003) 1749.
87. J. Scheers, P. Zawadzki, M. Kalita, P. Johansson, G.Z. Zukowska, W. Wieczorek, P. Jacobson, Communicated on 10th International Symposium on Polymeric Electrolytes, Brazil, 2006.
88. D. Fateaux, J. Electrochem. Soc. 135 (1989) 2231.
89. B. Scrosati, R. Neat, in: B. Scrosati (Ed.), Application of Electroactive Polymers, Chapman and Hall: London, 1993, Chap. 6.
90. D. Aurbah, A. Zaban, Y. Gofer, Y. Ein-Eli, I. Weissman, O. Chusid, O. Abramson, B. Markovsky, J. Power Sources 54 (1995) 76.
91. D. Aurbah, A. Zaban, Y. Ein-Eli, I. Weissman, O. Chusid, B. Markovsky, M.D. Levi, E. Levi, A. Schechter, E. Granot, J. Power Sources 86 (1997) 91.
92. E. Peled, J. Electrochem. Soc. 126 (1979) 2047.
93. E. Peled, in: J.P. Gabano (Ed.), Lithium Batteries, Academic Press: London, 1983, Chap. 3.
94. W.Wieczorek, J.R. Stevens, J. Phys. Chem. B 101 (1997) 1529.
95. R. Borkowska, A. Reda, A. Zalewska, W. Wieczorek, Electrochim. Acta 46 (2001) 1737.
96. L. Sannier, A. Zalewska, M. Marczewski, H. Marczewska, W.Wieczorek, Electrochim. Acta (in press)
97. J. Przyluski, M. Siekierski, W.Wieczorek, Electrochim. Acta 40 (1995) 2101.
98. M. Marcinek, M. Ciosek, G. Zukowska, W. Wieczorek, K.R. Jeffrey, J.R. Stevens, Solid State Ionics 171 (2004) 69.
99. M. Ciosek, L. Sannier, M. Siekierski, D. Golodnitsky, E. Peled, B. Scrosati, S. Głowinkowski, W. Wieczorek, Electrochim. Acta (in press)
100. J.R. Mac Callum, A.S. Tomlin, C.A. Vincent, Eur. Polym. J. (1986) 787.
101. R.M. Fuoss, F. Accascina, Electrolytic Conductance, Interscience: NY, 1959.
102. M.H. Cohen, D. Turnbull, J. Chem. Phys. 31 (1959) 1164.
103. G. Adam, J.H. Gibbs, J. Chem. Phys. 43 (1965) 139.
104. S.D. Druger, M.A. Ratner, A. Nitzan, Solid State Ionics 9–10 (1983) 1115.
105. M. Marcinek, A. Zalewska, G. Zukowska, W. Wieczorek, Solid State Ionics 136–137 (2000) 1175.
106. B.A. Boukamp, Solid State Ionics 10 (1986) 31.
107. J.R. Macdonald, Solid State Ionics 133 (2000) 79.
108. K. Funke, B. Roling, M. Lange, Solid State Ionics 105 (1998) 195.
109. K. Funke, D. Wilmer, Solid State Ionics 136–137 (2000) 1329.
110. K. Funke, R.D. Banhatti, S. Bruckner, C. Cramer, D. Wilmer, Solid State Ionics 154–155 (2002) 65.
111. A. Jonscher, Dielectric Relaxation in Solids, Chelsea Dielectric Press: London, 1983.
112. D.P. Almond, A.R. West, Solid State Ionics 23 (1987) 27.
113. D.P. Almond, A.R. West, Solid State Ionics 18–19 (1986) 1105.

114. K. Such, J.R. Stevens, W. Wieczorek, Z. Florjańczyk, M. Siekierski, J. Polym. Sci. B 32 (1994) 2221.
115. J. van Heumen, W. Wieczorek, M. Siekierski, J.R. Stevens, J. Phys. Chem. 41 (1995) 15412.
116. Z. Florjanczyk, W. Krawiec, W. Wieczorek, M. Siekierski, J. Polym. Sci. Polym. Phys. 33 (1995) 629.
117. M. Siekierski, W. Wieczorek, Solid State Ionics 60 (1993) 67.
118. W. Wieczorek, Solid State Ionics 53–56 (1992) 1064.
119. J. Fleig, J. Maier, J. Electrochem. Soc. 145 (1998) 2081.
120. J. Fleig, J. Maier, Electrochim. Acta 41 (1996) 1003.
121. J. Fleig, J. Maier, J. Electroceram. 1 (1997) 73.
122. J. Flieg, J. Maier, J. Am. Ceram. Soc. 82 (1999) 3485.
123. R.T. Coverdale, H.M. Jennings, E.J. Garboczi, Comp. Mat. Sci. 3 (1995) 465.
124. R.T. Coverdale, E.J. Garboczi, B.J. Christensen, T.O. Manson, H.M. Jennings, J. Am. Ceram. Soc. 76 (1993) 1153.
125. D.J. Frank, C.J. Lobb, Phys. Rev. B 37 (1988) 302.
126. D. B. Gingold, C.J. Lobb, Phys. Rev. B 42 (1990) 8220.
127. J.C. Maxwell, A Treaty on Electricity and Magnetism, Vol. 1, Clerendon: London, 1892, Chap. 11.
128. D.A.G. Bruggeman, Ann. Phys. (Leipzig) 24 (1935) 636.
129. R. Landauer, J. Appl. Phys. 23 (1952) 779.
130. C.W. Nan, Prog. Mater. Sci. 37 (1993) 1.
131. C.W. Nan, D.M. Smith, Mat. Sci. Eng. B 10 (1991) 99.
132. M. Nakamura, Phys. Rev. B 29 (1984) 3691.
133. W. Wieczorek, M. Siekierski, J. Appl. Phys. 76 (1994) 2220.
134. W. Wieczorek, A. Zalewska, M. Siekierski, J. Przyluski, Solid State Ionics 86–88 (1996) 357.
135. M. Siekierski, W. Wieczorek, J. Przyluski, Electrochim. Acta 43 (1998) 1339.
136. Z. Xu, P.M. Hui, Z.Y. Li, J. Appl. Phys. 90 (2001) 365.
137. M. Acharyya, B.K. Chakrabarti, J. Phys. (Paris), I5 (1995) 153.
138. M. Acharyya, B.K. Chakrabarti, Physica A 224 (1996) 254.
139. S. Jiang, J.B. Wagner, Jr., J. Phys. Chem. Solids 56 (1995) 1113.
140. E.C. Yeh, M. Chiou, K.Y. Hsu, Thin Solid Films 297 (1997) 88.
141. S. Kirkpatrick, Rev. Mod. Phys. 45 (1973) 574.
142. J. Bigalke, Physica A, 272 (1999) 281.
143. D.G. Han, G.M. Choi, Solid State Ionics 106 (1998) 71.
144. J.J.S. Andrade, Y. Shibusa, Y. Arai, A. Siqueira, Synthetic Met. 68 (1995) 167.
145. C. Pennetta, L. Reggiani, E. Alfinito, G. Trefan, J. Phys. Cond. Mater. 14 (2002) 2371.
146. O. Stenull, H.K. Janssen, Phys. Rev. E 6405 (2001) 65105.
147. J. Wang, Y. Wang, Z. Li, Phys. Lett. A 244 (1998) 30.
148. M.Siekierski, K.Nadara, Electrochim. Acta 50 (2005) 3796.
149. M.Siekierski, K.Nadara, P.Rzeszotarski, J. New Mat. Electrochem. Syst. 9 (2006) 375.
150. M. Siekierski, K. Nadara, J. Power Sources (in press)
151. M.Siekierski, K. Nadara, ECS Transactions 2(27) (2007) 49.
152. K.Nadara, M.Siekierski, Solid State Ionics 176 (2005) 1781.
153. M.Siekierski, P.Rzeszotarski, K.Nadara, Solid State Ionics 176 (2005) 2129.
154. J.B. Kerr, S.E. Sloop, G. Liu, Y.B. han, J. Hou, S. Wang, J. Power Sources 110 (2002) 389.
155. P. Johanson, P. Jacobson, J. Power Sources 153 (2006) 336.
156. P. Johanson, P. Jacobson, Solid State Ionics 170 (2004) 73.
157. G.J. Kearley, L. van Eijck, A.A. Van Well, Physica B 350 (2004) e987.
158. P. Johanson, P. Jacobson, Electrochim. Acta 50 (2005) 3782.
159. P. Johanson, Electrochim. Acta 48 (2003) 2291.
160. P.C. Redfern, L.A. Curtiss, J. Power Sources 110 (2002) 401.
161. N.R. Dhumal, S.P. Gejji, Chem. Phys. 323 (2006) 595.
162. P. Johanson, P. Jacobson, Electrochim. Acta 48 (2003) 2279.
163. A. Liivat, A. Aabloo, J.O. Thomas, J. Comput. Chem. 26 (2005) 716.

164. X. Wang, P.B. Balbuena, J. Phys. Chem. B 108 (2004) 15694.
165. P.B. Balbuena, E.J. Lamas, X. Wang, Electrochim. Acta 50 (2005) 3788.
166. L. van Eijck, A.S. Best, J. Stride, G.J. Kearley, Chem. Phys. 317 (2005) 282.
167. H. Kasemagi, M. Klintenberg, A. Aabloo, J.O. Thomas, Solid State Ionics 147 (2002) 367.
168. H. Kasemagi, M. Klintenberg, A. Aabloo, J.O. Thomas, Solid State Ionics 147 (2004) 249.
169. O. Borodin, G. Karlson, G.D. Smith, Meeting Abstracts 2004, p. 383.

Proton-Conducting Nanocomposites and Hybrid Polymers

Y.D. Premchand[1], M.L. Di Vona[2], and P. Knauth[1]

1 Introduction

This chapter is about proton-conducting nanocomposites and hybrid polymers. Before beginning to treat the different examples from literature, we must first define what we understand by the terms 'nanocomposite' and 'hybrid polymer'. Their definitions are neither simple nor unanimous. A useful criteria for hybrid materials classification is based on their chemical nature: Class I where organic and inorganic components are dispersed and held together only by weak forces, such as Van der Waals interactions, and Class II where the organic and inorganic moieties are linked through strong bonds, such as covalent bonds [1]. In this context, Van der Waals interactions are considered to include permanent dipole interactions (Keesom forces, including also hydrogen bonds), interactions between permanent and induced dipoles (Debye forces) and interactions between induced dipoles (London forces).

Class I hybrid materials and composites differ from each other in respect to the dimension of dispersion. However this difference is minimal when we consider 'nanocomposites'. A nanocomposite is a material with nanometric domains of two coexisting phases without mutual solubility. In the following we will use the two terms, Class I and nanocomposite, as interchangeable.

To clarify our definition, let us take the example of the most widely employed proton-conducting polymer today: Nafion. Nafion at high degree of humidification is itself a fascinating material, presenting hydrophilic and hydrophobic nanodomains and could, in a sense, already be considered as a nanocomposite. Figure 1 shows this microstructure schematically: one observes nanometric channels in the structure containing water molecules and dissociated sulfonic acid groups. Polymer domains are situated between these hydrophilic regions, where the hydrophobic perfluorated alkane chains are placed. No strong bonds exist between the two regions giving a relatively labile structure. The water containing domains are

[1] Université de Provence, UMR 6121 CNRS, Centre St Jérôme, F-13397 Marseille Cedex 20, France

[2] Università di Roma Tor Vergata, Dip. Scienze e Tecnologie Chimiche, I-00133 Roma, Italy

P. Knauth and J. Schoonman (eds.), *Nanocomposites: Ionic Conducting Materials and Structural Spectroscopies.*

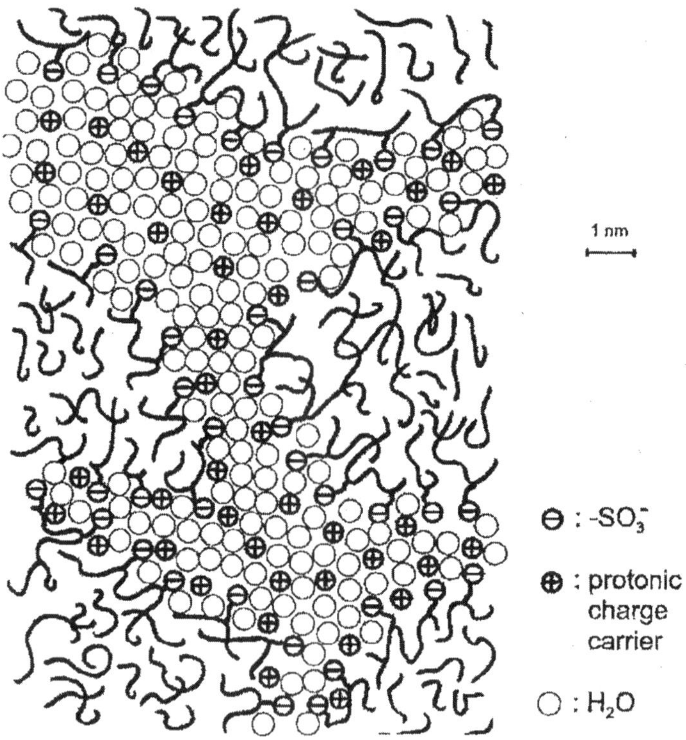

Fig. 1 Schematic representation of the microstructure of Nafion [3]

responsible for the relatively high conductivity of Nafion membranes under high humidification and at temperatures below 80°C. If one goes above this temperature, the conductivity drops sharply. There is still some debate about the origin of this effect: loss of water or change in polymer morphology [2, 3]. Whatever the origin, the technical need for permanent humidification and relatively low temperature of application for Nafion membranes is clearly established and limits the technological usefulness of these membranes, since fuel cell operation at higher temperature (typically 120°C) and under low relative humidity (such as 25% RH) is desirable.

An overview of recent literature in the domain of proton-conducting nanocomposite and hybrid materials shows mainly two categories of materials. The largest amount of literature has appeared on Nafion-related systems, initially with addition of binary oxides. The second largest group is hydrocarbon-based systems, generally aromatic but also some non-aromatic. This separation forms the basic structure of this chapter. It also includes an overview of synthesis and characterization procedures for these kinds of material as well as models used for microscopic description of transport phenomena. For further information on modeling of ionic conducting polymers, the reader is referred to the previous chapter of this book.

2 Synthesis

In the light of the general concepts described in the introduction, it may be presumed that the introduction of inorganic components within a polymer matrix intended for proton-conducting nanocomposite and hybrid membranes is accomplished through

1. Direct embedding
2. Covalent bonding directly to a polymeric matrix
3. Covalent bonding to supporting particles that can further bond covalently or embed physically into a polymer matrix.

2.1 Nanocomposites

Nafion is the widely preferred polymer component for the preparation of inorganic/organic nanocomposite membranes. Other polymers besides Nafion include aromatic hydrocarbon polymers, such as sulfonated polyetheretherketone (SPEEK), polybenzimidazole (PBI), non-aromatic hydrocarbon polymers like poly(ethylene oxide) (PEO) and many others. Solid inorganic components used are binary oxides such as SiO_2, TiO_2 and ZrO_2, or inorganic proton conductors like heteropolyacids. Recasting a bulk mixture of powdered or colloidal inorganic components with a polymer solution, and in-situ formation of inorganic components within a polymer membrane or in a polymer solution are the two preferred synthesis routes for achieving nanocomposite and hybrid proton-conducting polymer membranes using the above set of organic and inorganic components [4].

2.1.1 Recasting

One of the simplest methods for the preparation of nanocomposite membranes is 'Solution-Casting' [5] using suitable organic solvents, preferably with high boiling point, followed by solvent elimination. It enables direct incorporation of nanosized inorganic materials into a polymer matrix and a number of nanocomposite membranes with Nafion and SPEEK as the backbone polymers have been prepared using this procedure [6]. In a typical experimental procedure, appropriate amount of inorganics are first mixed with the polymer solution under vigorous stirring. The membrane is then obtained by film casting and heated until all the solvent evaporates. For bulk mixing, the inorganic components should be prepared in the form of powders or dispersions. Silica, titania, zirconia, silicotungstic acid, phosphomolybdic acid, zirconium phosphate or phosphonate and silica supported inorganic acids are some of the inorganic components introduced by bulk mixing. The size and dispersion of the solid particles are of particular importance in this case to obtain uniform and nonporous membranes.

Major disadvantages of the recast procedure are possible in-homogeneities of the composite membrane and the formation of pores around the oxide particles, especially for larger size additives. An appropriate method to avoid these difficulties is to use colloidal suspensions of inorganic additives; for example, a colloidal dispersion of hydrous oxides can be obtained by the hydrolysis of metal alkoxides in aqueous solution. According to this procedure, the aqueous dispersions of oxides are stratified over the polymer solution in the organic solvent. As water is evaporated, the oxide nano-particles transfer to the organic phase. This method differs from the conventional sol–gel technique in the sense that it does not involve condensation reactions that ultimately lead to the formation of an inorganic network within polymers. Instead the inorganic particles will remain as separate entities or as bulk in the resulting nanocomposite membranes. Parameters of interest are the type of inorganic filler, its surface area and size, and possible chemical surface treatment [7].

2.1.2 Sol–Gel

In situ generation of inorganic species within a polymer membrane by sol–gel process is a versatile strategy for the preparation of proton-conducting nanocomposite materials [8]. It enables the synthesis at nano- to sub-micrometer scales at low temperatures and the membranes obtained by this procedure are generally homogeneous. Ideally, the process begins with the infiltration of a molecular precursor solution into the polymer matrix at ambient temperatures. Hydrolysis of the infiltrated inorganic precursor then occurs due to the nucleophilic attack of the water present in the membrane on the inorganic atoms and the membranes are subsequently treated with necessary reactants to complete the condensation reactions. If the reaction is an acid-catalyzed hydrolysis, and the polymer happens to be ionomer like Nafion, the pendant $SO_3^-H^+$ group will itself act as a catalyst, and the need for external reactants do not arise. The original morphology of unfilled polymer membranes will persist even after the invasion by sol–gel derived phase that is the membrane in itself will act as a template for the whole sol–gel process. Introduction of the ORMOSIL phase within monomer membranes are feasible through in-situ acid-catalyzed co-polymerization of tetraethylorthosilicate (TEOS) and organically modified silane monomers [9]. One major difficulty associated with the impregnation of polymer membranes with inorganic precursors is maintaining concentration gradient of precursor solutions; it virtually limits the incorporation of inorganic components into the polymer matrix to a certain level.

2.2 *Hybrid Polymers*

Synthesis of organic/inorganic hybrid polymers via the formation of covalent bonds between totally aromatic polymers and inorganic clusters is another effective approach in the development of proton-conducting membranes for fuel cell applications. This approach requires molecular precursors that contain a hydrolytically stable chemical bond between the element that will form the inorganic network during sol–gel processing and the organic moieties. Alternatively, organic components can be chemically modified to covalently attach them to the inorganic network [10], e. g. silylation. The organic components after silylation are easily accessible to common hydrolysis and polycondensation reactions. This approach has been followed in the synthesis of an organic inorganic polymer known as Silylated-SPEEK (SiSPEEK) [11]; the primary aim is to synthesize a polymer containing both $-SO_3H$ functions and silicon moieties. The order of the synthesis steps is

- Sulfonation
- Silylation
- Sol–gel process

A novel, efficient and experimentally simple method for the introduction of silicon functional groups into polymeric carbon frameworks is to combine in one macromolecule the features of a cross-linked polymer and the presence of covalent organic–inorganic, C–Si, bonds [12]. Cross-linking can be obtained during the sulfonation step by formation of SO_2 bridges among the repeating unit of PEEK. The synthetic route takes the following steps:

- Direct sulfonation of PEEK with chlorosulfonic acid ($ClSO_3H$); this will result in SOPEEK – sulfochlorinated PEEK (Scheme 1)
- Silylation – this will introduce covalently linked silicon moieties in SOPEEK
- Hydrolysis of sulfochlorinated PEEK in order to obtain the desired $-SO_3H$ functions (SOSiPEEK, Scheme 2)

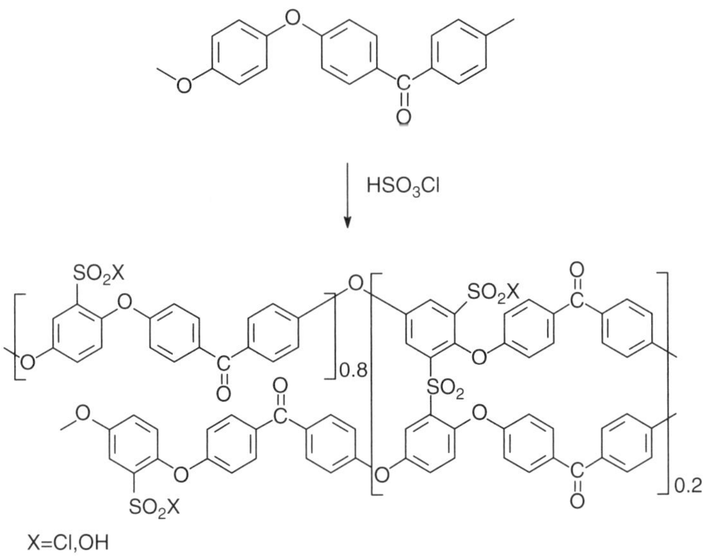

Scheme 1 Synthesis of SOPEEK

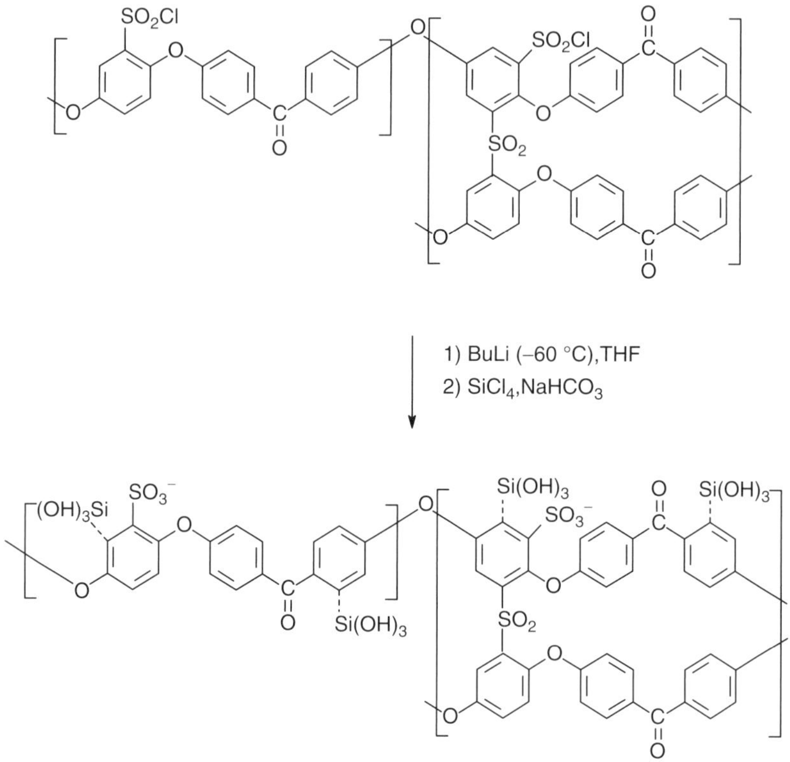

Scheme 2 Synthesis of SOSiPEEK

3 Characterization

The distinctive qualities or properties of a material are manifold, physical, chemical or electrical for that matter [13]. Characterization entails establishing these properties on the basis of chemical composition, structure and other features that influence them. The major features that decide on the ability of a nanocomposite or hybrid polymer to function as a proton exchange membrane within a fuel cell are (a) proton conductivity, (b) impermeability to reactants and (c) chemical and thermal stabilities at operation environment. Proton conduction in turn is dependent on level of hydration in those nanocomposite and hybrid membranes based on polymers that require water for proton transport. Therefore, the amount of water taken up by the nanocomposite membranes immersed in liquid water or exposed to water vapour of variable pressure is an important parameter in the development of the Class I hybrid membranes for fuel cell applications. Besides these features, the nature of acid-bearing host polymer and the acid content (often referred to as the ion exchange capacity or equivalent weight) of such membranes also plays a profound role in the membranes' ability to serve as a polymer electrolyte and so does the ability of the material to phase separate into hydrophobic and hydrophilic domains – especially in materials in which water content and corresponding number of water molecules per acid sites are low. Still other properties such as structure, morphology and mechanical strength may also exert a strong influence on their fuel cell performance. Membrane electrode assembly (MEA) testing is the principal means of characterization pertaining to its electrochemical performance. A brief discussion on each of these features and the principles behind various experimental techniques employed for their characterization is presented in this section.

3.1 Chemical and Electrochemical Properties

3.1.1 Proton Conduction and Diffusion

Proton conduction is fundamental for any fuel cell electrolyte and is usually the first characteristic considered when evaluating membranes for potential fuel cell use. Resistive loss is proportional to the ionic resistance of the membrane and high conductivity is essential for the required performance especially at high current density. At a molecular level, the proton transport in hydrated polymeric matrices is in general described on the basis of either of the two principal mechanisms: (a) diffusion mechanism [14] – through water as 'vehicle' – and (b) hopping mechanism (Grotthuss transport) [15, 16]. The prevalence of one or the other mechanism depends on the hydration level of the membrane. On the other hand, the mechanism of proton transport within nanocomposite and hybrid systems based on the aforementioned membranes is a much more complex process as it involves both the surface and chemical properties of the inorganic and organic phases [17]. Although the exact role of inorganic components in stabilizing the proton transport properties of nanocomposites based on Nafion and other polymers is still under discussion, it may be presumed that the primary function of the nano-particles is to stabilize the polymer morphology with increasing temperature. If the inorganic additive happens to be an alternative proton transporter like heteropolyacids, their contribution to the transport processes has also to be analysed. Proton conductivity improvements would, however, depend upon whether the fraction of bulk water and the bulk proton concentrations are increased as a result of the inorganic additives or not. Figure 2 is a summary of these transport mechanisms. For more details on different transport models please refer to the section on Models.

Polymers with chemically bonded –SO_3H groups are single-species conducting electrolytes, where the charged specie involved in the conduction is the proton. For this kind of systems, the conductivity equation can be written as [19]

$$\sigma = Fuc, \tag{1}$$

where F is the Faraday constant (96,500 C/mol), u is the proton mobility and c is the proton concentration.

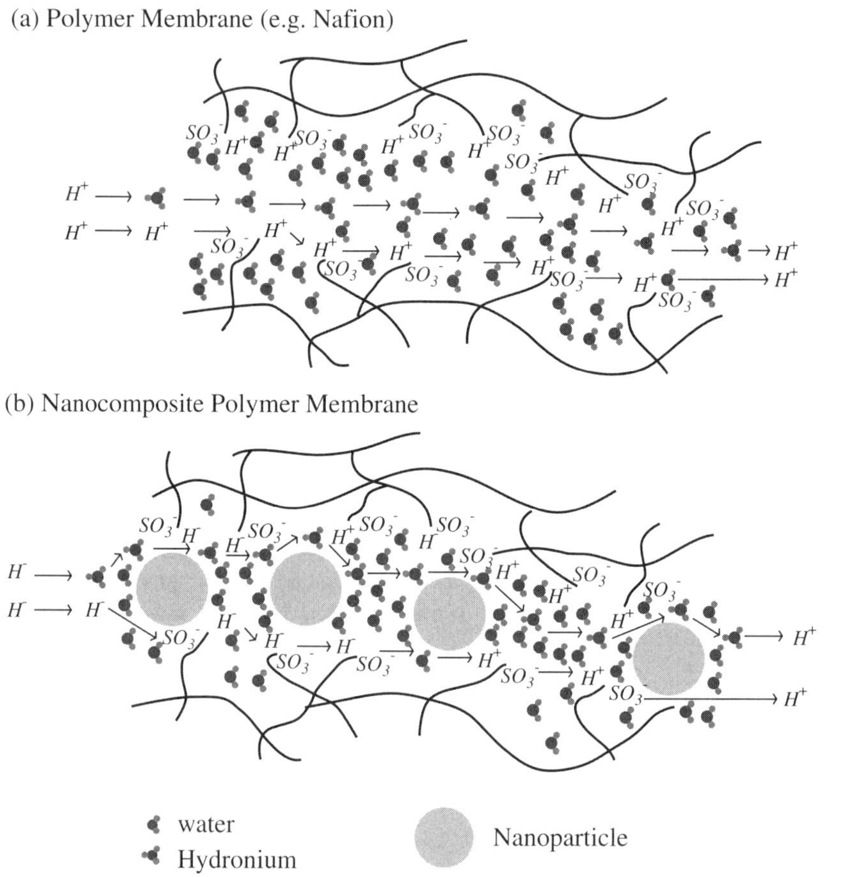

Fig. 2 Proton transport in different membrane configurations: (**a**) proton transport in Nafion membranes and (**b**) proton transport in polymer/nano-particle composite membranes [18]

It is useful to obtain the relationship between proton conductivity, mobility and diffusion coefficient. The transport equation takes the general form of a linear relationship between a flux J and a driving force, the gradient of the electrochemical potential η:

$$J = -\beta \nabla \eta = -\beta \nabla \mu - \beta F \nabla \phi. \tag{2}$$

This equation expresses the fact that both chemical (μ) and electrical (φ) potentials can act as driving forces. If uncharged particles are concerned, only the chemical potential term is relevant and the process is reduced to pure diffusion:

$$J = -\beta \nabla \mu = -\frac{RT\beta}{c} \nabla c. \tag{3}$$

A comparison with Fick's first law permits to identify

$$\beta = \frac{Dc}{RT}. \tag{4}$$

Conversely, if there is no noticeable chemical potential variation, the electrical field remains the only driving force. This is excellently fulfilled for materials with high charge carrier concentrations, such as metals, fast ion conductors or highly doped systems.

$$J = -\beta F \nabla \phi. \tag{5}$$

Converting the particle flux into a current density:

$$i = FJ = -\beta F^2 \nabla \phi. \tag{6}$$

This is Ohm's law; the prefactor corresponds to the electrical conductivity, using (4):

$$\sigma = \beta F^2 = F^2 \frac{Dc}{RT}. \tag{7a}$$

This relationship is known as Nernst-Einstein equation. Using (1), one obtains the relation between diffusion coefficient and mobility:

$$u = \frac{FD}{RT}. \tag{7b}$$

The Nernst-Einstein equation is particularly useful for determining the mobility of charged species and the diffusion coefficient from electrical conductivity data, or vice versa.

Ionic conduction in a solid material is due to a thermally activated process. Proton conductivity obeys either a simple Arrhenius law or a Vogel-Tamman-Fulcher (VTF) equation. At temperatures below the glass transition temperature, T_g, the conductivity obeys generally an Arrhenius-type law [20]:

$$\sigma = \frac{A}{T} \exp\left(-\frac{E_a}{RT}\right), \tag{8}$$

A is a constant proportional to the number of charge carriers, E_a is the activation energy and R is the ideal gas constant. Above T_g, the conductivity usually follows the VTF formula [19, 21]:

$$\sigma = \frac{A}{\sqrt{T}} \exp\left(-\frac{B}{T - T_0}\right), \tag{9}$$

B is a pseudo activation energy related to the segmental motions of the polymer and T_0 is called equilibrium glass transition temperature. The VTF behaviour describes conductivity in polymers where the segmental motions contribute to the proton transport. As a consequence, amorphous polymers exhibit a higher conductivity above glass transition temperature than below.

Typically, three distinct conductivity behaviours can be observed in polymer electrolytes [19]:

1. VTF behaviour is observed above glass transition temperature (T_g) of the polymers
2. Arrhenius-type behaviour at low temperatures and VTF at high temperatures
3. Arrhenius-type behaviour throughout the whole temperature range with a clear decrease of the activation energy around T_g. This is the behaviour of polymers with an extremely rigid structure; thus, the contribution of segmental motions is low

Supplementary information on the electrical properties of polymer electrolytes can be found in the first chapter of this book.

3.1.2 Water Uptake Measurement

The performance of a membrane is dependent on proton conductivity, which in turn often depends on its water content. High proton conductivity is supported by high level of water uptake; at the same time, it is also a sign of low-dimensional stability as water influences the polymer microstructure and mechanical properties. Since water is also known to assist the mass transport of methanol and oxygen through the membrane, the water uptake measurements could serve as a quantitative measure of membrane performance for DMFC application as well. Gravimetric technique has been widely used for this purpose. Water uptake measurements due to gravimetric techniques are generally done by double weighing. 'Wet' weights (W_{wt}) of the membranes are first measured after equilibrating with water at different temperatures or upon exposure to water vapour at various pressures. The membrane samples are then dried at a temperature above the boiling point of water for a particular period of time and their 'Dry' weights (W_{dry}) measured. The total water content (W_{uptake}%) is determined from the difference between the wet and the dry mass of the membranes as follows:

$$\text{Water uptake} = \frac{W_{wet} - W_{dry}}{W_{dry}} 100\% \tag{10}$$

If t_w and t_d are the thicknesses of wet and dry membranes, respectively, the membrane swelling ratio may be calculated as

$$\text{Swelling ratio} = \frac{t_w - t_d}{t_d} 100\% \tag{11}$$

The membrane water content parameter, 'λ', the number of moles of water per mole of acidic group can be calculated:

$$\lambda = \frac{N_{H_2O}}{N_{SO_3^-}}, \tag{12}$$

$$\lambda = \frac{(W_{wet} - W_{dry})}{IEC\ 18\ W_{dry}}, \tag{13}$$

IEC is the ion exchange capacity of the membrane.

Water uptake curves are generally represented as number of water molecules per sulfonic acid group 'λ' versus thermodynamic water activity or relative humidity, which is the ratio of water partial pressure and saturation partial pressure of water (Fig. 3). The factors that affect the extent of the water uptake of a membrane are temperature, ion-exchange capacity, pre-treatment of membrane, the physical state of absorbing water, whether it is in liquid or vapour phase, and the elastic modulus of the membrane. A major objective in the nanocomposite and hybrid membrane research is to determine which chemical attributes of the composite membranes improve water level at elevated temperature and thereby improve PEMFC performance.

3.1.3 Ion Exchange Capacity

IEC or EW is the measure of relative concentration of acid groups within polymer electrolyte membranes. Proton conductivity and water uptake both rely heavily on the

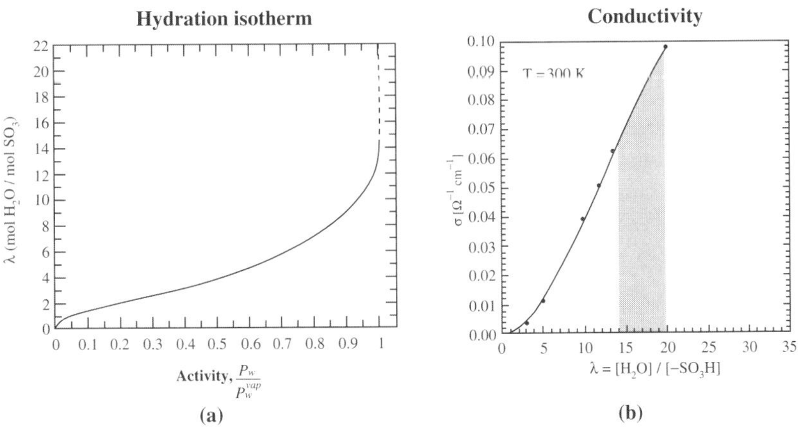

Fig. 3 (**a**) Equilibrium water-uptake or isotherm curve for Nafion at 300 K. The shaded line corresponds to coexistence with liquid water (λ = 22) (For details see Chap. 6.2). (**b**) Conductivity vs. water-uptake curve at 300 K. The shaded area corresponds to coexistence with liquid water (λ = 22) [16]

concentration of ion conducting units (most commonly sulfonic acid) in the polymer membrane. The ion content is characterized by the mass of dry membrane per molar equivalents of ion conductor and is expressed as EW with units of grams of polymer per equivalent or as IEC with units of milliequivalents per gram (mequiv. g^{-1} or mmol g^{-1}) of polymer. Varying the ion content of the membrane can control both its water uptake and conductivity. While it is desirable to maximize the conductivity of the membrane by increasing its ion content (decreasing equivalent weight), other physical properties must be considered. Too many ionic groups will cause the membrane to swell excessively with water, which compromises mechanical integrity and durability.

IEC of membranes is determined by titration at room temperature. The membranes in the acidic forms (H^+) are first converted to the sodium forms by immersing the membranes in NaCl solutions to exchange the H^+ ions for Na^+ ions. The exchanged H^+ ions within the solutions are titrated with 0.01 N NaOH solutions. IEC values may be calculated from the titration result using the formula

$$\mathrm{IEC} = \frac{\text{Consumed ml of NaOH} \times \text{Molarity of NaOH}}{\text{Weight dried membrane}} (\text{mequiv. g}^{-1}). \quad (14)$$

3.1.4 MEA Testing

The development of new or modified PEMs with improved characteristics for fuel cell applications requires a quantitative determination of their electrochemical performance under relevant fuel cell conditions. The most straightforward approach for this is to construct a membrane electrode assembly (MEA) and measure the cell parameters in a single cell configuration. A membrane electrode assembly includes an anode, a cathode, a membrane disposed between the anode and the cathode and an extended catalyst layer between the membrane and the electrodes. The conversion efficiency of MEA depends on many factors including type and thickness of both membrane and gas-diffusion material, nature of binder used in the electrodes and the binder to catalyst ratio apart from the operation conditions (temperature, pressure, flow rates, humidification of reactant gases). In addition, the ability to collect data from an operating electrochemical system can be alluring.

Single cell testing is relatively straightforward and operation conditions can be accurately monitored as it allows specific control over humidity, reactant flow and temperature. Cell performance is often described by the polarization curve, i. e., cell voltage vs. current density. A typical curve is shown in Fig. 4. In general three main polarization losses can be identified: (a) activation overpotentials, arising from charge transfer and other reaction kinetics; (b) ohmic losses, arising from the electrical resistances of the cell materials and interfaces and (c) mass transport overpotentials, arising from the limitations of mass transport. At low current densities, the shape of the curve is primarily determined by activation polarization, which gives it the characteristic logarithmic shape. It plays an important role if the reaction rate on the electrode surface is restricted by sluggish electrode kinetics. Similar to a

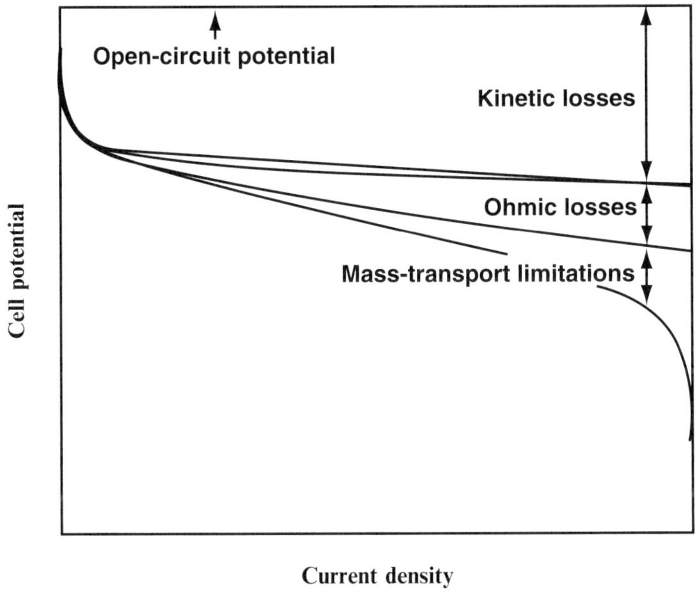

Fig. 4 Example of a polarization curve showing the typical losses in a polymer–electrolyte fuel cell

chemical reaction, the electrochemical reaction has to overcome an activation barrier. This barrier usually depends on the electrode material (electrocatalyst). When pure hydrogen is used as fuel, the activation losses of the anode are negligible, because the rate of the hydrogen oxidation reaction is orders of magnitude higher than the rate of the cathode reaction. Hence, the main source of activation overpotential is the cathode, which means the oxygen reduction. When current density increases, the shape of the curve becomes approximately linear, reflecting the effect of *ohmic losses*. This is caused by both the resistance due to the migration of ions within the electrolyte and the resistance due to the flow of electrons. It can be expressed by the product of cell current (I) and the overall cell resistance (R, including electronic, ionic and contact resistances). When current density is increased further, the curve begins to bend down due to mass transport overpotentials, which result from limitations in the availability of reactants at the catalyst surfaces. The main source of losses is the cathode side again, because the diffusivity of oxygen is significantly lower than that of hydrogen, due to the larger molecular size of oxygen.

Fuel cell efficiency on the other hand is directly proportional to the power density (in W cm^{-2}), which can be linked directly to the chemistry of the polymer membrane. Higher achievable power density directly translates to smaller, thus less expensive fuel cells. Thus, a swift comparison of the obtained data against those obtained with unmodified membranes will provide useful information on the influence of inorganic phase on the nanocomposite efficiency. Their effectiveness as a catalyst binder may be evident from an investigation on the interfacial effects of membrane on electrodes and catalysts. In the case of Class I membranes intended for high temperature operations, apart from I–V measurements, the methanol crossover flux vs. methanol feed concentration can be collected; the suitability of the membrane for DMFC applications may be accessed from these data. The long time stability of the membranes against different operating conditions may also be studied. In short, membrane electrode assembly (MEA) testing will be of great advantage to fine tune the hybrid membrane properties in order to give them commercial viability. However, a major technical challenge is up-scaling single cell performance to multi-cell stacks.

3.2 Physical Characterization

3.2.1 Structure and Microstructure

Structure decides macroscopic functions; consequently, a better understanding of the behaviour of any solid material is facilitated by the knowledge of its structure. Most of the prominent proton-conducting polymers considered for fuel cell applications are hydrated acidic polymers, which typically phase separate into a percolating network of hydrophilic functional domain and hydrophobic polymer-rich structural domain. Water separated morphology and their hierarchical microstructure are therefore of great importance for the fundamental understanding of proton conduction mechanism, conductivity dependence on solvent (water and methanol) content and stability. It is acutely demonstrated by the fact that the difference in the water uptake and transport properties of Nafion and SPEEK membranes arises due to the difference in their microstructure [3]. Small-angle X-ray and neutron scattering (respectively SAXS and SANS) are the most prevalent experimental technique for microstructure analysis. Backbone hydrophobicity, backbone flexibity and acid group (or side chain) characteristics have been identified as the major factors influencing microstructure of polymer electrolyte membranes. On the other hand, the morphological structure of hydrated membranes is mainly decided by the nature of water-filled network through which proton transport occurs [22]. A successful evaluation of structure may therefore demand the characterization of water within the membranes as well, since they are inter-related. The structural features that determine the behaviour of water within the membrane are water domain size, shape and nature of domain itself. Nuclear magnetic resonance (NMR), quasi-elastic neutron scattering, dielectric spectra studies and differential scanning calorimetry are some of the common techniques used to probe the nature of water within the membranes. Eventually, it is feasible to develop a simple and more self-consistent PEM microstructural model from the correlation of the data obtained from microstructure studies with water diffusion coefficients derived from pulsed field gradient (PFG)-NMR measurements [23]. In fact, a satisfactory structural model taking into account all the necessary features relevant to fuel cell properties is not yet available, basically due to the complexity of membrane structures arising from phase separations when hydrated [24]. Still certain generalizations about the structure are useful in setting a basis to understand the structural changes in Nafion and other membranes upon introduction of inorganic phases [25].

In SAXS and SANS experiments, fluctuations of the electron density give rise to characteristic patterns due to the scattering of incident X-rays. The scattered intensity $I(\mathbf{q})$ oscillates with increasing wave vector ($\mathbf{q}$):

$$\mathbf{q} = \frac{4\pi}{\lambda}\sin\theta, \tag{15}$$

where λ and θ are the wavelength and scattering angle, respectively. Bragg spacing d is related to q as

$$d = \frac{2\pi}{q}. \tag{16}$$

Scattering maximum, often known as 'ionomer peak', results due to the scattering of the ionic group clusters in low dielectric polymer matrices. The origin of this scattering maximum has been extensively studied, but it is still a subject of controversy. Various models have been proposed to interpret the SAXS observations, and they can be generally divided into intra-particle models and inter-particle models. The intra-particle models attribute the ionomer peak' to the interference within the ionic cluster, implying that the scattering maximum is related to the internal structure of the cluster. On the other hand, the inter-particle models attribute the 'ionomer peak' to the interference between different ionic clusters, implying that the center-to-center distance between two clusters is the Bragg spacing. The inter-particle models are now commonly being accepted. Water content and equivalent weight are the two parameters that decide the inter-ionic cluster distance in hydrated membranes. Since hydrated silica- and titania-containing membranes support larger water clusters than unmodified membranes, the inter-ionic cluster distances obtained from SAXS, in comparison with water diffusion coefficients, can be effectively interpreted to evaluate the microstructural changes in Nafion and SPEEK-based composite membranes [26].

3.2.2 Hydrophilicity

Proton conductivity, and many other properties of the membrane (e. g., mechanical strength and water retention), is a delicate balance between hydrophobicity and hydrophilicity and, in particular for conductivity, the molecular state of water in the polymer. The hydrophobicity/hydrophilicity of nanocomposite membranes in a more common way is expressed in terms of wettability of the membrane, which can be quantified by contact angle measurements. Experimental measurements are usually made by first submerging the membrane sample in room temperature deionized water for several minutes, spreading the saturated membrane flat on a clean glass microscope slide, and finally removing excess surface water by wiping with a Kimwipe. The contact angle is then immediately measured. Contact angle measurements on polymer surfaces are not only influenced by the interfacial tensions according to Young's equation but also by many other phenomena, such as surface roughness, chemical heterogeneity, sorption layers, molecular orientation, swelling and partial solution of the polymer or low-molecular constituents in the polymer material. These effects have to be considered when contact angle measurements are used to calculate the surface tension of solid polymers.

3.2.3 Thermal Stability

One of the major issues to be addressed in the development of proton-conducting nanocomposite and hybrid membranes for fuel cell applications is their high temperature stability. It arises primarily due to the fact that the sulfonic acid side chains in backbone polymers such as Nafion and other sulfonated hydrocarbon membranes undergo desulfonation with increase in temperature. While the sulfonic acid groups in Nafion are stable up to a temperature of 280°C in air [27, 28], the degradation temperature of SPEEK is reported to be in the range of 240–330° C. Desulfonation is in general studied by means of thermogravimetric analysis (TGA), differential thermal analysis (DTA), Fourier-transform infra-red spectroscopy (FT-IR) and TGA-mass spectrometry (MS). In Nafion-based composites this decomposition behaviour is attributed to the loosening of sulfonic acid groups present in the unmodified Nafion membrane [29]. It was also observed that the temperature at which this decomposition occurs shifts with the nature of inorganic additive within the pores of Nafion membrane. For example, sharp thermal degradation of the unmodified Nafion occurs at about 325°C, whereas for Nafion–ZrO_2, and Nafion–SiO_2 sol–gel membranes, degradation temperature shifts to about 360°C and 470°C, respectively [30]. The TiO_2 incorporated membranes on the other hand are reported to show not much improvement in thermal degradation temperature as compared with Nafion. A systematic investigation on the thermal behaviour of nanocomposite and hybrid membranes would thus give more insights into their stability at high temperatures. Although informative, these thermal stability results can hardly be used to predict the long-term durability of these membranes.

3.2.4 Mechanical Properties

Fuel cells operating above 100°C possess many technical advantages such as smaller heat exchangers, easier integration with reformers and many others. The thermo-mechanical stability of the polymer 'backbone' in the nanocomposite and hybrid membranes at higher temperatures are therefore of paramount importance for better consumption as well as from the point of view of producing economical fuel cells. Mechanical integrity of membranes as mounted in cells, and under the perturbation of pressure gradients, swelling-dehydration cycles, mechanical creep and the onset of brittleness and tear resistance are the other relevant factors that need to be taken into consideration. Stress–strain behavior at varying load and at different temperature is generally analyzed. The information from dynamic mechanical analysis (DMA) can be used to identify thermomechanical transitions that are assigned to morphological features in the microphase separated morphology. For viscoelastic materials, like Nafion, stress consists of an elastic and a viscous component. The elastic component or storage modulus accounts for how the material behaves like an ideal solid and the viscous component or loss modulus accounts for how the material behaves like an ideal fluid. The phase angle shift between stress and strain is represented by δ, which varies between 0 (100% elastic) and 90 (100% viscous):

$$\frac{E'}{E''} = \tan(\delta), \tag{17}$$

E' is the storage modulus (elastic component) and E'' is the loss modulus (viscous component) of the material. The glass transition temperature (T_g) of a material can be taken as either the peak of the loss modulus versus temperature curve or the peak of the tan (δ) versus temperature curve.

4 Nafion-Based Systems

Nafion is a perfluoro-sulfonic acid ionomer membrane possessing a polytetrafluoroethylene (PTFE) backbone known as Teflon and regularly spaced long perfluorovinyl ether pendant side chains terminated by a sulfonate ionic group. Its general chemical structure is given below as Fig. 5.

Nafion is currently the most widely used and studied polymer electrolyte membrane in low temperature fuel cell applications. The qualities that make it a prototypical material for polymer exchange membrane are its conductivity as high as 0.1 S cm^{-1} at 80°C under fully hydrated conditions, coupled with good water uptake, IEC, low gas permeability and excellent electrochemical stability [31, 32]. However, proton conductivity of Nafion decreases drastically when temperature is higher than 80–90°C. The elevated temperature also causes shrinkage of the membrane, which in turn makes a poor contact between the membrane and the electrodes. Hence it cannot be used for fuel cell operations at temperatures above 100°C under atmospheric pressure. However, for practical applications, operation at temperatures near and above the normal boiling point of water is essential both for hydrogen and methanol fuelled cells. When hydrogen is used as a fuel, an increase in the cell temperature above 100°C produces enhanced CO tolerance of Pt catalyst, faster reaction kinetics, easier water management and reduced heat exchanger requirement. In the case of fuel cells using methanol as fuel, high temperature operation provides many advantages, such as improved kinetics at the

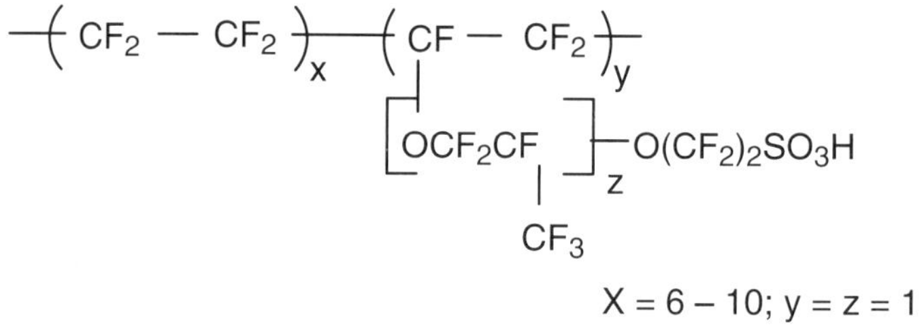

Fig. 5 Nafion perfluorinated ionomer

surface of the electrode, which is especially important in methanol and CO-containing reformat feeds, fast transport of protons across the PEM, efficient heat and water management, and also opening a new possibility of integrating fuel cells with methanol reformer, which can result in compact fuel cell systems. Hence, the stability of the Nafion membrane becomes more crucial for the performance of the PEMFC at high temperature. Thus, there is a strong incentive to develop alternate polymer electrolytes that can work above 100°C under low relative humidity (RH). Therefore, increasing the operation temperature of proton-conducting membranes is a key issue in the development of PEMFC technology.

Much of the early interest in these directions was simulated by the pioneering works of Watanabe et al. [33–35] who modified Nafion PEMs by incorporating nanosized particles of SiO_2, TiO_2, Pt, Pt-SiO_2 and Pt-TiO_2 for humidification requirements and by Malhotra and Datta [36], who proposed the impregnation of conventional Nafion membranes with inorganic solid acids with the objective of improving water retention as well as providing additional acidic sites. Their approaches showed the early promise for developing PEMs that function adequately at temperatures above 120°C under low relative humidity (RH) conditions. Several approaches have been attempted to improve the performance of Nafion at temperatures near and above the normal boiling point of water. They include incorporation of (a) hydrophilic oxides such as SiO_2; (b) inorganic acidic oxides such as ZrO_2/SO_4^{2-} to increase the concentration of acid sites; (c) bifunctional particles having both hydrophilic and proton-conducting properties such as zirconium phosphates, metal sulphates and silica supported inorganic proton conductors; (d) inorganic proton conductors such as heteropolyacids to enhance further proton conductivity using inorganic-assisted proton transport and (e) layered additives such as Montmorillonite (MMT) clays to reduce methanol permeability.

4.1 Binary Oxide Additives

The synergic merge of two different materials is known to produce distinct characteristics associated with each of the individual components. Incorporation of hygroscopic and nanosized inorganic binary oxides, such as SiO_2, TiO_2 and ZrO_2, within Nafion membrane are therefore intended to (a) maintain the usefulness of the membrane at high temperatures under low relative humidity, (b) increase their resistance to fuel crossover by creating barriers in the flow channels and (c) improve thermo-mechanical properties. In effect, the fuel cells employing binary oxide-modified composite membranes are found to be characterized by lower level of fuel crossover and are able to work up to 145°C [37]. Reasons for the improved fuel cell performances are still speculative: water sorption by the inorganic fillers [38–40] or specific chemical interactions existing between polymer sulfonate groups and the metal oxide surface [41]. Further investigations therefore need to be carried out in order to understand the function of inorganic fillers in composite solid oxide/polymer membranes rather than on the beneficial effects of incorporating fillers within a membrane. Special attention should be given to the physicochemical characteristics of the various oxides, in particular, to the interfacial chemistry that occurs between the oxide particles and the Nafion membrane, along with a consideration of how this interface affects elevated-temperature fuel cell dynamics. The improvement of the membrane performances seems to be only very little dependent on the nature of the filler, while it is strongly dependent on the size and distribution of the inorganic particles inside the membrane [7].

A major disadvantage of Nafion/binary oxide composite membranes, in spite of the better fuel cell performance at high temperature, is their reduced proton conductivity compared with that of unmodified membrane. The long-term stability of these membranes is also still in question.

4.1.1 Recast Membranes

Among the binary oxides, nanosized TiO_2 has been extensively studied as a filler material for Nafion [42–44]. To begin with the influence of particle size and crystallographic phase (anatase or rutile) on the electrochemical performance of recast membranes containing various amounts of TiO_2 in DMFC was analyzed. The best electrochemical performance was found to be exhibited by TiO_2 particles having anatase structure and the smallest particle size. As a follow up of these studies, the electrochemical performance of the nanocomposite membranes have also been analysed in terms of filler specific surface area and oxygen functional groups on the filler surface. These studies reveal that a large number of water molecules, promoting proton migration, are coordinated by the surface groups. This conclusion is in agreement with that of Chalkova et al. on similar materials [39]. Therefore, for an effective use of solid oxides, surface chemistry (solid oxide/water interface chemistry) should be understood in greater detail over a wide range of temperatures and increased electrochemical performance may be achieved by appropriate tailoring of the surface characteristics of the inorganic filler.

4.1.2 Sol–Gel Membranes

Sol–gel technique was first adopted by Mauritz [8, 45–47] and his co-workers to introduce silicon oxide networks into the fine hydrophilic channels of Nafion membranes.

Their performance against recast membranes were assessed by Adjmein et al. [48] in hydrogen/oxygen proton-exchange membrane fuel cells (PEMFCs) at different temperatures ranging from 80 to 140°C. At 130°C under a pressure of 3 atm, composite membranes prepared using Mauritz's synthetic procedure delivered a current density at least six times higher than that of unmodified membranes due to their better water management. The benefit of these Class I membranes from their study appears to be stable operation against the conventional Nafion at a cell temperature of 130°C due to high rigidity, both tested under fully humidified conditions. The sol–gel prepared nanocomposite membranes (silica content as high as 20 wt%) have also been reported to exhibit much lower methanol permeation rate than an unmodified Nafion [49].

In recent years, a great deal of effort has been put forth by Datta and his co-workers on sol–gel synthesized Nafion/SiO_2, Nafion/TiO_2 and Nafion/ZrO_2 nanocomposite membranes, in particular on their relative performances against unmodified Nafion in terms of water uptake, proton conductivity at different relative humidity conditions (RHs), fuel cell performance and ion exchange measurements [30]. At temperatures of about 90°C and 120°C, all these Nafion–MO_2 sol–gel Class I membranes were found to display better water uptake at a given RH than unmodified Nafion membrane. The basic sorption trend at both temperatures is similar, with water uptake increasing from silica to titania to zirconia nanocomposites, and is in order of increasing acid strength. However, at these temperatures of interest, higher conductivity has been observed only in the case of Nafion–ZrO_2 sol–gel composite, the enhancement at 40% RH being 10% over Nafion. The difference in the nature of acidic sites offered by the inorganic fillers is thus presumed to be the reason behind this behaviour. Also, the sorption isotherm shape obtained for nanocomposite membranes were found to be similar to that of Nafion, with sharp increase in the amount of water uptake above water activity of 0.6. Hence, the basic mechanism of water sorption must be similar for all the oxide incorporated nanocomposite membranes. The difference is probably due to the change in acidity or active surface area of the membranes.

The Nafion/MO_2 (M = Si, Ti, Zr) nanocomposite membranes have also been characterized using TGA and DMA to determine degradation and glass transition temperatures (T_g). All the sol–gel synthesised nanocomposite membranes are found to exhibit better thermomechanical properties than Nafion. The degradation temperatures and T_g improved for all nanocomposites and there is ample proof that these membranes are tolerant to high temperature above 120°C.

4.2 Inorganic Superacids

Apart from the high surface to volume ratios inherent of nano-particles that make them attractive for a variety of applications, the ability to modify surfaces is important. While incorporation of nanosized acidic inorganics with higher surface areas enhances water uptake properties of Nafion membrane, active proton transport depends more on surface properties involved. Inorganic superacids and/or oxides with purposely modified surface acidity are therefore considered as potential candidate to enhance the conductivity of nanocomposite Nafion membranes. In particular, sulfated metal oxides, such as (SO_4^{2-} /TiO_2), (SO_4^{2-} / ZrO_2) and (SO_4^{2-} / Al_2O_3) have become subjects of more interest as they are known to be thermally more stable than other solid superacids [50]. The surface acidic strengths of different oxide/SO_4^{2-} type superacids have been compared from their Hammett acid constant (H_0) [51]; H_0 is an expression for the acidity of a medium, defined as $H_0 = K_{BH}^{+}[BH]^{+}/[B]$, where K_{BH}^{+} is the dissociation constant of the acid form of the indicator, and $[BH^{+}]$ and $[B]$ are the concentrations of the protonated base and the unprotonated base, respectively. It is evident from these studies that the acid strength of the Al_2O_3 /SO_4^{2-} system changed considerably with H_2SO_4 surface content $-7 < H_0 < -10$ (1%), $-10 < H_0 < -14$ (2–3%), $H_0 < -14$ (4–9%), while for TiO_2 /SO_4^{2-} preparations it was high and less dependent on catalyst composition: $H_0 < -14$ (≥1%) and finally for SiO_2 /SO_4^{2-} it was low with $-0.75 < H_0 < -3.50$ (1–9%).

Sulfated zirconia (SO_4^{2-} /ZrO_2) is the strongest superacid ($H_0 = -16$) among all the known superacids so far [52]. It retains the sulfuric acid groups, responsible for proton conduction, until about 500°C. Hara and Miyayama studied the proton conductivity of sulfated zirconia (SO_4^{2-} /ZrO_2) prepared by three different methods [53]. Thampan et al. [54] first intermingled sulfated zirconia into Nafion for higher temperature composite PEMs and investigated the effects of particle size, chemical treatment and additive loading on membrane performance. In comparison with commercial Nafion, the nanostructured (SO_4^{2-} /ZrO_2) /Nafion PEM exhibited an increase in IEC, in water sorption and enhanced conductivity, but hydrolysis during fuel cell operation might represent a problem of this kind of additive.

4.3 *Ormosils*

The addition of binary oxides, such as SiO_2, TiO_2 and ZrO_2, improves the properties of a Nafion membrane in many ways, thermo-mechanical stability, water uptake capacity and reduction of methanol crossover to name a few, but most often these advantages could not be transformed into desired level of improvement in their performance when used inside a membrane–electrode assembly (MEA). It is mainly because the chemically inert nano-oxides within the Nafion matrix wither down the proton conductivity of the composite membranes. An exemplary way to prevent the conductivity loss without compromising much on the high water uptake ability achieved is to tether the inorganic oxides with proton-conductive organic functional groups such as $-SO_3H$, –SH etc., for example, sulfonic acid functionalized silica. ORMOSILs or organic-modified silicates are inorganic/organic hybrid monomers incorporating a series of organically functional groups that are covalently bounded to silica matrix. Sol–gel processes enable the preparation of the monomers at low temperatures [55], the inorganic skeleton can be modified by starting from many different silicon, main group, or transition metal alkoxides that predominantly influence the mechanical, optical and thermal properties, whereas the organic part is modified by selecting certain types out of an immense number of available organo-(alkoxy)silanes as well as organic monomers or pre-polymers, which are responsible for the resulting flexibility and processability [56]. These tuneable features of ORMOSILs facilitate the preparation of a family of nanocomposite membranes whose structure, morphology and functionality can be varied.

4.3.1 Recast Membranes

Significant conductivity enhancement can be achieved by using sulfonated organically modified silane monomers possessing hydrophilic –Si–OH and proton conductive $-SO_3H$ functional groups such as sulfonated phenylethylmethoxysilane as inorganic additive to Nafion [57]. At 80°C and 100% relative humidity, the composite membrane demonstrated conductivity values at least six times higher than that of commercial Nafion 117 under the same conditions. However, attempts to prepare high conducting Nafion/bifunctional organosilica composite membranes with sulfonated diphenyldimethoxy-silane (S-DDS) [58] monomer having hydrophilic SO_3H functional group or mercaptopropyl-methyldimethoxysilane (MPDS) [59] monomer grafted with thiol group –SH using similar synthetic strategy were not fruitful. Even though these membranes could not produce proton conductivity values as high as bifunctional sulfonated phenyethylsilica incorporated nanocomposite membranes, their single cell DMFC performance were significantly up to the level of Nafion 117 membrane under the same fuel cell conditions.

4.3.2 Sol–Gel Membranes

In-situ generation of ORMOSIL phase within nanophase-separated morphology of Nafion membranes is feasible due to the incorporation of organic groups within a silicate structure via co-polymerization of Si alkoxides (monomer) and organically modified silane monomers (co-monomer). Mauritz and his co-workers first prepared a Nafion/ORMOSIL nanocomposite by in-situ co-polymerization of tetraethoxysilane (TEOS) and diethoxydimethylsilane (DEDMS) using Nafion membrane as template [60]. They later followed it with a series of TEOS-$R'_nSi(OR)_{4-n}$ co-monomers [61]. R′ in the co-monomer is the organic group joined to the Si atom via covalent Si–C bonds, which are hydrolytically stable, while R denotes the organic group attached to oxygen atom. TEOS will act as a glass network former and inserted $R'_n\mathrm{Si(OR)}_{4-n}$ units will function either as bridging ($n = 2$) or end capping ($n = 3$) network modifiers. Depending upon the co-monomer feed ratio, the ORMOSILs will range from being rigid to flexible or from hydrophilic to hydrophobic; the structural topology of the networks are controlled in this way.

4.4 Heteropolyacids

One of the most promising strategies to obtain Nafion-based nanocomposite membranes with improved thermo-mechanical properties as well as high proton conductivity is incorporation of bifunctional particles having both hydrophilic and proton-conducting properties such as heteropolyacids, zirconium phosphates and metal sulphates. [62–64] Heteropolyacids (HPA) are an example of a class worth mentioning as they demonstrate strong acidity and reasonably high proton conductivity in their hydrated crystalline forms [65]. Typically HPAs, such as $H_3PW_{12}O_{40}\cdot nH_2O$ (PTA), $H_3PMo_{12}O_{40}\cdot nH_2O$ (PMA) and $H_4SiW_{12}O_{40}\cdot nH_2O$ (STA) exist in a series of hydrate phases with the degree of hydration varying from 6 to 30 molecules of water (waters of hydration) per HPA molecule. The exact number of waters of hydration depends upon the temperature and relative humidity of the environment. In dehydrated state, they exist as stable $[PM_{12}O_{40}]^{3+}$ anion clusters, known as Keggin unit [66] and their molecules (about 1 nm in diameter) can be regarded as nano-particles. If the HPA particle size is sufficiently small, the existing waters of hydration of the additive can form a bridge between shrunken clusters, thereby providing a pathway for proton hopping from one cluster to another. In this manner the activation energy for hopping may be reduced. Owing to these characteristics, HPA are suitable membrane fillers for increasing the protonic conduction and improving the hydrophilic character of the membranes [67].

Nafion/HPA nanocomposite membranes can be obtained through simple impregnation [36] of preformed membranes with a heteropolyacid solution as well as by mixing a Nafion solution with an appropriate amount of a heteropolyacid followed by casting [68, 69]. In comparison with unmodified Nafion, Nafion membranes impregnated with PTA show stronger electrochemical performances attributable to the presence of PTA, which provides high proton concentrations and improved water retention [36]. Similarly, substantial increase in water uptake and high proton conductivity can also be accomplished with recast Nafion membranes loaded with $H_3PW_{12}O_{40}\cdot 29H_2O$, $H_3PMo_{12}O_{40}\cdot 29H_2O$ and $H_4SiW_{12}O_{40}\cdot 29H_2O$ [67]. Water uptake in the Nafion 117 incorporated with STA increases to 60% from the normal value of 27% for Nafion 117 and with PMA it reaches a maximum of 95%. Also the fuel cell performance of heteropolyacid-loaded recast membranes at 80°C is far better than unmodified Nafion with the PMA-Nafion witnessing a maximum current density value of 940 mA cm^{-2} as compared to 640 mA cm^{-2} for Nafion.

Two major factors limit the performance of Nafion/HPA composite membranes: (1) the extremely high solubility of the HPA additive and (2) the large particle size of the inorganic additive within the membrane matrix, which in turn results in ineffective bridging between the ionic domains. Stabilization (i. e. limiting the solubility and leaching of the HPA additive) can be achieved by heat treatment [65]. 'Heterogenizing' [70] or confining the HPA additive into support materials is the most effective way to increase the additive stability. The presence of a second dopant namely SiO_2 increases the crystallinity of Nafion–silica-immobilized phosphotungstic acid, PTA/SiO_2, or silicotungstic acid, STA/SiO_2, recast membranes thereby improving the initial degradation temperature of the Nafion membrane.[71, 72] Superior stabilization over a range of membrane compositions can be achieved by using in-situ sol–gel approach rather than recasting [73, 74]. The sol–gel membranes are much more robust, and demonstrate low hydrogen crossover for membranes that are approximately 25 μm thick. The decomposition onset temperature of the composite membrane was extended by 30°C over Nafion. The conductivities of the composite membranes at 120°C and 35% relative humidity were on the order of 1.6×10^{-2} S cm^{-1}. An interesting recent addition in this direction is sulfonic acid functionalized heteropolyacid–silica nano-particles as fillers [75]. The resulting nanocomposites have been claimed as thermally stable up to 290°C as also their DMFC performance at high temperatures, but the reported power density values remain relatively small.

4.5 *Layered Additives*

Diffusion of small molecules through membranes with impermeable objects is a classic problem in transport phenomena. Investigations by means of various objects having different size and shapes indicate that decrease in permeability is governed mainly by aspect ratio and concentration (volume fraction) of the objects used. Upon uniform dispersion and complete exfoliation within a continuous polymer electrolyte membrane, high aspect ratio (length to width) nanofillers are therefore expected to significantly decrease the methanol crossover due to the resulting longer diffusion pathways [76]. This is the primary motivation for exploring inorganic fillers with high aspect ratio such as layered silicates, phosphates and phosphonates for preparing polymer nanocomposite membranes for DMFC applications. In addition to low methanol permeability, these Class I membranes even at a loading of less than 5 wt% would exhibit better mechanical properties than unmodified membranes [77].

Montmorillonite (MMT) clay or layered alumino-silicate is among the most widely used nano-fillers in polymer nanocomposites for DMFC application because of its (a) high IEC allowing modification of the inter-layer spacing to achieve better compatibility with host polymer matrices; (b) high aspect ratio giving better reinforcement effect [78]; clay particle length had a direct effect on the permeability of water. Longer, more tortuous pathways were generated by introducing clay with a higher aspect ratio, and thus the permeability of water was more effectively reduced [79]; (c) large surface areas (220–270 m^2 g^{-1}) and (d) high potential for large-scale commercial use. Thus membrane electrode assemblies using recast nanocomposite membrane loaded even with 3 and 5 wt% of MMT are found to show improved performance at 125°C [80]. Nafion/MMT nanocomposite membranes with 1 wt% MMT demonstrating increase in power density even at concentrated methanol feed have also been reported [81].

However, it is only when this nano-filler is completely exfoliated within the polymer that its impact on polymer properties is maximum. For this purpose, the nano-fillers have been made organophilic by exchanging the native inorganic counter-ions with bulky organic ones [82], e. g., alkyl ammonium cations. Not only the affinity of the filler to the polymer matrix is improved, but the inter-layer distance will also be increased by ion exchanging. Incidentally, addition of the nanoclays also decreases the ionic conductivity of the membrane, although this decrease is largely lower compared to the methanol permeability decrease.

In addition to MMT, the use of organo-clays (MMTs modified with an organic sulfonic acid group: HSO_3–MMT) has therefore been lately drafted into Nafion nanocomposite membrane systems [83]. They manifestly provide high barrier properties, an increase in mechanical strength as well as thermal stability when the organically modified clay is exfoliated in the matrix polymers. The relative permeabilities of methanol and water reduce up to 90% and 80%, respectively, relative to pristine Nafion. Proton conductivity of the composite membranes is found to be lowered only slightly from that of Nafion. The combination of these effects has been found to significantly improve the performance of a DMFC made with Nafion/HSO_3–MMT composite membranes. The encouraging results may be attributed to two factors: (a) the relatively high proton conductivity of HSO_3–MMT compared to other inorganic fillers employed previously and (b) the highly anisotropic morphology of the MMT platelets that could be more effective in blocking the passage of methanol than materials of other geometries. More details on the theoretical modelling of polymer–clay nanocomposites can be found in the fifth chapter of this book.

5 Hydrocarbon-Based Systems

Development of proton-conducting nanocomposite and hybrid polymers based on non-fluorinated hydrocarbons is another active area in the inorganic/organic polymer electrolyte membrane research [84]. Hydrocarbons as backbone polymers are less expensive, commercially available, more stable and their structure permits functionalization, i. e., the introduction of polar sites as pendant groups for increasing proton conductivity and water uptake. Functionalization can be made through acid doping, chemical grafting of protogenic groups or direct sulfonation by electrophilic substitution on the polymer backbone of these hydrothermally stable polymers. A synergetic composite membrane with the functionalized polymers will be more effective for at least one property (either conductivity or methanol crossover) than the polymer only membrane. In this way, hydrocarbon polymers are rendered fuel cell active as well as an attractive prospect for possible high temperature operation.

Most attractive of the new generation hydrocarbon polymers are arylene main chain polymers, such as polyethersulfones (PES), polyetherketones (PEK) with varying number of ether and ketone functionalities (such as PEEK, PEKK, PEKEKK etc.), poly(arylenethers), polyesters and polyimides (PI). Because of the aromatic rings, these hydrocarbon polymers as backbones allow the introduction of sulfonic moieties to render the desired level of conductivity for fuel cell application and also possess excellent chemical resistance and good mechanical properties. From a chemical point of view, the good oxidation resistance of aromatic hydrocarbons is due to the fact that the C–H bonds of the benzene ring have typical bond strength of around 435 kJ mol^{-1}, compared with aliphatic C–H bond strengths, around 350 kJ mol^{-1}. This section is restricted mainly to the advances in the research on SPEEK as a matrix for the incorporation of inorganic proton conductors or inorganic oxides as well as on polybenzimidazoles (PBI), another family of aromatic polymers often considered for this purpose.

5.1 SPEEK-Based Systems

PEEK is a semicrystalline thermoplastic polymer with an aromatic, non-fluorinated backbone, in which 1,4-disubstituted phenyl groups are separated by ether (–O–) and carbonyl (–CO–) linkages. It is known for its good thermal stability, excellent mechanical properties, broad chemical resistance and low cost. Sulfonation brings in proton conductivity as an additional feature and is sufficient to meet the requirements needed for applications in fuel cells. The chemical structures of PEEK and sulfonated PEEK are shown in Scheme 3.

Proton conduction in sulfonated PEEK is water assisted; consequently, the hydration level is a crucial factor for better electrochemical performance. Water uptake increases with degree of sulfonation (DS = number of $-SO_3H$ groups per repeat unit), thereby improving the conductivity of the hydrated membrane. However, highly polar water molecules act as a plasticizer, undermining the electrostatic interactions between SPEEK molecular chains and causes membrane swelling. Highly sulfonated PEEK swells rather strongly in water and becomes soluble if the sulfonation degree is high enough. It makes the long-term stability of highly sulfonated PEEK membranes questionable. Introduction of inorganic components into SPEEK will modulate the solubility, compensate for the mechanical behaviour, assist the improvement of the thermal stability, reduce methanol permeability and enhance the proton conductivity when solid inorganic proton conductors are used.

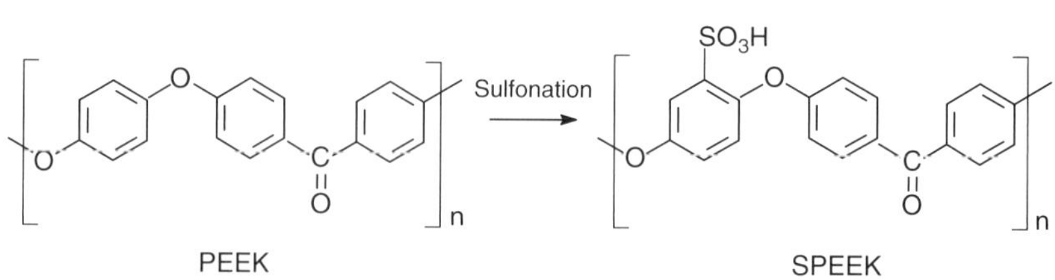

Scheme 3 Chemical structures of PEEK before and after sulfonation

5.1.1 Binary Oxides

The deployment of nanosized binary oxide materials (SiO_2, TiO_2 and ZrO_2) as inorganic fillers to obtain SPEEK-based Class I membranes has several attributes of interests including decreased water swelling, reduced permeability towards methanol and improved morphological stability without compromising proton conductivity under high degree of sulfonation. Studies carried out in these directions on recast SPEEK/SiO_2 membranes (silica loading up to 20 wt%) obtained by bulk mixing finely powdered amorphous silica with polymer solution indicate that the best balance of electrical and mechanical characteristics can be accomplished with a silica content of 10 wt% [85]. However, attempts to prepare nanocomposite membranes by in-situ generation of the silica phase within SPEEK membranes were not successful, due to the creation of large particles, auto-aggregation of the silica phase as well as by formation of cavities in the polymer matrix during the hydrolysis of silanes [86]. It is also very hard to achieve homogeneous dispersion of SiO_2 particles within the SPEEK matrix by sol–gel technique and the level of adhesion between the inorganic domains and the polymer matrix in the composite membranes obtained are generally very low (Fig. 6). Proper control at atomic level and dispersion of the inorganic component is feasible either by covalently bonding the inorganic component to the polymer matrix [41] or by using organic functionalized silica [87]. While non-functionalized silica into SPEEK membranes causes worsening of the mechanical properties of the composites, hybrid membranes and membranes containing organic-functionalized silica exhibit the same and even higher

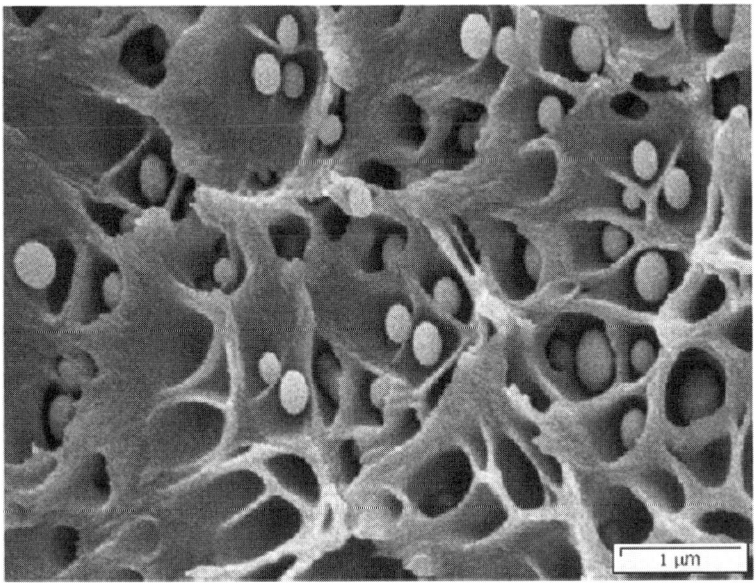

Fig. 6 Scanning electron micrograph of a SPEEK membrane modified with hydrolyzed TEOS [86]

mechanical modulus than that of unmodified SPEEK. On the other hand, hydrolysis of silanes and $Ti(OEt)_4$ in the polymer solution allows the synthesis of nanocomposite SPEEK/TiO_2 membranes having reasonable stability and reduced methanol and water permeability without any need for modifications. The distribution of inorganic phase within the obtained membranes is homogeneous and the observed proton conductivity of a particular membrane with 16 wt% TiO_2 is 1.5×10^{-2} S cm^{-1}, about one-third of the value of the unmodified membrane. By the same synthetic approach, but with added acetyl acetone as a chelating agent to avoid precipitation of the inorganic phase, nanocomposite SPEEK/ZrO_2 membranes with a wide variety of physical/chemical properties using a systematic variation of the inorganic content from 2.5 to 12.5 wt% of ZrO_2 SPEEK DS 0.87 have also been prepared, their properties investigated [88] and fuel cell performances evaluated [89]. It is clear from these studies that the zirconium oxide network within SPEEK polymer apart from being advantageous in improving barrier properties of the composite membranes leads to considerable reduction of water swelling and proton conductivity as well. It also enables the preparation of membranes with improved morphological stability for DMFC application, even though the SPEEK polymer possesses high degree of sulfonation. MEA tests indicate that SPEEK-based hybrid membrane with 7.5 wt% ZrO_2 presents the highest open-circuit potential due to its better ratio between methanol crossover and ohmic resistance in comparison with the other studied membranes.

5.1.2 Ormosils

Proton conductivity losses occurring due to the incorporation of inorganic fillers can be evaded by using organically modified silanes as filler for SPEEK matrixes. Diphenylsilanediol (DPDO) is an organically modified alkoxysilane with an aromatic structure having a decomposition temperature around 142°C. Organic moieties in DPDO are phenyl rings, a highly stable chemical entity that also plays a prominent role in the structure of PEEK and allows sulfonation with ease.

Sulfonated diphenylsilanediole (SDPDO) can be homogeneously dispersed into a SPEEK (DS as high as 0.9) matrix without any difficulty because of its lack of reactivity towards condensation due to the steric hindrance of the phenyl groups [90] (Scheme 4). Structural affinities between SDPDO and SPEEK would also enhance the composite morphological stability especially by shortening the distance between sulfonic groups, thus helping proton transfer and improving electrical properties. Furthermore, SDPDO is not soluble in water and its presence could be expected to modify SPEEK mechanical and solubility properties as well. SDPDO incorporated SPEEK membranes in effect are characterized with reduced water swelling, increased thermal stability from 250°C (pure SPEEK) to 350°C (composite), proton conductivity values as high as 0.1 S cm^{-1} at 120°C and proton diffusion coefficients higher than those reported for Nafion.

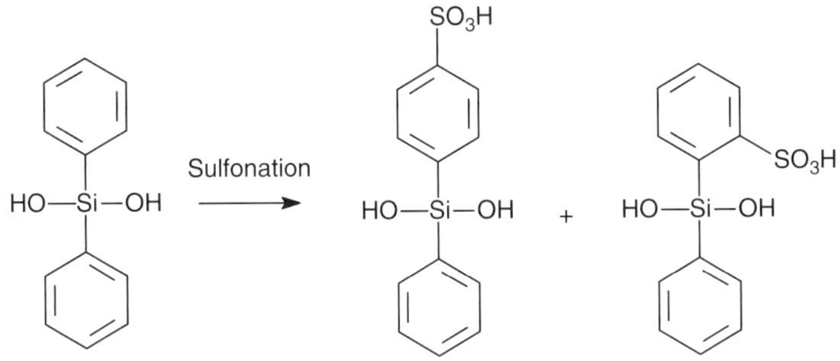

Scheme 4 Chemical structures DPDO and sulfonated DPDO

5.1.3 Heteropolyacids

HPA especially molybdophosphoric acid $H_3PMo_{12}O_{40}{\cdot}29H_2O$ (MPA) and tunstophosphoric $H_3PW_{12}O_{40}$ $29H_2O$ (TPA) incorporated SPEEK membranes exhibit conductivity as high as 10^{-2} S cm^{-1} at ambient temperature and up to a maximum of about 10^{-1} S cm^{-1} above 100°C (DS varying from 0.7 to 0.8) [91]. SPEEK/PMA and SPEEK/TPA membranes are also characterized by higher glass transition temperature compared to pure sulfonated polymer and even greater hydration at room temperature (up to five times for TPA-loaded SPEEK with DS = 0.8). Despite these improvements, most often HPAs get dissolved in water present within the membrane because of their highly soluble nature and washed out from the membrane during fuel cell operation. This problem can be minimized by the incorporation of a second inorganic additive [92] such as ZrO_2 or poly(silsesquioxane) resins or by using mesoporous molecular sieves incorporated withHPA [93], which in turn can be grafted into the SPEEK polymer.

5.1.4 Layered Additives

Laponite and Montmorillonite (MMT) are well-known layered silicates composed of silica tetrahedral and alumina octahedral sheets. When introduced, these layered inorganic materials are found to considerably improve the properties of SPEEK polymers. In particular, a SPEEK membrane with 10 wt% laponite was found to exhibit higher conductivity than unmodified SPEEK membrane [94]. Figure 7 presents the relative methanol flux of SPEEK/laponite 10 wt% composite membrane in comparison with unmodified SPEEK and Nafion-115 as a function of time [95]. The stationary methanol permeation rate decreases from 100% for Nafion 115 membrane to 25% for the SPEEK/Lapo10 composite membrane. SPEEK membrane without any modification displays much reduced relative methanol permeability (70%) than Nafion 115, possibly due to the difference in their microstructure. In fact, SPEEK membranes possess much narrow channel than Nafion-115 as their hydrophilic/hydrophobic separation in less pronounced. On the other hand, even lower methanol

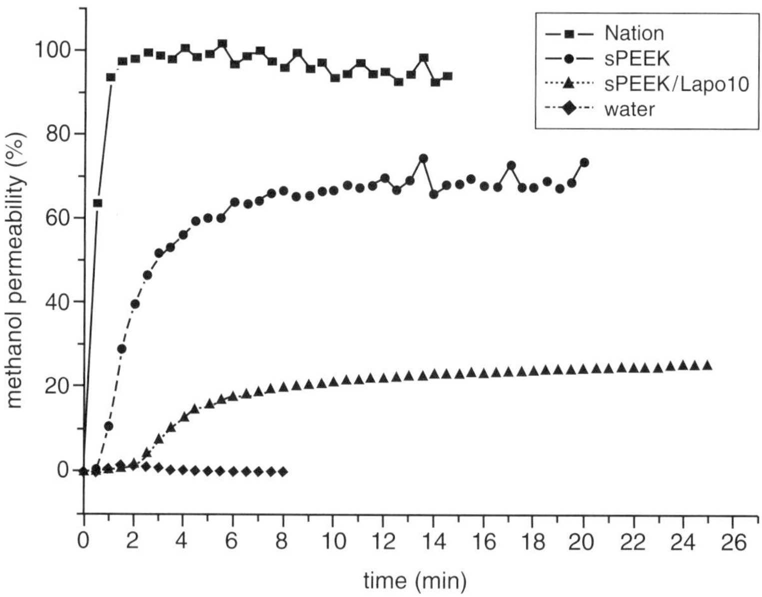

Fig. 7 Methanol permeability of Nafion 115, SPEEK and SPEEK composite membranes [95]

permeability in composite membranes is due to the microsized dispersion of the clay that prevents methanol from transferring through the membrane.

α-Zirconium phosphate (α-ZrP) is another well-known layered inorganic additive preferred to achieve reduced permeability towards methanol species in polymer electrolyte membranes. Conductivity of SPEEK membranes incorporated with nanosized α-ZrP increases marginally with α-ZrP loading up to 30 wt% [96]. Hydrogen fuel cells based on these nanocomposite SPEEK-ZrP membranes show improved performance when temperature is increased from 80 to 100°C. Organic modification of the layered additives will increase the inter-layer distance, acidic surface area and room temperature proton conductivity at least by two or three orders in comparison with that of the original material. Incorporation of layered additives intercalated with organic moieties may enable the preparation of membranes with improved relationship between proton conductivity and permeability towards methanol (sorption and diffusion) [97].

5.2 *PBI-Based Nanocomposites*

Polybenizimidazoles are a large family of amorphous thermoplastic polymers with high glass transition temperature (T_g = 425–436°C), good chemical resistance and excellent textile fibre properties. Commercially available polybenzimidazole is poly-[2,2′-(*m*-phenylene)-5,5′-bibenzimidazole] or simply 'PBI' synthesised from diphenyl-*iso*-phthalate and tetraaminobiphenyl, the chemical structure of which is presented in Fig. 8.

PBI easily reacts with strong acids such as H_2SO_4 or H_3PO_4, by undergoing hydrogen bond interaction or proton transfer reactions because of the presence of imidazole rings in its structure. The resultant solid membranes show remarkably high ionic conductivity (>10^{-3} S cm^{-1}) at room temperature and they form a new family of proton conductors know as PBI-acid complexes. For reviews covering the vast literature on PBI polymer and the polymer acid complexes based on PBI the reader is referred elsewhere [98, 99]. Proton conduction in acid-doped PBI is mainly influenced by the fraction of acid termed as 'doping level'. High acid doping levels result in higher conductivity. However, a very high acid doping level deteriorates the mechanical properties of the acid doped polymer membrane, especially at temperatures above 100°C. The endurance of PBI-acid membranes is therefore a critical issue in the development of PBI as polymer electrolyte membranes. An alternate option, both from the point of view of their conductivity and long-term endurance is the preparation of composite membranes using solid inorganic proton conductors.

HPAs in general are considered to be the best among solid proton-conducting inorganic compounds and their incorporation into sulfonated and perflurosulfonated conventional membranes constitute a significant research trend. This approach can also been extended to PBI membranes, but the conductivities achieved with the direct incorporation of HPAs are not high enough to be considered for application in a PEMFC. It is because HPAs interact very feebly with PBI; consequently, it is very difficult to anchor the HPAs within the polymer. HPA immobilization is feasible only via the dispersion of HPAs entrapped on high-surface-area inorganic fillers such as SiO_2.

PBI composites with silica-immobilized phosphotungstic acid, PTA/SiO_2, or silicotungstic acid, STA/SiO_2, are thermally stable up to 400°C [100, 101]. Highest conductivity achieved is 3×10^{-3} S cm^{-1} with a PTA/SiO_2 loading of 60 wt% at 100°C under a RH of 100% and is stable over the temperature range 100–150°C. The highest conductivity in the case of PBI–STA/SiO_2 composite membrane is due to the membrane having STA/SiO_2 weight ratio of 45/65. On the other hand, the conductivity of a membrane loaded with only 50 wt% STA was five orders of magnitude lower, thus indicating that hydrated silica provides the main pathway for

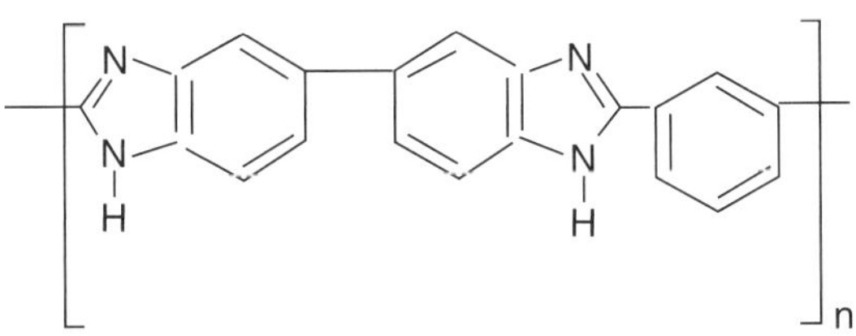

Fig. 8 Poly[2,2'-(*m*-phenylene)-5,5'-bibenzimidazole], PBI

proton conduction. An interesting observation about these PBI–heteropolyacid/SiO_2 membranes is, although their conductivities are still too low for fuel cell applications, they are only weakly dependent on relative humidity.

6 Hybrid Polymers

Hybridization is an important and evolutionary route for obtaining strong polymer–filler interaction in polymer composites so as to overcome the problems related to their poor mechanical properties and morphological stabilities. It involves bridging two generically different constituent materials, an organic hydrocarbon polymer and inorganic clusters, at the molecular level. A major appeal of such materials is linked to the extraordinary properties the synergistic effect carrying to the hybrid materials, which are normally not achievable by physical mixing of macroscopic phases, as usually made in classical composites (Nanocomposites or Class I hybrids). Two main types of hybrid materials can be individuated: inorganic/organic and organic/inorganic. In the former case, organic moieties are bound to an inorganic polymeric matrix, generally polysiloxanes, generating organically modified silanes (ORMOSILs) [102]. The latter are characterized by a high content of organic networks similar to carbon-based polymers (organic inorganic polymers, OIP). Modulation of the properties of the final materials can be achieved by proper dosage of the inorganic and organic components.

Hybrids that attract much interest in the field of PEM are those containing Si–O–Si links. Structural units that contain Si–O–Si network spanning a whole range from inorganic glass type to compositions with a high content of organic structures, similar to carbon-based polymers are feasible (Fig. 9). Their properties depend on various parameters such as the degree of phase dispersion, the relative amount of organic and inorganic components, the molar ratio of water to silane as well as the molecular weight of the polymer.

Silicones or polysiloxanes, chemical formula $[R_2SiO]_n$, where R represents organic groups such as methyl, ethyl or phenyl, are examples of inorganic/organic polymers with Si–O bonds. These materials consist of an inorganic silicon-oxygen backbone (…–Si–O–Si–O–Si–O–…) with organic side groups attached to the silicon atoms, which are four coordinate. In some cases organic side groups can be used to link two or more of these –Si–O– backbones together. By varying the –Si–O– chain lengths, side groups and crosslinking, silicones can be synthesized with a wide variety of properties and compositions. SPEEK membranes incorporating polysilsequioxane network show remarkable reduction in methanol and water permeabilities while retaining the proton conductivity of the pristine sulphonated polymer.

Preparation and characterization of organic/inorganic hybrid polymers where Si atoms are covalently bound to the SPEEK backbone is another point of interest. Inorganic component as $-Si(OH)_3$ moieties can be introduced directly into sulfonated aromatic polymer backbone SPEEK (SiSPEEK) [11] or into an aromatic polyphenylsulfone (PPSU) polymer backbone followed by sulfonation (SiSPPSU)

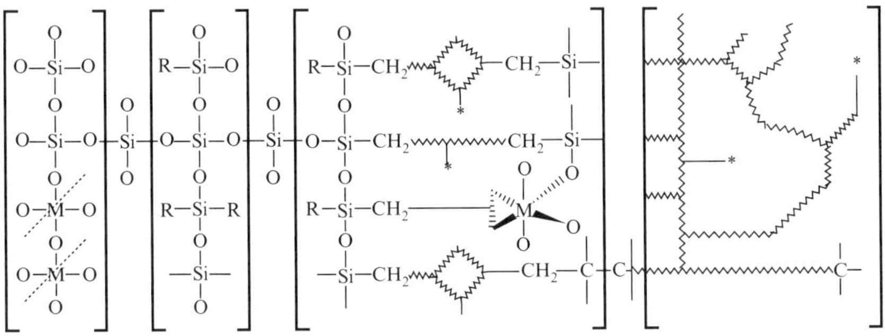

Fig. 9 Structural units of hybrids from inorganic/organic polymers (IOP) to organic–inorganic polymers (OIP) [102]

[103] or into sulfochlorinated PEEK (SOPEEK – a covalently cross-linked sulfonated polyetheretherketone) to obtain SOSiPEEK [12, 104]. Characterisation of these hybrid materials prepared with high degree of sulfonation (DS > 0.9) indicates high room-temperature conductivity values, enhanced water uptake capability, reduced solubility and high thermal stability with respect to the unmodified membranes. Figure 10 shows the TG curves of hybrid SOSiPEEK samples along with TG curves of PEEK and SOPEEK. Unmodified PEEK shows the highest thermal stability (decomposition temperature around 480°C), followed by SOSiPEEK/50 (1:0.50 per monomeric unit), SOSiPEEK/25 (1:0.25 per monomeric unit) and SOPEEK in that order. SOPEEK is clearly the less stable material, with weight loss between 250 and 300°C, corresponding to the decomposition of sulfonic groups. Complete pyrolysis occurs at about 400°C. Introduction of silicon increases the thermal stability, leading to products insoluble in water, thus allowing their electrochemical characterization and use in wet conditions.

An entirely new family of organic/inorganic hybrid macromolecules is constituted of SiO_2 linked with non-aromatic polyethers: polyethylene oxides (PEO), polypropylene oxide (PPO) and polytetramethylene oxide (PTMO) [105]. 'Hybrid precursors' first obtained by end-capping the organic polymers with alkoxysilanes are subsequently hydrolyzed and condensed to obtain highly flexible organic/inorganic hybrid materials. Hybrids are also feasible from the hydrolysis and condensation of precursors containing polydimethylsiloxane (PDMS) and the hydrocarbon monomers [106]. Sol–gel processing of end groups (OH) of organic parts with reactive inorganic moieties allows the formation of hybrids at molecular scale in the second procedure. The structure of these hybrid membranes is considered to be an interpenetrated network of nanosized silica skeleton and polymer phase, in which each silica domain seems to have a distance of few nm by chemically bound interior polymer chain. In the presence of water vapour, the membranes become proton conducting by doping with acidic moieties such as MDP or phosphotungstic acid (PWA), which probably is entrapped within the silica domains [107]. Accordingly, strong interactions between inorganic silica framework and the polyanion prevent the PTA from leaching out.

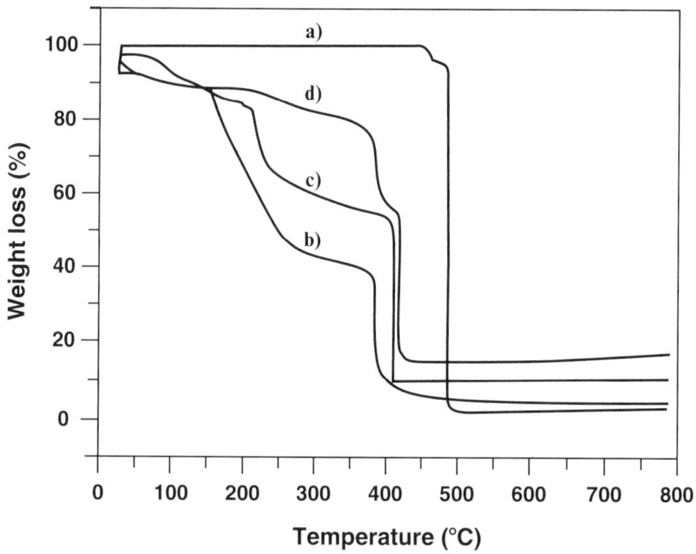

Fig. 10 Thermogravimetric curves for (**a**) PEEK, (**b**) SOPEEK, (**c**) SOSiPEEK/25 and (**d**) SOSiPEEK/50 [104]

7 Models

7.1 *Microstructural Models*

Nafion is an ionomer containing electrically neutral repeating units as well as a distinct fraction of ionic units. Nanophase separation occurs due to the de-mixing of the molecular and ionic units, while any macrophase separation is prevented by the chemical connectivity of the components. The exact structure is not yet clearly known and has been explained on the basis of several models, describing the way in which the molecular/ionic units organize themselves within the ionomer matrix. These models include the Mauritz-Hopfinger Model [108], the Yeager Three-Phase Model [109], the Eisenberg-Hird-More Model of Hydrocarbon Ionomers [110] and the Gierke Cluster Network Model [111]. In the forefront of them all was the model due to Gierke [112], which presumed the nanophase phase separated domains as a network of ionic clusters – spherically inverted micelles interconnected by short narrow channels and embedded in a fluorocarbon medium. Yeager's Three-Phase structural model [109] on the other hand, described Nafion as a three phase material consisting of a fluorocarbon region (A), an interfacial zone (B) and an ionic cluster region (C). These regions are depicted in Fig. 11. Region A consists of the fluorocarbon backbone and is quite hydrophobic. Region C consists of clusters of pendant sulfonate groups. This region is quite hydrophilic, most of the absorbed water and counter ions exist in this region. Region B is an interfacial region containing the

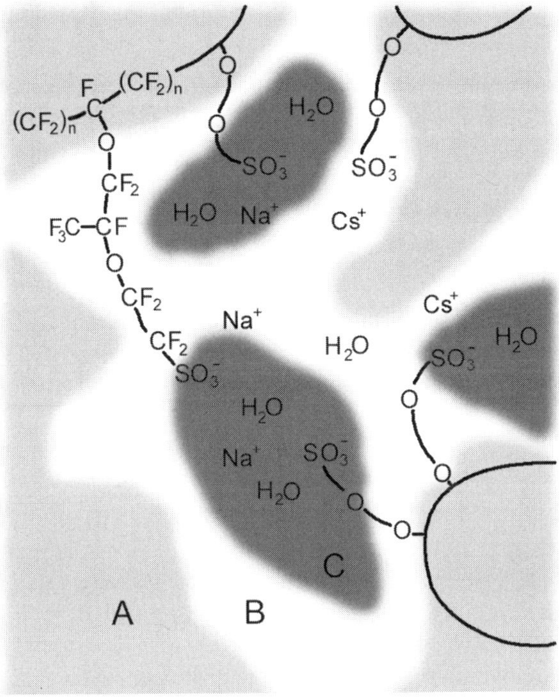

Fig. 11 Yeager's three-phase model of Nafion; a fluorocarbon region (A), an interfacial zone (B) and an ionic cluster region (C) [109]

pendant side chain material and sulfonate groups that are not clustered. Hence, only part of the absorbed water and counter ions would be present in this region. Gebel and Lambard [113] later confirmed the phase separation between the water domain and the perfluorinated matrix in a swollen ionomer membrane with their small-angle X-ray and neutron scattering (SAXS and SANS respectively) experiments, while the formation of clusters was confirmed by atomic force microscopy (AFM) studies (Fig. 12) [114, 115].

In a slightly different perspective, Eikerling et al. [116] extended Gierke's cluster-network model by assuming the existence of channels and a random network of pores that are filled with either bulk-like water or bound water, as impregnation by liquid water is easier than condensation. They used effective medium theory to predict conductivity results from impedance data and were able to demonstrate the importance of the connectivity of the pores and the coodination of the water in the pores to the overall conductivity of the membrane. A number of other models that attempt to describe the structure of the hydrophilic and hydrophobic domains of the Nafion membrane were also proposed over the years, but many of them were faced with one or other of the existing challenges.

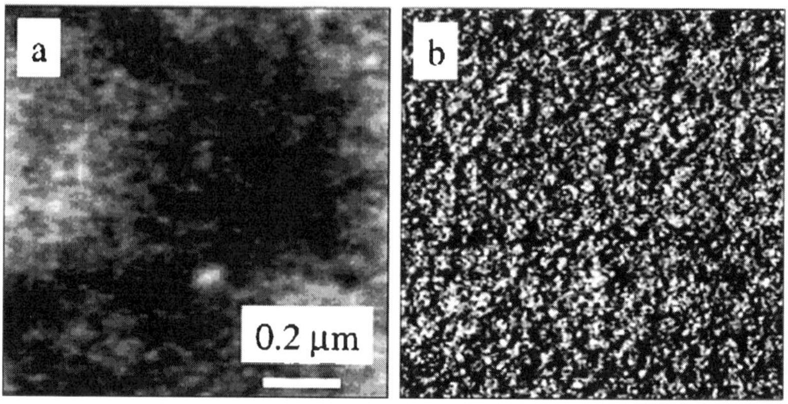

Fig. 12 AFM images of Nafion 115 obtained using the same cantilever under identical imaging conditions. (**a**) Tapping-mode AFM topography image under ambient conditions. (**b**) A phase image corresponding to part a (Z-scale, 25 nm and 10°, respectively) [115]

Nevertheless, it is a recognized fact now that Nafion combines in one macromolecule, the high hydrophobicity of the backbone with the high hydrophilicity of the sulfonic acid functional groups, which results in a constrained hydrophobic/hydrophilic nanoseparation. The sulfonic acid functional groups aggregate to form a hydrophilic domain that is hydrated upon absorption of water. It is within this continuous domain that ionic conductivity occurs: protons dissociate from their anion ($-SO_3^-$) and become hydrated by water. The morphology of water filled channels in Nafion has already been presented schematically as Fig. 1.

Most comprehensive of all the morphological structures proposed, when Nafion changes from dry state to swollen state, is the one due to Gebel [117]. In Fig. 13 is presented the schematic representation of structural evolution. In the dry state, isolated spherical ionic clusters are formed with a diameter close to 1.5 nm and an inner-cluster distance close to 2.7 nm. The absorption of water molecules induces modification of the cluster structure in spherical water domains with the ionic groups at the polymer–water interface in order to minimize the interfacial energy.

The diameter of water pools is about 2 nm and the inter-aggregate distance is roughly 3 nm, indicating that they are still isolated as revealed by the low value of ionic conductivity. As the membrane absorbs more water, the cluster swells and the diameter of it increases from 2 to 4 nm but relatively small increase in the inter-cluster distance leads to percolation. In this process, the number of ionic groups per cluster increases, and consequently the total number of clusters in membrane decreases. The high increase in ionic conductivity for a water volume fraction 'φ' larger than 0.2 reveals the percolation of the ionic aggregates in the membrane. When the water volume fraction is between $\varphi = 0.2$ and 0.5, the structure is formed of spherical ionic domains connected with cylinders of water dispersed in the polymer matrix. The diameter of ionic domain increases from 4 to 5 nm. At φ larger

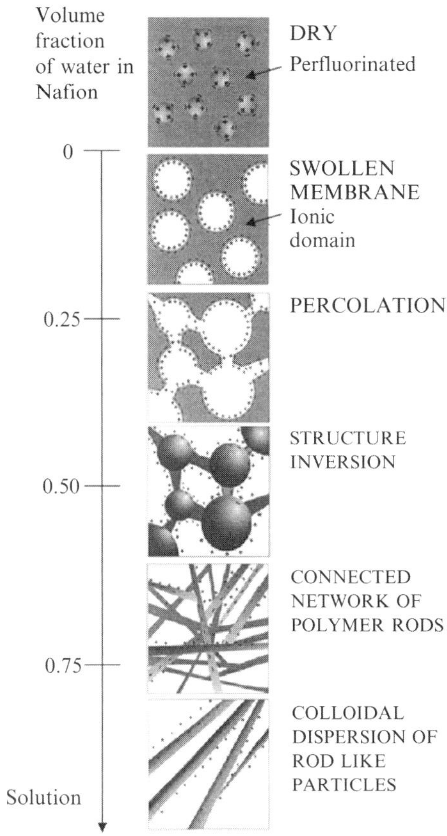

Fig. 13 Schematic representation of structural changes with water content in Nafion [117]

than 0.5, a structural inversion occurs and the membranes correspond to a connected network of rod-like polymer aggregates. For $\varphi = 0.5$–0.9, the rod-like network swells and the radius of the rod is about 2.5 nm. The structure of the highly swollen membrane would be very close to that of the Nafion solution.

Nafion solution aggregates into rods with the hydrophobic backbone making up the core of the rod and the hydrophilic sulfonate groups pointed toward the solvent, making rod-like fringes [118]. Specific chemical interaction between the polymer sulfonic acid groups and the various oxide surfaces has therefore been advocated as a vital first step towards making a successful elevated-temperature composite membrane in a recent Nafion–metal oxide nanocomposite model put forth by Bocarsly and his co-workers [38]. According to their model, the Nafion rods during the recasting process orient themselves around the metal oxide particles because of the electrostatic and chemical interactions and, after evaporation of the solvent, form homogeneous composite membranes. If the Nafion–metal oxide interaction is too weak, the particles will precipitate out of solution and an inhomogeneous material will form. This latter condition is observed when large metal oxide particles are employed. Likewise, if the selected oxide interacts very weakly with the polymer surface, a homogeneous membrane will not form. It has therefore been postulated that the metal oxide particles act to crosslink the Nafion polymer chains, which in turn results in the improved performances of the Nafion–metal oxide nanocomposite membranes during high-temperature PEMFC operations. The discussion about the origin of property improvement by oxide additives on polymer membranes was previously also made in the case of lithium-ion conducting polymers. Here, space charge effects but also influence on morphology and specific polymer structure, including glass transition temperature and degree of crystallinity (see chapter by Wieczorek and co-workers in this book).

7.2 *Thermodynamic Models*

The main focus of these models is how the membrane structure changes as a function of water content λ (the number of water molecules per sulfonic acid group of the Nafion membrane), which in turn could be determined by thermodynamics. Fig 14 is a schematic of the state of a membrane for λ in the range 1–14. The dis-

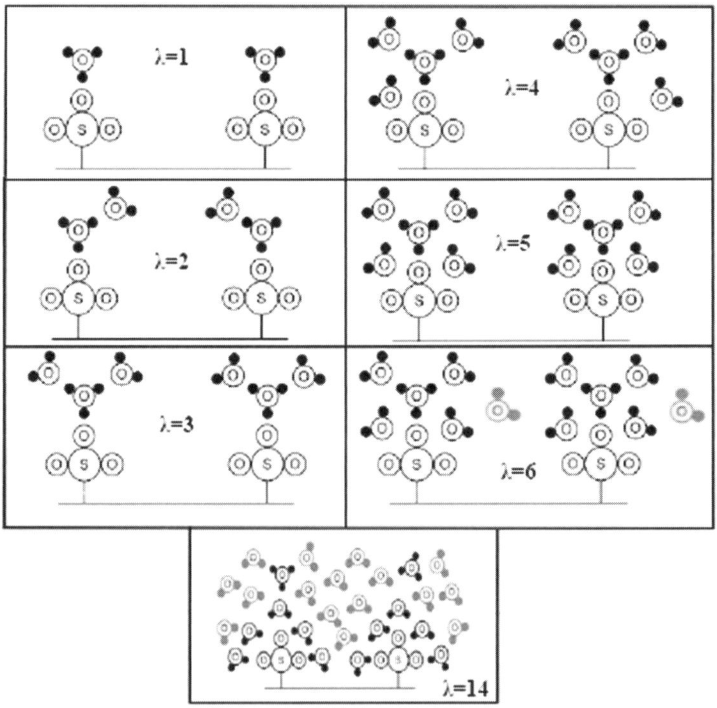

Fig. 14 Schematic hydration diagram for Nafion. Free water is shown in gray [122]

tance between sulfonate groups will be somewhat less than in an average membrane as sulfonate heads cluster together: transport is possible even at lower water contents ($\lambda < 2$). λ value of 2 corresponds to the percolation threshold [119]. In the range 2–5, the clusters continue to grow while protons are mobile over the entire cluster. Many different research groups have studied the dependence of conductivity on water content [119–121]. All these studies indicate that the membrane exhibits low conductivity for λ less than 5, but as it approaches 5 the conductivity increases by about 2 orders of magnitude. The extreme variation in conductivity in the range of $\lambda = 2$–5 highlights how significant the formation of a continuous phase is.

The number of water molecules forming the primary hydration shell for Nafion is expected to lie in the range 4–6 [121]. Molecular dynamic simulations indicate that 5 waters form the primary hydration shell for the sulfonate head, and additional water molecules are not as strongly bound and thus form a free phase [123, 124]. For $\lambda \geq 6$, counter ion clusters coalesce to form larger clusters, and eventually form a continuous phase with properties that approach those of bulk water [119]. This is supported by measurements that show that water mobility and water self-diffusion values approach the bulk water values [125]; the mobility of protonic charge carriers approaches the value in bulk water as well [3]. The free water phase is shielded from the sulfonate heads by the strongly bound water molecules of the primary hydration shell [3].

In saturated conditions, $\lambda = 14$, the clusters are filled with water and have a spherical shape of about 4 nm in diameter, an energetically favoured state (Fig. 15). The membrane acid groups are hydrated and dissociate, creating charged groups

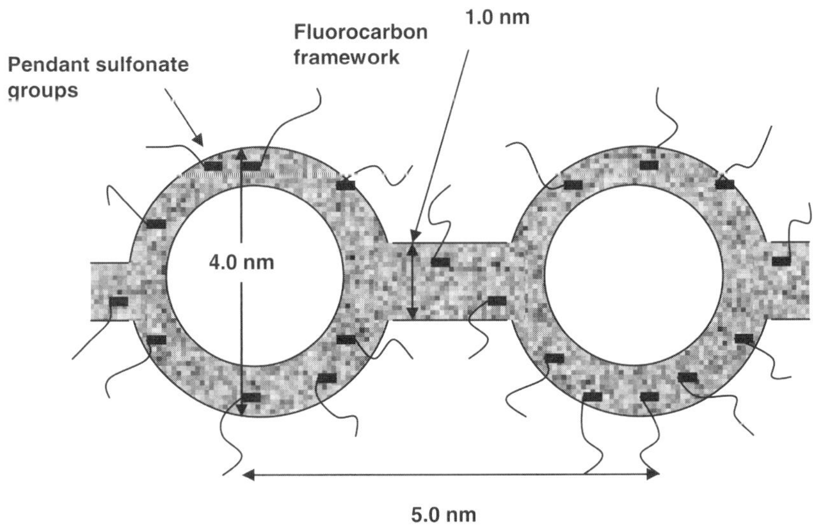

Fig. 15 Hsu and Gierke representation of Nafion with $\lambda = 14$ [126]

that participate in coulombic repulsion. These interactions are opposed by the work required to deform the polymer matrix. Hence, there is a balance between the surface or electrostatic energy and the elastic or deformation energy [126, 127]. The clusters are connected by short pathways, collapsed channels, of around 1 nm in diameter. They were determined by Hsu and Gierke to be transient with stability on the order of ambient fluctuations [126].

In general, water uptake in Nafion is measured by equilibrating the membrane with water either by immersing the membrane into water, or by bringing it into contact with water vapour. Both methods are widely followed and, thermodynamically, must give identical results for polymer solutions. However, the amount of water uptake in Nafion, with respect to the two modes differs substantially – the number of water molecules absorbed per acid site (λ) in Nafion is 22 for the liquid phase sorption, whereas in the vapour phase sorption it is 14 [116]. Apparently, a highly hydrated membrane, i. e. under water-equilibration conditions, would exist as a two-phase regime; when removed and exposed to saturated water vapour, it would emerge into a thermodynamically more stable membrane with λ dropping from 22 to 14. This phenomenon, known as Schroeder's paradox [128] (See Fig. 3), is apparently not uncommon in polymer systems and is an important feature in the development of any model where the membrane is not either fully hydrated or dehydrated. As recently shown by Alberti [129], the complicated hydration behaviour is related to the fact that an ionomer, being a semi-crystaline material, can easily give rise to metastable phases. The structural conformations of these metastable phases depend on their water content that, in turn, depends on the previous thermal treatments and on the temperature of liquid water in which the ionomer has been equilibrated. By using appropriate thermal treatments in air and/or in liquid water, samples of Nafion 117 with a continuous change of structural conformations corresponding to a change of water-uptakes (determined in liquid water at 20°C) between 8 and 55 were easily obtained The memory of the previous treatments decreases with increasing temperature and completely disappears at about 140°C. It was also observed that the changes of the structural conformation obtained with increasing the temperature were not reversible and this is in agreement with the large hysteresis loops found at low temperature. The conclusion was that models based on thermodynamic equilibrium fail to represent the complex ionomer properties. In this sense, Schroeder's paradox is not really a paradox since thermodynamic equilibrium cannot be assumed in metastable phases.

7.3 Transport Models

Transport of protons within polymer electrolyte membranes is a phenomenon of great interest in the design and development of new materials. The complex nature of proton transport is related to its unique nature. It is the only ion that possesses no electronic shell. As a result, it can strongly interact with the electron density of its environment. In the conducting polymeric matrix, characterized by the presence of electronegative oxygen atoms mainly due to the sulfonic sites, the proton could interact with the two nearest neighbour oxygen atoms. This results in the formation of O–H bonds. For medium distances between two oxygen atoms (2.5–2.8 Å), the proton may be involved in two bonds, a short with a proton donor and a long, weak bond with a proton acceptor. This is the case of an asymmetrical hydrogen bond, which is directional in character. For shorter oxygen separation (~2.4 Å), a symmetrical hydrogen bond may be formed where proton is involved in two equivalent bonds.

In general, proton transfer phenomena in polymer electrolyte membranes follow two basic mechanisms. The most trivial case requires the translational dynamics of bigger species; this is the *vehicle mechanism* [130]. In this mechanism the proton diffuses through the medium together with a 'vehicle', water. The counter diffusion of unprotonated water molecules allows the net transport of protons. The observed conductivity, therefore, is directly dependant on the rate of vehicle diffusion and it can be expressed as $D_{H2}O$.

In the other principal mechanism, the vehicle shows pronounced local dynamics but reside on their sites. The protons are transferred from one water molecule to the other by hydrogen bonding. Simultaneous reorganization of the proton environment, consisting in reorientation of individual water molecule dipoles, then leads to the formation of an unprotonated path. This mechanism is known as the *Grotthuss mechanism* [15, 125]. The rate of proton transport and reorganization of its environment affects directly the mechanism.

The proton transport can be described on the basis of the discussed mechanism depending on hydration level [16, 126]. Depending on the hydration level of the membrane, one or the other mechanism occurs. At low water content, (in Nafion $\lambda = 2–3$), the relative population of hydrogen bonds is low as well and the conductivity follows the vehicle mechanism. In such conditions, no free water is present neither in the clusters nor in the collapsed channels and the only complex that can be formed and transported is a hydronium ion (H_3O^+). Thus, proton transport is due to the hydronium ion that hops from acid side to acid side through the clusters and across the collapsed channels. When the water content increases, the channels form continuous water pathways from cluster to cluster and from one side of the membrane to the other. Thus, the membrane liquid phase is well inter-connected and the effect of the sulfonic sites on the free water is reduced due to shielding. Larger structures, such as Zundel ($H_5O_2^+$) and Eigen ($H_9O_4^+$) ions, are present and protons migrate by hopping from one structure to the other [125]. The transport occurs by continuous formation of hydrogen bonds between the proton, its environments and water molecules. Regarding the influence of temperature, it is well known that the Grotthuss mechanism is less influenced than the vehicular proton transport mechanism. This results in a higher activation energy value for the proton conduction, determined from an Arrhenius plot, in the case of vehicular mechanism.

Apart from the two principal mechanisms discussed above, the proton transport in a membrane has also been assumed to occur via (a) surface mechanism where proton transport proceeds along the array of acid groups (i. e., via structure diffusion) over the interface and (b) bulk mechanism where the protons are transported with the Grotthuss mechanism in Eikerling's [131] phenomenological model. Proton mobility through the surface water is considerably smaller than that in the bulk due to the strong electrostatic attraction of SO_3 groups; subsequently, the measured proton conductivity of Nafion was considered as a weighted average of surface and bulk conductivities depending upon the radial distribution of protons and water content in Nafion [132]. However, in spite of being assumed to be via the Grotthuss mechanism, proton diffusion in bulk water would also undergo traditional mass diffusion, known as vehicular diffusion [128, 133]. Proton conductivity within a pore (σ_p) has therefore been assumed to be the contributions of proton conductivity from surface hopping, Grotthuss, and vehicular diffusion mechanisms, respectively in the proton transport model proposed by Choi et al. [134] to describe proton diffusion within Nafion at various hydration levels. It also incorporates the effects of water content, structural variables such as porosity, tortuosity, the ratio of diffusion coefficients, distributions of protons and diffusion coefficients for the proton conduction processes.

In a similar vein, proton transport model proposed by Choi et al. [17] to describe diffusion in nanocomposite Nafion/$(ZrO_2/SO_4)^{2-}$ membrane distinguishes the surface and bulk mechanisms of proton transport in the nanocomposite membrane in which the proton conduction depends on the water content, diffusion coefficients at the surface and bulk regions in the membrane, and concentration and distribution of protons. Whilst the nanocomposite model considers the inorganic additive within the nanocomposite framework (Fig. 16) as an additional dust species immobilized within the polymer matrix similar to a dusty-fluid model (DFM) and views

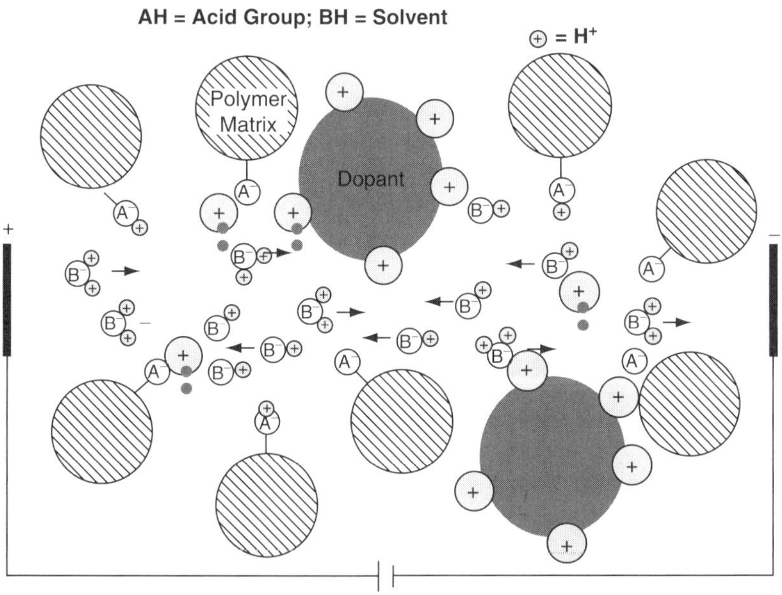

Fig. 16 A dusty-fluid model depiction of the PEM describing proton conductivity through the Nafion polymer matrix and the superacidic dopant. The framework treats the Nafion matrix as large dust particles through which the current carrying ions must traverse [17]

the obstruction presented by the nanocomposite matrix to proton diffusion as an additional frictional interaction with large immobile dust or gel particles. The transport of proton occurs via a sluggish hopping process through the membrane surface, and relatively fast structural and ordinary mass diffusion of hydronium ions in the bulk of the membrane pores.

The surface diffusion of proton, which takes place dominantly under low-humidity environments, is slow due to high coulombic interaction around the acid surface, while the transport of protons in the bulk water is relatively fast and occurs via Grotthuss and vehicular mechanisms. As a result, the sol–gel incorporation of ZrO_2 /SO_4^{2-} into Nafion increases the amount of water uptake and provided additional acid sites for proton diffusion, which ultimately results in higher proton conductivity compared to that of the host membrane.

A number of other physical and quantitative mathematical models have also been developed to describe the conductivity in Nafion membranes during the last decades. They are based on statistical mechanics [135, 136] molecular dynamics [137–139] and macroscopic phenomena [140], applied to the microstructure of the membranes. Despite these efforts, a comprehensive transport mechanism in PEMs as well as nanocomposite and hybrid membranes has not yet been advanced due to their complex nanostructure and inhomogeneous nature when hydrated.

8 Conclusions

The preceding chapter presents the status of research on proton-conducting nanocomposite and hybrid polymers. There appears at this point a clear general need for more fundamentals, especially a better understanding of the interface structure (between polymer and inorganic compound in composites, but also between different polymers in blends) and of the interaction between protons and/or water molecules and the secondary phase. Modelling appears as a crucial tool for better understanding, but it is still at an early stage of development, as shown in the next chapter of this book. Some special problems are still unresolved: for example, the breakdown of Nafion at higher temperatures is still not understood and different explanations (loss of water or morphological change) are competing. From a preparative point of view, the synthesis of hybrid materials and the use of sol–gel technique must be continued and generalized, because, so far, these techniques appear the most suitable for obtainment of non-porous membranes. An important aspect for the practical use of the membranes concerns mechanical properties, but the experimental studies are still relatively few, especially exploring the effects of humidity and thermal pre-treatments on mechanical properties. Considering the future importance of improved energy conversion systems and PEM fuel cells, synthesis and characterization of new proton-conducting polymer membranes will certainly remain a worthwhile topic for the years to come.

References

1. C. Sanchez, F. Ribot, *New J. Chem.* **18** (1994) 1007.
2. M. Casciola, G. Alberti, M. Sganappa, R. Narducci, *J. Power Sourc.* **162** (2006) 141.
3. K.D. Kreuer, *J. Membr. Sci.* **185** (2001) 29.
4. Q. Li, R. He, J.O. Jensen, N.J. Bjerrum, *Chem. Mater.* **15** (2003) 4896.
5. R.B. Moore, C.R. Martin, *Anal. Chem.* **58** (1986) 2569.
6. G. Alberti, M. Casciola, *Annu. Rev. Mater. Res.* **33** (2003)129.
7. G. Alberti, M. Casciola, M. Pica, T. Torpanelli, M. Sganappa, *Fuel Cells* **5** (2005) 366.
8. K.A. Mauritz, *Mater. Sci. Eng. C* **6** (1998) 121.
9. S.K. Young, W.L. Jarett, K.A. Mauritz, *Polymer* **43** (2002) 2311.
10. G. Schottner, *Chem. Mater.* **13** (2001) 3422.
11. M.L. Di Vona, D. Marani, A. D'Epifanio, E. Traversa, M. Trombetta, S.Licoccia, *Polymer* **46** (2005) 1754.
12. M.L. Di Vona, D. Marani, C. D'Ottavi, M. Trombetta, E. Traversa, I. Beurroies, P. Knauth, S. Licoccia, *Chem. Mater.* **18** (2006) 69.
13. M. Valcarcel, *Principles of Analytical Chemistry*, Springer-Verlag, Berlin (2000).
14. A.Z. Weber, J.B.R. Breslau, I.F. Miller, *InEng. Chem. Fundam.* **10** (1971) 554.
15. C.J.T. de Grotthuss, *Ann. Chim.* (Paris) **58** (1806) 54.
16. K.D. Kreuer, S. Paddison, E. Spohr, M. Schuster, *Chem. Rev.* **104** (2004) 4637.
17. P. Choi, N.H. Jalani, R.J. Datta, *J. Electrochem. Soc.*, **152** (2005) A1548.
18. W.H.J. Hogarth, J.C. Diniz da Costa, G.Q.(Max) Lu, *J. Power Sourc.* **142** (2005) 223.
19. P.G. Bruce, *Solid State Electrochemistry*, Cambridge University Press, Cambridge (1995).
20. A.G. Guzman-Garcia, P.N. Pintauro, M.W. Verbugge, R.F. Hill, *AIChE J.* **36** (1990) 1061.
21. Y. Yang, P.N. Pintauro, *AIChE J.* **46** (2000) 1177.
22. M.A. Hickner, B.S. Pivovar, *Fuel cells* **5** (2005) 213.
23. S.J. Paddison, *Annu. Rev. Mater. Res.* **33** (2003) 289.
24. B. Yang, A. Manthiram, *J. Power Sourc.* **153** (2006) 29.
25. L.A.S. De, A. Prado, H. Wittich, K. Schulte, G. Goerigk, V.M. Garamus, R. Willumeit, S. Vetter, B. Ruffmann, S.P. Nunes, *J. Polym. Sci. B: Polym. Phys.* **42** (2003) 567.
26. A.S. Aricò, V. Baglio, V. Antonucci, I. Nicotera, C. Oliviero, L. Coppola, P.L. Antonucci, *J. Membr. Sci.* **270** (2006) 221.
27. J. Surowiec, R. Bogoczek, *J. Thermal Anal.* **33** (1988) 1097.
28. S.R. Samms, S. Wasmus, R.F. Savinell, *J. Electrochem. Soc.* **143** (1996) 1498.
29. T. Kyu, M. Hashiyama, A. Eisenberg, *Can. J. Chem.* **61** (1983) 680.
30. N.H. Jalani, K. Dunn, R. Datta, *Electrochim. Acta* **51** (2005) 553.
31. S. Banerjee, D.E. Curtin, *J. Fluorine Chem.* **125** (2004) 1211.
32. K.A. Mauritz, R.B. Moore, *Chem. Rev.* **104** (2004) 4535.
33. M.Watanabe, H. Uchida, Y. Seki, M. Emori, P. Stonehart, *J. Electrochem. Soc.* **143** (1996) 3847.
34. M. Watanabe, H. Uchida, Y. Seki, M. Emori, *J. Phys. Chem. B* **102** (1998) 3129.
35. H. Uchida, Y. Ueno, H. Hagihara, M. Watanabe, *J. Electrochem. Soc.* **150** (2003) A57.
36. S. Malhotra, R. Datta, *J. Electrochem. Soc.* **144** (1997) L23.
37. P.L. Antonucci, A.S. Aricò, P. Creti, E. Ramunni, V. Antonucci, *Solid State Ionics* **125** (1999) 431.
38. K.T. Adjemian, S.J. Lee, S. Srinivasan, J. Benziger, A.B. Bocarsly, *J. Electrochem. Soc.* **149** (2002) A256.
39. E. Chalkova, M.B. Pague, M.V. Fedkin, D.J. Wesolowski, S.N. Lvov, *J. Electrochem. Soc.* **152** (2005) A1035.
40. N.H. Jalani, R. Datta, *J. Membr. Sci.* **264** (2005) 167.
41. K.T. Adjemian, R. Dominey, L. Krishnan, H. Ota, P. Majsztrik, T. Zhang, J. Mann, B. Kirby, L. Gatto, M. Velo-Simpson, J. Leahy, S. Srinivasan, J.B. Benziger, A.B. Bocarsly, *Chem. Mater.* **18** (2006) 2238.

42. S.P. Nunes, B. Ruffmann, E. Rikowski, S. Vetter, K. Richam, *J. Memb. Sci.* **203** (2002) 215.
43. V. Baglio, A.S. Aricò, A. Di Blasi, V. Antonucci, P.L. Antonucci, S. Licoccia, E.Traversa, F.S. Fiory, *Electrochim. Acta* **50** (2005) 1241.
44. Z. Liu, B. Guo, J. Huang, L. Hong, M. Han, L.M. Gan, *J. Power Sourc.* **157** (2006) 207.
45. K.A. Mauritz, I.D. Stefanithis, S.V. Davis, R.W. Scheetz, R.K. Pope, G.L. Wilers, H. Huang, *J. Appl. Polym. Sci.* **55** (1995) 181.
46. Q. Deng, R.B. Moore, K.A. Mauritz, *J. Appl. Polym. Sci.* **68** (1998) 747.
47. K.A. Mauritz, J.T. Payne, *J. Membr. Sci.* **168** (2000) 39.
48. K.T. Adjemian, S. Srinivasan, J. Benziger, A.B. Bocarsly, *J. Power Sourc.* **109** (2002) 356.
49. N. Miyake, J.S. Wainright, R.F. Savinell, *J. Electrochem. Soc.* **148** (2001) A905.
50. F. Lonyi, J. Valyon, J. Engelhardt, F. Mizukami, *J. Catal.* **60** (1996) 279.
51. M. Marczewski, A. Jakubiak, H. Marczewska, A. Frydrych, M. Gontarz, A. Snieguia, *Phys. Chem. Chem. Phys.* **6** (2004) 2513.
52. M. Misono, T. Okuhara, N. Mizuno, *Successful Design of Catalysts*, T. Inui, Ed., p. 267, Elsevier, Amsterdam (1988).
53. S. Hara, M. Miyayama, *Solid State Ionics* **168** (2004) 111.
54. T.M. Thampan, N.H. Jalani, P. Choi, R. Datta, *J. Electrochem. Soc.* **152** (2005) A316.
55. C.H. Brinker, G. Scherer, *Sol–gel Science*; Academic Press, London (1990).
56. G. Schottner, *Chem. Mater.* **13** (2001) 3422.
57. H. Wang, B.A. Holmberg, L. Huang, Z. Wang, A. Mitra, J.M. Norbeck, Y. Yan, *J. Mater. Chem.* **12** (2002) 834.
58. C. Li, G. Sun, S. Ren, J. Liu, Q. Wang, Z. Wu, H. Sun, W. Jin, *J. Membr. Sci.* **272** (2006) 50.
59. S. Ren, G. Sun, C. Li, Z.Liang, Z.Wu, W. Jin, X. Qin, X. Yang, *J. Membr. Sci.* **282** (2006) 450.
60. Q. Deng, R.B. Moore, K.A. Mauritz, *Chem. Mater.* **7** (1995) 2259.
61. S.K.Young, W.L. Jarrett, K.A. Mauritz, *Polymer* **43** (2002) 2311.
62. C. Yang, S. Srinivasan, A.S. Aricò, P. Cretì, V. Baglio, V. Antonucci, *Electrochem. Solid State Lett.* **4** (2001) A31.
63. G. Alberti, M. Casciola, R. Palombari, *J. Membr. Sci.* **4** (2000) 172.
64. G. Alberti, M. Casciola, L. Massinelli, B. Bauer, *J. Membr. Sci.* **185** (2001) 73.
65. P. Colomban (Ed.), *Chemistry of Solid State Materials. 2. Proton Conductors–Solids, Membranes and Gels–Materials and Devices,* Cambridge University Press, Cambridge, UK (1992).
66. J.F. Keggin, *Nature* **131**(1933) 908.
67. V. Ramani, H.R. Kunz, J.M. Fenton, *J. Membr. Sci.* **232** (2004) 31.
68. B. Tazi, O. Savadogo, *Electrochim. Acta* **45** (2000) 4329.
69. B. Tazi, O. Savadogo, *J. New Mater. Electrochem. Systems* **4** (2001)187.
70. Y. Wu, X. Ye, X. Yang, X. Wang, W. Chu, Y. Hu, *Ind. Eng. Chem. Res.* **35** (1996) 2546.
71. P. Staiti, A.S. Aricò, V. Baglio, F. Lufrano, E. Passalacqua, V. Antonucci, *Solid State Ionics* **145** (2001) 101.
72. Z.-G. Shao, P. Joghee, I.-M. Hsing, *J. Membr. Sci.* **229** (2004) 43.
73. V. Ramani, H.R. Kunz, J.M. Fenton, *J. Membr. Sci.* **279** (2006) 506.
74. W. Xu, T. Lu, C. Liu, W. Xing, *Electrochim. Acta* **50** (2005) 3280.
75. H.J. Kim, Y.G. Shul, H. Hana, *J. Power Sourc.* **158** (2006) 137.
76. M. Alexandre, P. Dubois, *Mater. Sci. Eng.* **28** (2000) 1.
77. B.J. Chisholm, R.B. Moore, G. Barber, F. Khouri, A. Hempstead, M. Larsen, E. Olson, J. Kelley, G. Balch, J. Caraher, *Macromolecules* **35** (2002) 5508.
78. G.S. Borchardt, *Minerals in Soil Environments*, J.B. Dixon, S.B.Weed, Eds., Soil Science Society of America, Madison, WI (1989); Chapter 14, pp. 675.
79. K.Yano, A. Usuki, A. Okada, *J. Polym. Sci. A: Polym. Chem.* **35** (1997) 2289.
80. D.H. Jung, S.Y. Cho, D.H. Peck, D.R. Shin, J.S. Kim, *J. Power Sourc.* **118** (2003) 205.
81. M-K. Song, S-B. Park, Y-T. Kim, K-H. Kim, S-K. Min, H-W Rhee, *Electrochim. Acta* **50** (2004) 639.

82. J.M. Thomassin, C. Pagnoulle, D. Bizzarri, G. Caldarella, A. Germain, R. Jerome, *e-Polymer* **018** (2004).
83. C.H. Rhee, H.K. Kim, H. Chang, J. Sung Lee, *Chem. Mater.* **17** (2005) 1691.
84. M.A. Hickner, H. Ghassemi, Y.S. Kim, B.R. Einsla, J.E. McGrath, *Chem. Rev.* **104** (2004) 4587.
85. B. Bonnet, D.J. Jones, J. Roziere, L. Tchicaya, G. Alberti, M. Casciola, L. Massinelli, B. Baner, A. Peraio, E. Ramunni, *J. New Mater. Electrochem. Syst.* **3** (2000) 87.
86. S.P. Nunes, B. Ruffmann, E. Rikowski, S.Vetter, K. Richau, *J. Membr. Sci.* **203** (2002) 215.
87. C.S. Karthikeyan, S.P. Nunes, K. Schulte, *Macromol. Chem. Phys.* **207** (2006) 336.
88. V.S. Silva, B. Ruffmann, H. Silva, Y.A Gallego, A. Mendes, L.M. Madeira, S.P. Nunes, *J. Power Sourc.* **140** (2005) 34.
89. V.S. Silva, J. Schirmer, R. Reissner, B. Ruffmann, H. Silva, A. Mendes, L.M. Madeira, S.P. Nunes, *J. Power Sourc.* **140** (2005) 41.
90. S. Licoccia, M.L. Di Vona, A. D'Epifanio, D. Marani, M. Vittadello, J.R.P. Jayakody, S.G. Greenbaum, *J.Electrochem. Soc.* **153** (2006) A1226.
91. S.J.M. Zaidi, S.D. Mikhailenko, G.P. Robertson, M.D. Guiver, S. Kaliaguine, *J. Membr. Sci.* **173** (2000) 17.
92. M.L. Ponce, L. Prado, B. Ruffmann, K. Richau, R. Mohr, S.P. Nunes, *J. Membr. Sci.* **217** (2003) 5.
93. S.M.J. Zaidi, M.I. Ahmad, *J. Memb. Sci.* **279** (2006) 548.
94. C.S. Karthikeyan, S.P. Nunes, K. Schulte, *Eur. Poly. J.* **41** (2005) 1350.
95. J.-H. Chang, J.H. Park, G.-G. Park, C.-S. Kim, O.O. Park, *J. Power Sourc.* **124** (2003) 18.
96. D.J. Jones, J. Rozière, *J. Membr. Sci.* **185** (2001) 41.
97. V.S. Silva, S. Weisshaar, R. Reissner, B. Ruffmann, S.Vetter, A. Mendes, L.M. Madeira, S.P. Nunes, *J. Power Sourc.* **145** (2005) 485.
98. J. Rozière, D.J. Jones, *Annu. Rev. Mater. Res.* **33** (2003) 503.
99. Q. Li, R. He, J.O. Jensen, N.J. Bjerrum, *Fuel Cell* **4** (2004) 147.
100. P. Staiti, S. Freni, S. Hocevar, *J. Power Sourc.* 79 (1999) 250.
101. P. Staiti, M. Minutoli, *J. Power Sourc.* **94** (2001) 9.
102. K.H. Haas, K. Rose, *Rev. Adv. Mater. Sci.* **5** (2003) 47.
103. M.L. Di Vona, A. D'Epifanio, D. Marani, M. Trombetta, E. Traversa, S.Licoccia, *J. Membr. Sci.* **279** (2006) 186.
104. D. Marani, M.L. Di Vona, E. Traversa, S. Licoccia, I. Beurroies, P.L. Llewellyn, P. Knauth, *J. Phys. Chem. B* 110 (2006) 15817.
105. I. Honma, S. Nomura, H. Nakajima, *J. Membr. Sci.* **185** (2001) 83.
106. J.-D. Kim, T. Mori, I. Honma, *J. Electrochem. Soc.* **153** (2006) A508.
107. H. Nakajima, S. Nomura, T. Sugimoto, S. Nishikawa, I. Honma, *J. Electrochem. Soc.* **149** (2002) A953.
108. K.A. Mauritz, C.J. Hora, A.J. Hopfinger, *Ions in Polymers*, A. Eisenberg, Ed., ACS Advances in Chemistry Ser. No. 187, American Chemical Society, Washington, DC, (1980), pp. 124.
109. H.L. Yeager, A. Steck, *J. Electrochem. Soc.* **128** (1981) 1880.
110. A. Eisenberg, B. Hird, R.B. Moore, *Macromolecules* **23** (1990) 4098.
111. T.D. Gierke, G.E. Munn, F.C. Wilson, *J. Polym. Sci. Polym. Phys. Ed.* **19** (1981) 1687.
112. T.D. Gierke, *Am. Phys. Soc.* **26** (1981) 462.
113. G. Gebel, J. Lambard, *Macromolecules* **30** (1997) 7914.
114. P. Zammaretti, A. Fakler, F. Zaugg, K. Spichiger, E. Ursula, *Anal. Chem.* **72** (2000) 3689.
115. J. James, M. Antognozzi, J. Tamayo, T.J. McMaster, J.M. Newton, M.J. Miles, *Langmuir* **17** (2001) 349.
116. M. Eikerling, A.A. Kornyshev, U. Stimming, *J. Phys. Chem.*, **101** (1997) 10807.
117. G. Gebel, *Polymer* **41** (2000) 5829.
118. S.H. Jiang, K.Q.Xia, G. Xu, *Macromolecules* **34** (2001) 7783.
119. T.A. Zawodzinski, T.E. Springer, J. Davey, R. Jestel, C. Lopez, J. Valerio, S.Gottesfeld, *J. Electrochem. Soc.* **140** (1993) 1981.
120. Y. Sone, P. Ekdunge, D. Simonsson, *J. Electrochem. Soc.* **143** (1996) 1254.

121. P. Choi, R. Datta, *J. Electrochem. Soc.* **150** (2003) E601.
122. J. Fimrite, H. Struchtrup, N. Djilali, *J. Electrochem. Soc.* **152** (2005) A1804.
123. J. Elliot, S. Hanna, A.M.S. Elliot, G.E. Cooley, *Phys. Chem. Chem. Phys.* **1** (1999) 4855.
124. A. Vishnyakov A.V. Neimark, *J. Phys. Chem. B* **104** (2000) 4471.
125. K.D. Kreuer, *Solid State Ionics* **136** (2000) 149.
126. W.Y. Hsu, T.D. Gierke, *J. Memb. Sci.* **13** (1983) 307.
127. K.A. Mauritz, C.E. Rogers, *Macromolecules* **18** (1985) 483.
128. P.V. Schroeder, *Z. Phys. Chem.* **45** (1903) 75.
129. G. Alberti, R. Narducci, *Proceedings in Advances in Materials for Proton Exchange Membrane Fuel Cell Systems*, Asilomar, California, February 18–21 (2007).
130. A.Z. Weber, J. Newman, *J. Electrochem. Soc.* **150** (2003) A1008.
131. M. Eikerling, A.A. Kornyshev, A.M. Kuznetsov, J. Ulstrup, S. Walbran, *J. Phys. Chem. B* **105** (2001) 3646.
132. S.J. Paddison, R. Paul, *Phys. Chem. Chem. Phys.* **4** (2002) 1158.
133. A.A. Kornshev, A.M. Kuznetsov, E. Spohr, J. Ulstrup, *J. Phys. Chem.* **107** (2003) 3351.
134. P. Choi, N.H. Jalani, R. Datta, *J. Electrochem. Soc.* **152** (2005) E123.
135. S.J. Paddison, R. Paul, T.A. Zawodzinski, *J. Chem. Phys.* **115** (2001) 7753.
136. X.D. Din, E.E. Michaelides, *AIChE J.* **44** (1998) 35.
137. A. Vishnykov, V. Neimark, *J. Phys. Chem. B* **104** (2000) 444.
138. B.R. Breslau, I.F. Miller, *Ind. Eng. Chem. Fund.* **10** (1971) 554.
139. A.G. Guzman-Garcia, P.N. Pintauro, M.W. Verbugge, R.F. Hill, *AIChE J.* **36** (1990) 1061.
140. Y. Yang, P.N. Pintauro, *AIChE J.* **46** (2000) 1177.

Hybrid Metal Oxide–Polymer Nanostructured Composites: Structure and Properties

Alla Pivkina[1], Sergey Zavyalov[2], and Joop Schoonman[3]

1 Introduction

Within metal (metal oxide)/polymer nanocomposites, nanoparticles reveal specific interparticle interactions and interactions with the matrix they are dispersed in [1, 2]. Nanostructured anatase titanium dioxide has attracted widespread attention as a photo-electrode in an advanced regenerative dye-sensitised solar cell, referred to as the Grätzel cell [3]. It has been shown also that the nanostructured anatase material exhibits an enhancement factor of about 3×10^6 compared to the mean lithium-ion intercalation time of a dense layer of this Li-battery anode material [4].

Nanostructured materials comprising 3-d transition metal oxide nanoparticles or alloys have been investigated extensively for their potential application as anode materials in lithium-ion batteries. A serious drawback of such systems is the substantial volume change of the active phase (up to 300%) during the charge/discharge process, which leads to mechanical disintegration of the electrode. The use of polymeric matrix could stabilize the nanoparticles within nanocomposite.

This chapter presents an extensive study of the structure, morphology, electrical properties, oxidation kinetics and electrochemical parameters of metal (metal oxide)/polymer nano-composites.

[1] Semenov Institute of Physical Chemistry, Russian Academy of Science, Kosygin st. 4, 119991, Moscow, Russia

[2] Karpov Institute of Physical Chemistry, Vorontsovo Pole, 10, 103064 Moscow, Russia

[3] Delft University of Technology, Delft Institute for Sustainable Energy, P.O. Box 5045, 2600 GA Delft, The Netherlands

P. Knauth and J. Schoonman (eds.), *Nanocomposites: Ionic Conducting Materials and Structural Spectroscopies.*

2 Experimental Aspects

2.1 Formation of Nanocomposites

The flux of metal atoms in vacuum (Pd, Sn, Al, Ti, Zn), evaporated from a bulk sample, condenses onto a cooled substrate together with the monomer. The condensate consists of nanoparticles of the metal and the monomer (Fig. 1). Upon heating the substrate to ambient temperature, the monomer polymerises to poly-*para*-xylylene. The structure thus obtained is a porous matrix with dispersed nanoparticles in it. The properties of these nanocomposites containing metal and/or metal-oxide nanoparticles in the polymeric matrix are presented. Manipulating the synthesis conditions, i. e., the distance between the vapour source and the substrate, the tilt angle of the beam and the deposition time, allowed for optimising the deposition regime. Measuring the electrical resistance of the condensate and composite permitted the control of the film formation in relation to the oxidation behaviour.

2.2 PPX Vacuum Co-Deposition

The schematics of the *para*-xylylene monomer polymerisation [5] is presented in Fig. 2. The monomer beam was introduced from the source consisting of the zone of evaporation of di-*para*-xylylene and its pyrolysis zone. Di-*para*-xylylene was introduced into the evaporation zone, which then evaporated (without destruction) in the temperature range 350–400 K. Then the molecules of di-*para*-xylylene reached the pyrolysis zone with a temperature of 930 K. Under these conditions the C–C bond

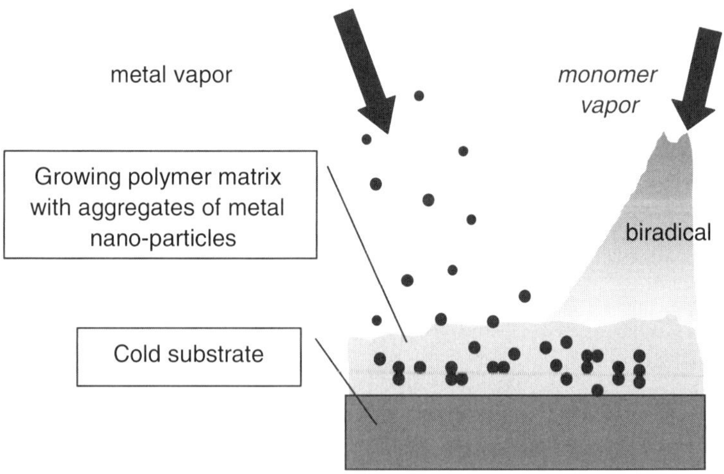

Fig. 1 Schematics of the cold-wall vacuum co-deposition process

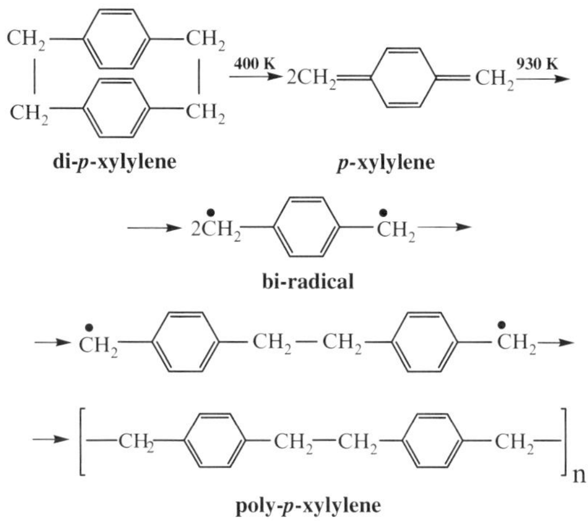

Fig. 2 Pathway for pyrolysis of di-*para*-xylylene to form PPX

shows destruction with almost 100% output of bi-radical. The monomer thus obtained condenses onto the cooled substrate. With heating up to room temperature, the condensed monomer polymerises into poly-*para*-xylylene as indicated in Fig. 2.

Several types of substrates have been used in the experiments reported here:

1. A polished quartz substrate of size 5 × 5 mm and 1 mm thick with Pt-contacts for electrical measurements
2. A polished NaCl single-crystalline substrate of the same size and thickness for TEM analysis
3. A polished quartz substrate of size 10 × 20 mm and 2 mm thick for optical and AFM investigations. To obtain identical samples both types of substrates were fixed close to each other onto a cooled surface of the sample holder
4. Al-foil and Cu-foil substrate for nanocomposites

2.3 Methods of Nanocomposite Analysis

Oxidation kinetics during air exposure after vacuum synthesis was measured using the data acquisition board L-1250 connected to a PC. The temperature coefficient of the electrical resistance in vacuum (the slope of $R_v(T)/R_v(293\,K)$ vs. temperature) was measured after cooling of the synthesized composite from 293 K to 77 K.

The morphology of the nanocomposites was studied with Transmission Electron Microscopy (TEM JEM-2000 EX-II at 200 kV). Samples for TEM were prepared by standard procedures, including separation of the nanocomposite layer from the NaCl substrate in water and the film deposition onto a Cu grid for further detailed investigations. The metal content of the composites was calculated by atomic absorption analysis using a Perkin-Elmer 503 spectrometer.

The surface morphology, film thickness, lateral forces and spreading resistance were studied by AFM (P47-SPM-MDT, Russia, NT-MDT) with silicon cantilevers having a tip radius less than 10 nm and 20° apex angle (NSC11, Estonia, Mikromasch) and conductive cantilevers (silicon coated with Ti-Pt) having a tip radius of 40 nm and 30° apex angle (CSC21, Estonia, Mikromasch). Topographic and spreading resistance AFM images were obtained in tapping and contact modes in air. Additionally, applying simultaneous AFM imaging of surface topography, spreading resistance and lateral force allowed us to study the surface structure of thin films in more detail and to distinguish the conducting areas within the electrically inert polymeric matrix.

The optical spectra of the films were recorded with a spectrophotometer (Shimadzu UV-3100) in the wavelength range 200–2,000 nm.

Mechanical adhesion was studied by carrying out a pull-off test by measuring the minimum tensile stress necessary to detach a thin film in a direction perpendicular to the substrate (in accordance with standard ISO 4624:1978).

After long-term contacting with air (during 5 months) the electrical conductivity upon heating from 10 K to 300 K in a vacuum of 10^{-6} Torr has been measured by a two-probe DC technique (TR-8652). Prior to measurements, the sample on the quartz substrate has been kept in vacuum at room temperature during 2–3 days. The electrical current I was measured under controlled voltage U and a temperature T increment of 5 K.

The specific electrical conductivity has been calculated as follows:

$$\sigma = \frac{l}{dhU} I = \kappa I, \tag{1}$$

where d is the thin-film thickness, h the length of electrodes and l the distance between the electrodes.

For electrochemical measurements, a conventional two-electrode cell was employed using 1.4-cm-diameter electrodes. The deposited thin film was the working electrode and metallic lithium was used as the reference and counter electrode. The electrolyte consists of 1-M $LiClO_4$ in a 1:2 molar ratio of ethylene carbonate to propylene carbonate solvent. A porous polyethylene sheet was used as an electrolyte separator. The cells were sealed in a coin cell casing (Hohsen) in an Ar-filled glove box. Specific capacity and cycling measurements were performed at room temperature using a Maccor battery test system. The cells were cycled between 0.08 and 2.6 V vs. metallic lithium at a constant current in the range of 0.1–0.0005 mA. The typical initial open-circuit voltage for these cells was about 2.9 V.

To measure AC electrical conductivity of the synthesized thin film nanocomposites we used the impedance measurement system LCR-821 (INSTEK, Taiwan). Errors of the conductivity and capacity measurements were less than 0.05% in the frequency range 12 Hz–100 kHz. Measurements were performed under different conditions, i. e., (i) in vacuum at room temperature, (ii) in vacuum at temperature 77 K and (iii) in air at room temperature.

3 Results and Discussion

3.1 PPX Thin Films

Varying the polymer precursor deposition time, thin films of different thicknesses were synthesized. The atomic force microscopy study shows a marked variation in morphology and particle size, comprising thin films of different thickness. Thus, a thin film of 1–2 μm thickness (deposition time 30 min) is formed by outstretched spheroids with main diameters of 100 and 200 nm (Fig. 3). Adsorbed species of the precursor are hardly moving before being polymerised into the growing film. This results in the deposition of an amorphous film, as was shown by X-ray analysis. However, increasing the deposition time up to 90 min, i. e., a thin film thickness of up to 4–8 μm leads to polymer globules to grow and to be crystallized, as follows from the particle size and shape on the polymer surface (Fig. 4). In this case, the film contains large crystallites of 300–400 nm. Increasing the resolution allows distinguishing of nano-sized polymer 'threads' and globules having a size of 5–30 nm (Fig. 5) on the surface of these crystallines. The degree of crystallization and the resulting alignment of polymer chains vary with the process conditions of the vapour-phase deposition for thin films with poly(*p*-xylylene) as observed previously [6].

3.2 Pd/PPX Nanocomposites

Samples of nanoporous composites of a metal and the polymer PPX have been synthesised in the form of thin films. The AFM study of the Pd/PPX nanocomposites in the wide range of metal concentration (7–14%) reveals the Pd nanoparticles within the polymeric matrix to have a size of 7–30 nm (Figs. 6 and 7), while the

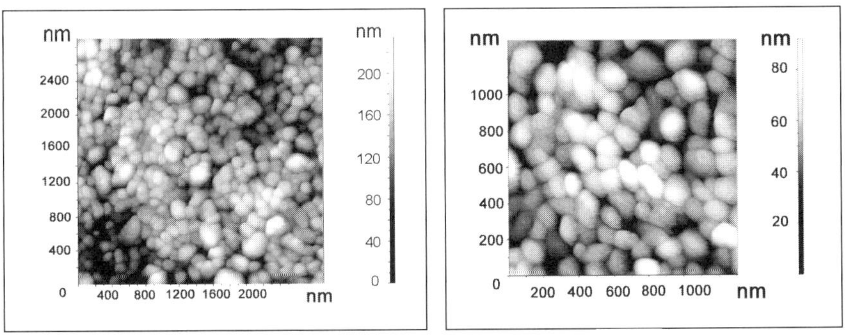

a) b)

Fig. 3 Surface topography of a pure PPX thin film, deposited during 30 min onto Al-foil. Film thickness is 2 μm. The polymeric globule size is 100–200 nm; (**a**) Scan size 2.8 × 2.8 μm and (**b**) Scan size 1.2 × 1.2 μm

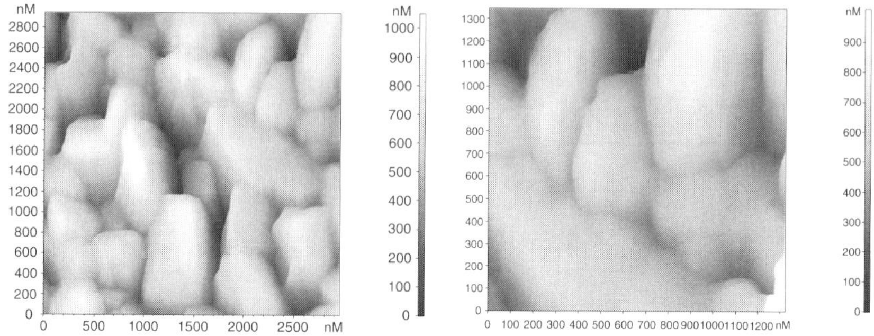

Fig. 4 Surface topography of a pure PPX thin film, deposited during 90 min onto Al-foil. Film thickness is 8 μm. The polymeric globule size is 300–800 nm with the crystallinity features: (**a**) Scan size 2.8 × 2.8 μm and (**b**) Scan size 1.2 × 1.2 μm

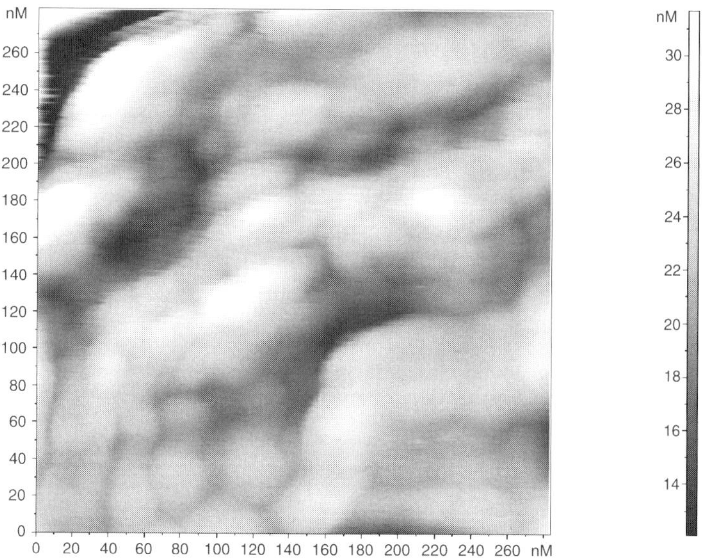

Fig. 5 Surface topography of a pure PPX thin film, deposited during 90 min onto Al-foil. Film thickness is 8 μm. Scan size 240 nm × 240 nm

polymer forms more or less spherical globules with diameters of 100 and 200 nm, i. e., of the same size of microstructure of the pure polymer thin film. It means that metal nanoparticle 'constellation' within a polymeric matrix is not influenced by the polymerization process, and Pd clusters and monomer are not interactive, being condensed onto substrate at temperature 77 K. Considering the high mobility of Pd clusters under these conditions, the 'constellation' of Pd nanoparticles already at the condensation stage could be suggested.

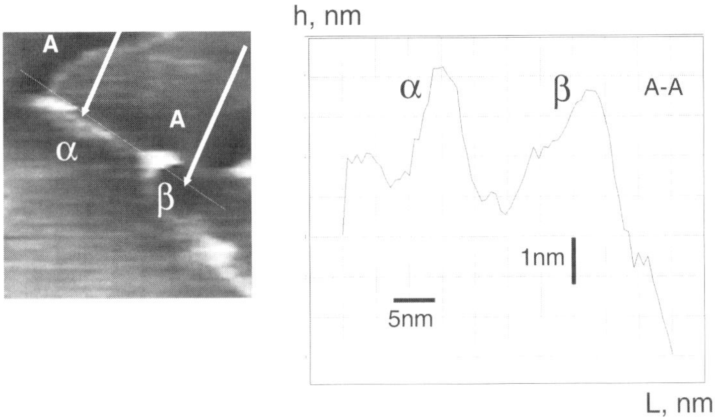

Fig. 6 AFM image of a Pd/PPX nanocomposite (**a**) Phase-contrast image: dark regions are polymer spherulits and light spots are Pd nano-particles situated on the boundaries between polymer globules and (**b**) Cross-section A-A. Profile maximums correspond to Pd particles embedded into the boundary surfaces between the polymeric globules

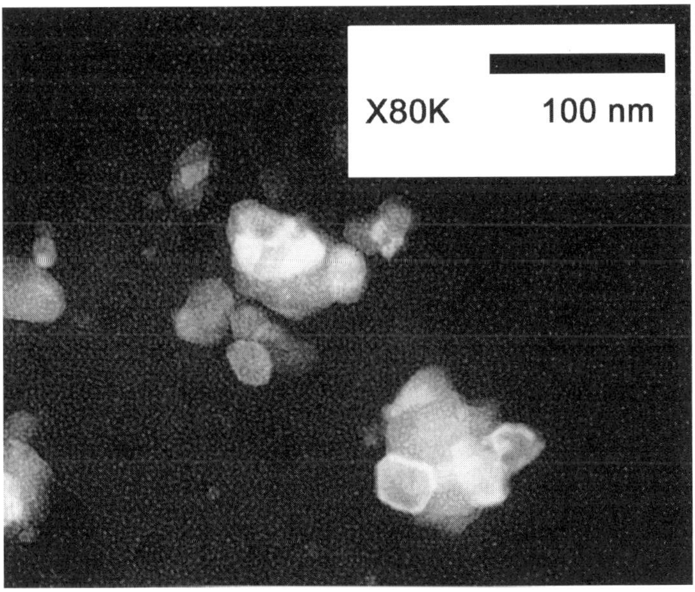

Fig. 7 Transmission-electron microscopy image of a Pd/PPX thin film, deposited during 30 min onto the polished NaCl single-crystalline substrate. Nanoparticles of palladium have a size of 5–30 nm

3.3 $Sn(SnO_2)$/PPX Nanocomposites

Freshly synthesized tin-containing samples exhibit a grey colour with a metallic lustre. On contacting with the ambient air during 2 min the samples became transparent for composites with Sn concentrations below the percolation threshold of 10 vol%, while the samples with a tin content beyond 10 vol% do not change their colour during several months (Fig. 8). Experimental data [7] reveal that for the metal content in as-prepared samples below or at the percolation threshold (samples 1 and 2, Table 1) the inorganic particles are isolated and the interparticle distance varies from 5 to 20 nm. Slightly above the percolation threshold (sample 3) the particles form continuous filaments of varying diameter, but the maximal diameter never exceeds that of the single metal nanoparticle. Beyond the percolation threshold (sample 4), the nanoparticles form aggregates located on the boundaries between

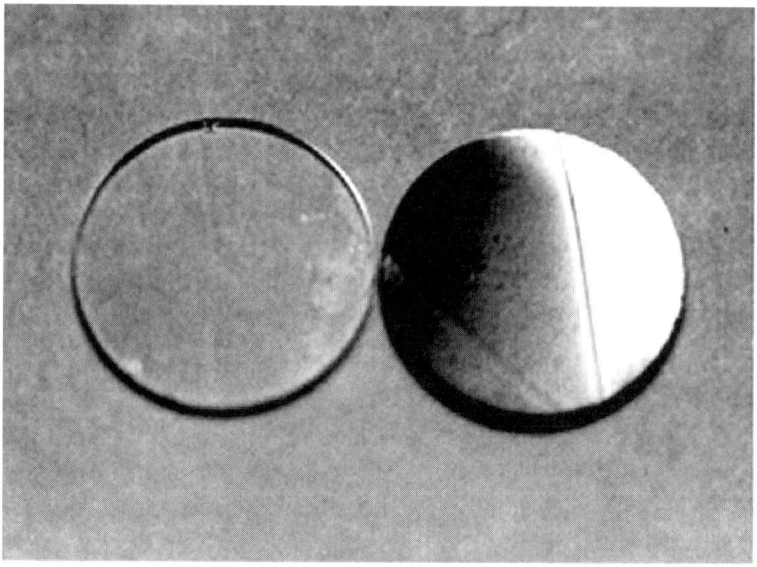

a) b)

Fig. 8 Sn-containing samples of nanocomposites after air exposure: (**a**) sample 1 (8 vol% of Sn) and (**b**) sample 3 (12 vol% of Sn)

Table 1 Properties of $Sn(SnO_2)$/PPX nanocomposites

Sample No.	Sn (vol%)	Resistivity in vacuum	Morphology
1	8	6 MOhm	Isolated particles, localization on the polymer spherulite surface
2	10	880 Ohm	
3	12	45 Ohm	Formation and growth of nanoparticle chains
4	16	13 Ohm	Chains exhibiting percolation

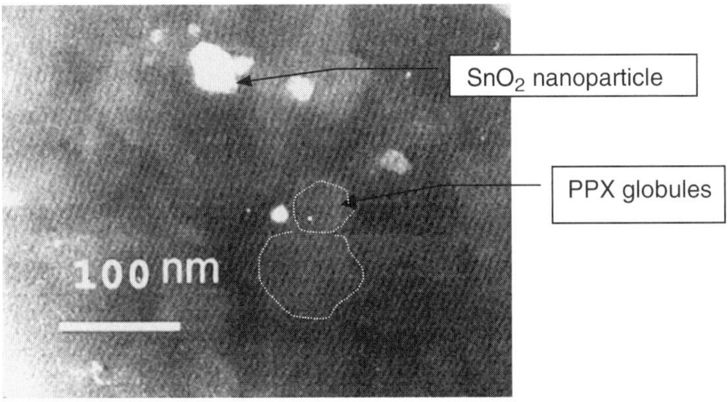

Fig. 9 TEM image of sample 1–SnO_2/PPX nanocomposite (8 vol%Sn)

the polymer spherulites. Hence, the metal–polymer and metal oxide–polymer nanocomposites are the components, wherein the inorganic particles form structured subsystems with respect to the polymeric matrix. In increasing the metal content, the nanoparticles localise along the borders of polymeric spherulites accompanied with the formation of conducting chain structures. Further increase in the metal content gives rise to aggregation of nanoparticles and their coalescence.

Analysis of the Sn(SnO_2)/PPX composites reveals that for the metal content in as-prepared samples below or at the percolation threshold the inorganic particles (SnO_2) are isolated and the interparticle distance varies from 5 to 100 nm (Fig. 9). Slightly above the percolation threshold the metal particles (Sn) form continuous filaments of varying diameter, but the maximal diameter never exceeds that of the single metal nanoparticle. Beyond the percolation threshold, the nanoparticles form aggregates located on the boundaries between the polymer globules. In increasing the metal content beyond the percolation threshold, the SnO_2 nanoparticles localize along the borders of polymeric globules accompanied with the formation of conducting chain structures. Further increase of the metal content gives rise to aggregation of nanoparticles and their coalescence.

3.4 Al(Al_2O_3)/PPX Nanocomposites

TEM analysis of the nanocomposite with an Al content beyond the percolation threshold reveals spherical pure metal nanoparticles with a mean diameter of about 10 nm (Fig. 10a), while below the percolation threshold the composite contains agglomerates of rhombohedral Al_2O_3 (corundum) with a mean size of 55 nm (Fig. 10c). A sample with a metal content just at the percolation threshold contains metal nanoparticles of 10 nm and alumina aggregates of 28 nm in diameter (Fig. 10b). The inorganic phase is homogeneously dispersed within the polymeric matrix in all of the investigated samples. It has been shown that the nanocomposite structure determines the oxidation behaviour of Al nanoparticles within the polymeric matrix under air exposure.

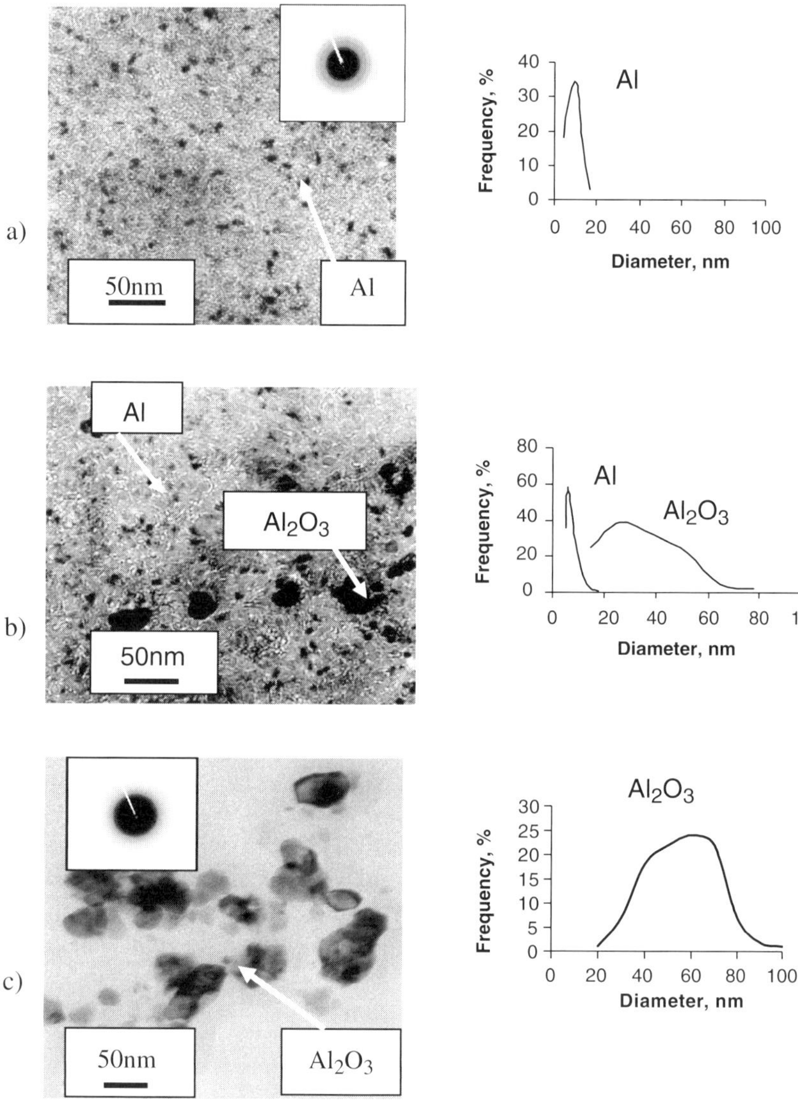

Fig. 10 TEM images of $Al(Al_2O_3)$/PPX nanocomposites with different contents of the inorganic phase: (**a**) sample 5 (12 vol%) (**b**) sample 6 (10 vol%) and (**c**) sample 7 (8 vol%)

Freshly synthesized Al-containing samples (Table 2) exhibit normally a dark colour with a metallic lustre in vacuum. On contacting the composite with the ambient air sample 7 became transparent, whereas samples 5 and 6 do not change their colour. Substantial differences in oxidation behaviour exist between the investigated samples. Figure 11a–c shows the resistivity change of freshly synthesized samples 5–7, if exposed to air at 1 atm. According to the above TEM results,

Table 2 Parameters of synthesized $Al(Al_2O_3)$/PPX nanocomposites

Sample number	Crystal phase	d_m (Al) (nm)	d_m (Al_2O_3) (nm)
5	Al	10	–
6	Al + Al_2O_3	6	28
7	Al_2O_3	–	55

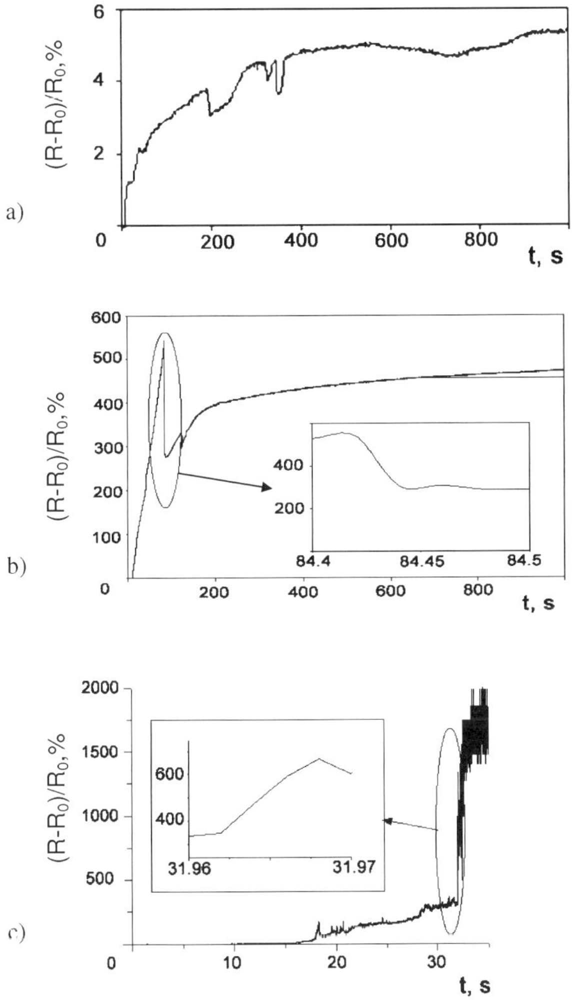

Fig. 11 Kinetics of oxidation during air exposure at room temperature of $Al(Al_2O_3)$/PPX nanocomposites with different contents of the inorganic phase: (**a**) sample 5 (12 vol%), (**b**) sample 6 (10 vol%) and (**c**) sample 7 (8 vol%)

nanoparticles in sample 5 are Al crystallites, which are clearly reflected in the minor increase of the electrical resistance (Fig. 11a) during air exposure (ΔR_{max} = 5.5%). However, the electrical resistivity of sample 7 increases dramatically during several seconds (Fig. 11c). In fact, nanoparticles in this sample comprise pure alumina dielectric material. The high resistivity of the nanocomposite is caused by the high resistivity of the alumina particles and large distance between them. TEM micrographs of sample 6 revealed alumina and aluminum crystallites within the polymeric matrix. The dramatic increase in resistivity is followed by a sharp decrease after 84 s of air exposure (Fig. 11b).

3.5 Ti(TiO_2)/PPX Nanocomposites

A series of samples of nanocomposites of Ti and PPX with different Ti content has been synthesized (Table 3). AFM analysis shows that the inorganic phase comprises nanoparticles of 10–20 nm in diameter, which are homogeneously distributed between the polymer globules (Fig. 12).

XRD studies show that synthesized composites do not contain any crystal phase, just an amorphous phase. Optical absorption measurements prove that synthesized nanocomposites are containing TiO_2 and Ti phases. For comparative analysis the pure Ti containing thin film was deposited onto the cold substrate (77 K) and onto the substrate at room temperature. The same result was obtained: XRD analysis

Table 3 Freshly-synthesized Ti(TiO_2)/PPX nanocomposites: inorganic phase content

Sample	Inorganic phase	Inorganic phase content (vol%)
8	TiO_2	7
9	TiO_2	9
10	TiO_2	10
11	Ti	11
12	Ti	14

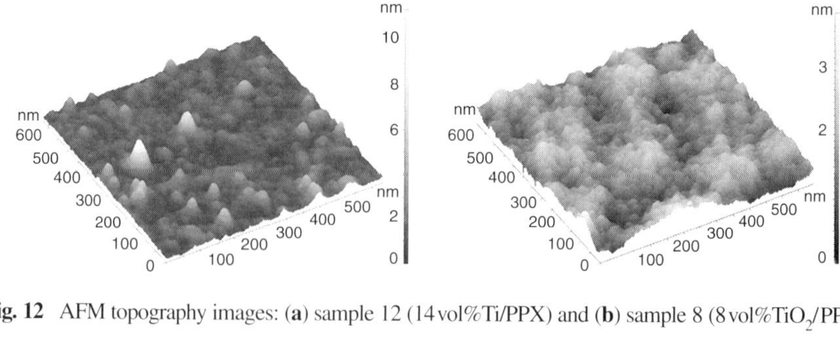

Fig. 12 AFM topography images: (**a**) sample 12 (14 vol%Ti/PPX) and (**b**) sample 8 (8 vol%TiO_2/PPX)

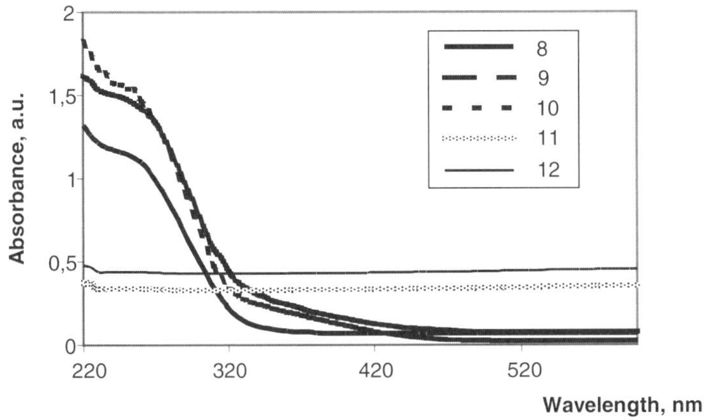

Fig. 13 UV-visible spectra of synthesized titanium/polymer nanocomposites: Samples 8, 9 and 10 contain amorphous particles of TiO_2, and samples 11 and 12 contain amorphous Ti

shows that the synthesized films only contain the amorphous phase. Kinetics of the electrical resistance grows with the air exposure of Ti/PPX nanocomposites (after synthesis under vacuum) is similar to the Al/PPX ones. For a metal content below the percolation threshold, the metal particles became isolators within several seconds, whereas for the samples beyond the threshold the observed resistance grows is several percents and takes hours.

The optical absorption spectra of five different samples are shown in Fig. 13. The polymeric matrix does not absorb in the range 200–600 nm as shown by Zavyalov [7]. According to data [8], the strongest absorption band of the nanocomposite samples 8–10 at 360 nm is assigned to TiO_2 particles in accordance with pure TiO_2. The optical spectra of samples 11 and 12 show the non-selective absorption over the entire wavelength span of 220–620 nm, which is typical for free electrons in the Ti metal in the composite films. Two types of inorganic filler are stabilised by the polymeric matrix, i. e., amorphous TiO_2 in samples 8, 9 and 10, and amorphous Ti in samples 11 and 12 (Table 3).

Simultaneously with topography acquisition under scanning of TiO_2/PPX nanocomposites, one can imagine some other characteristics of the investigated samples. Superposition of topography, lateral force and spreading resistance images allows to understand that the high- conductivity points are localized in between the polymeric globules (Fig. 14).

Figure 15a shows the electrical resistivity vs. time history for sample 8 with the metal content below the percolation threshold and for sample 11 above this threshold. The resistivity of sample 8 increases fast, i. e., for the 20 s of oxidation R grows three orders of magnitude, whereas the resistivity of sample 11 increases much slower, i. e., 1.5 times for 20 min. Kinetics of the curve for sample 8 (Fig. 15a) could be approximated by the dependency

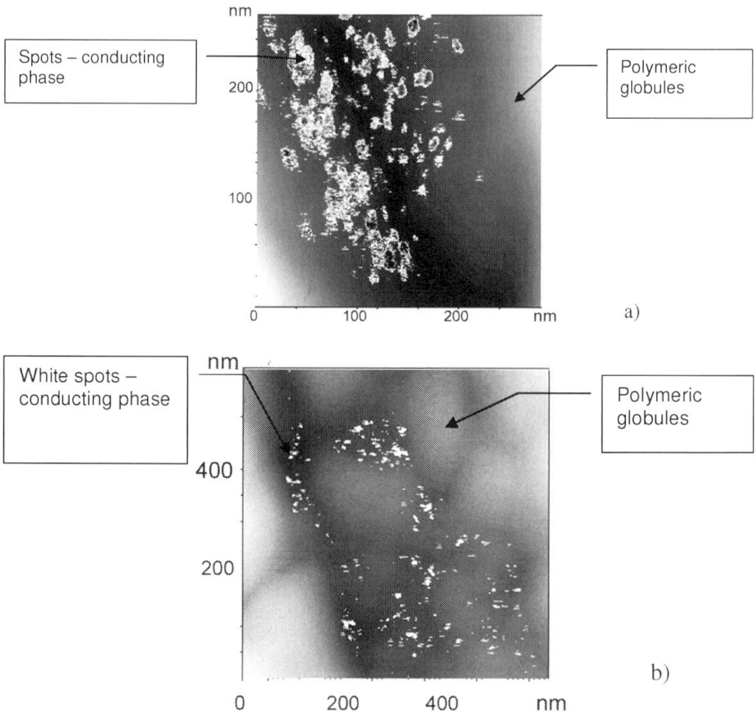

Fig. 14 AFM of sample 12 (14 Vol% TiO_2/PPX): superposition of topography and spreading resistance images. Scan sizes are (**a**) 250 × 250 nm and (**b**) 550 × 550 nm

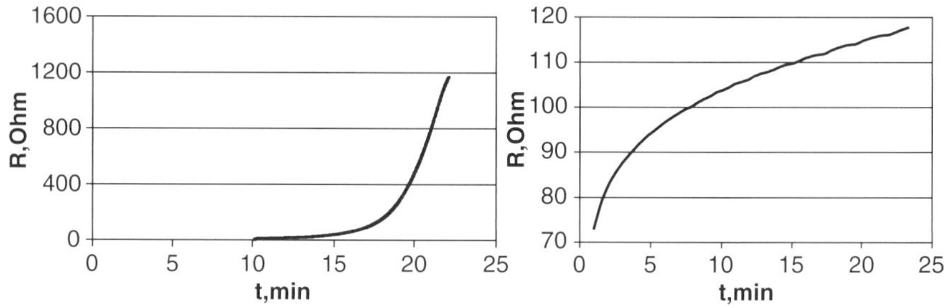

Fig. 15 Kinetics of oxidation of Ti(TiO_2)/PPX nanocomposites: (**a**) sample 8, experimental data follow the function ln(R) ~ 1/ln(t) and (**b**) sample 11, experimental data follow the function $R^{-0.5} \sim \ln(t)$

$$\ln(R) \sim 1/\ln(t) \tag{2}$$

and for sample 11 (Fig. 15b) the $R(t)$ dependency reveals

$$R^{-0.5} \sim 1/\ln(t) \tag{3}$$

The observed difference in the oxidation kinetics is a consequence of the different mechanisms of the charge transfer in the samples below and above the percolation threshold. For the samples below the percolation threshold, tunnel effects define the electrical conductivity, which is an exponential function of the distance between the metal nanoparticles. For the samples beyond the percolation threshold the charge transfer is realized via a grid of the metal 'nanowires', and the conductivity depends on the 'nanowire' cross-section mostly.

According to literature data [9, 10] for the early stages of the low-temperature metal oxidation, when the oxide layer thickness is below 2–3 nm, the oxide layer growth depends on the oxygen and metal ion diffusion and on the electron diffusion towards the reaction surface. The chemical potencial field is formed by the adsorbed oxygen on the oxide surface and by the induced oxygen activity at the metal–metal oxide boundary. With the metal content growth, the conditions of diffusion via the oxide layer are changing because of the change in the charge transfer mechanism within the ensemble of metal nanoparticles. As a result, the logarithmic oxidation law transforms to the inverse logarithmic one.

3.6 *Thin-Film Adhesion*

Thin-film adhesion is a very important property of materials for microelectronics and magnetic recording industries. In general, films that will strongly adhere to the substrate are desired. One should note that adhesion is not a constant, but rather a very complicated variable property, a concept very important for understanding length scale effects in small volumes [11].

For pure PPX films, irrespective of a metallic substrate foil material, i. e., aluminium or copper foil, the adhesion strength critically depends on the film thickness (h). Adhesion strength of thin films with $h < 6\,\mu m$ is not sensitive to sample thickness, whereas for thin films with $h > 6\,\mu m$ the adhesion strength substantially increased, as shown in Fig. 16.

In the case of a thin film, the yield stress is typically much higher than for a bulk material [12]. This is partly explained by the Hall–Petch type relationship between the film yield stress and thin-film grain size, d:

$$\sigma = \sigma_i + kd^{-n}, \tag{4}$$

where σ_i is the intrinsic stress, independent of the grain size d, and n is typically between 0.5 and 1 [11]. Since the grain size of a thin film may scale with the film thickness h, the latter can be used instead of the grain size as the scaling parameter [13]. Thus, the smaller the film thickness the larger the yield stress. Additionally, the structure of pure polymer thin films is considerably influenced by interfaces, i. e., the end group concentration decreases in three times with the film thickness growth from the growth surface to 400 nm, which in turn arises intrinsic stress with the film thickness decreasing [14]. The observed amorphous structure of thin films

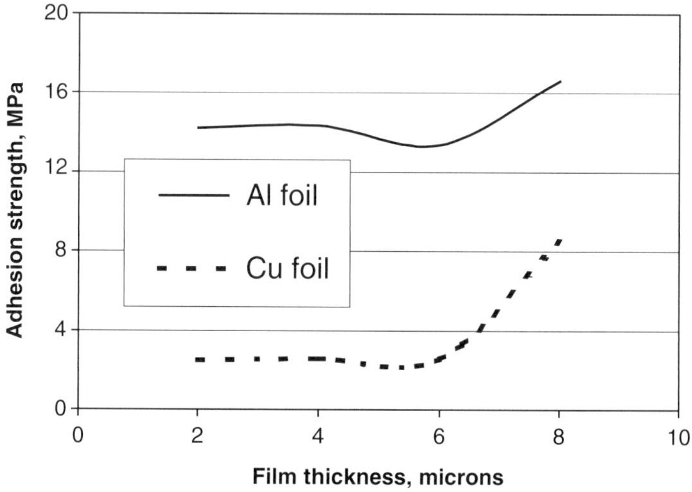

Fig. 16 Pull-off test results: adhesion strenght of pure polymer thin films deposited onto aluminium and copper foils as a function of the film thickness

of 2 μm thickness confirms the high intrinsic stress, whereas for thin films of 6–8 μm thickness crystallinity indications were observed. Increasing of the film thickness leads to the grain size growth, decreasing of the effects of interfaces and of the intrinsic stress. The resulting adhesion strength is remarkably increasing for films with a thickness of $h > 6\,\mu m$.

3.7 Electrical Resistance in Vacuum

The mechanism of the charge transfer processes within an ensemble of ultra-fine metal particles (in a composite with non-conducting particles) depends on the ratio of the particle's conductivity and the conductivity of the barrier regions between them and also on the ratio of the particle size and the interparticle distance. Figure 17 presents the temperature coefficient (the slope of $R(T)/R(20°C)$ vs. reciprocal temperature) of the electrical resistivity of as-prepared composites with metal contents ranging from 7 vol% to 12 vol%. These samples were deposited on quartz substrates with Pt-contacts. The composites with a lower metal content show a semiconductor-like negative temperature coefficient. This indicates a loss of metal-like contacts between metal particles. If the metal particle density increases, the temperature-coefficient increases. This is typical for porous films comprising islands of conducting material. Previously, we reported about a sign changing of the temperature coefficient of metal/polymer nanocomposites once the metal content is close to the percolation threshold. The composites with a metal content of 12 vol% reveal a positive temperature-coefficient, indicating an electrical conductivity

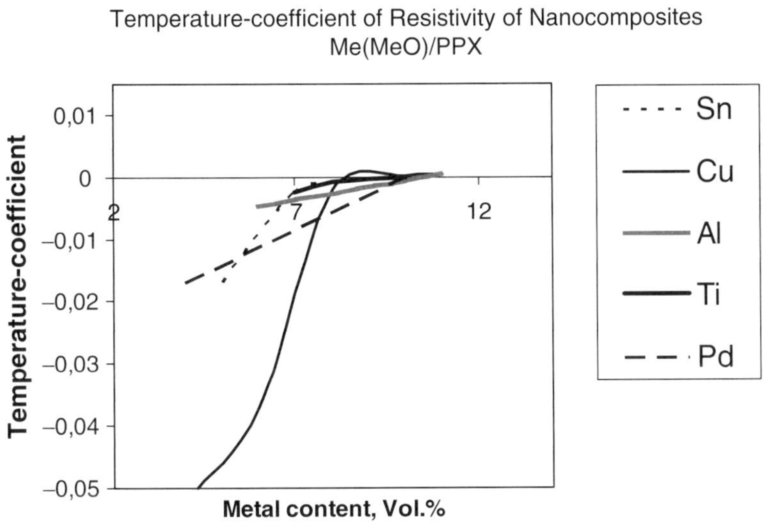

Fig. 17 Temperature-coefficient of the resistivity of synthesized metal (metal oxide)–polymer nanocomposites vs. metal content

determined by a continuous network with metal-like contacts between the metallic nanoparticles. In contrast, when the metal concentration is 8 vol%, the temperature-coefficient becomes semiconductor-like. This indicates a loss of metal-like contacts between particles of the metal phase. This is to be expected from percolation behaviour of the composites on the metal filler content. The percolation threshold can be determined by the variation in the temperature dependence of the electrical resistance, which is for the present case 10 vol% of metal. The synthesised nanocomposites demonstrate two types of electrical conductivity, i. e., the electrical conductivity in vacuum is limited by (i) the non-conducting polymer layer and (ii) the conductivity of the metal nanoparticles.

3.8 Impedance Spectroscopy

The charge transfer processes in as-synthesized samples are investigated by AC Electrochemical Impedance Spectrometry (EIS) in vacuum. Thin films of Pd/PPX nanocomposites with a metal content below the percolation threshold exhibit a linear frequency dependency of the admittance as presented in Fig. 18, which corresponds to a hopping mechanism of the electron transfer [15]. Thin films of Ti/PPX nanocomposites show more complex behaviour upon frequency variation. The frequency dependencies of the real and imaginary parts of the admittance ($Y^* = 1/Z^* = \mathrm{Re}Y + j\,\mathrm{Im}Y$) were analysed for samples with different Ti content, as shown in Fig. 19. For the low-frequency region samples could be modelled as a quasi-linear RC-circuit, where the real and imaginary parts of the admittance increase linearly with frequency. For the high-frequency region the imaginary

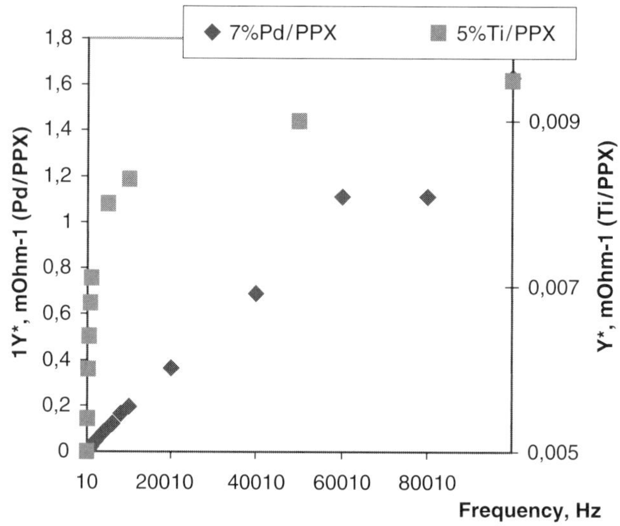

Fig. 18 Frequency dependences of the admittance (Y^*) of 7% Pd/PPX (**a**) and 5% Ti/PPX (**b**) nanocomposites in vacuum at room temperature

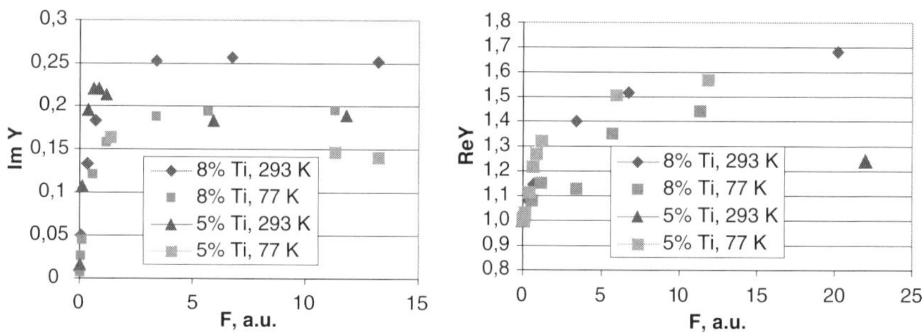

Fig. 19 Normalized frequency dependences of the admittance (Y^*) for Ti/PPX nanocomposites in vacuum at different temperatures: (**a**) real part of admittance and (**b**) imaginary part of admittance

part indicates Constant-Phase Element behaviour, due to a frequency-dependent capacitance, which is illustrated in Fig. 20.

For the high-frequency region, when the applied voltage period became smaller than the relaxation time of the surface statement, the polymeric matrix is no longer involved in the charge transfer process. Thus, the imaginary part of the admittance decreases with increasing frequency, which is the reason of the experimentally observed dispersion.

Figure 20 shows that when the Ti content within Ti/PPX nanocomposite is increasing above the percolation threshold, the capacity is decreasing up to zero.

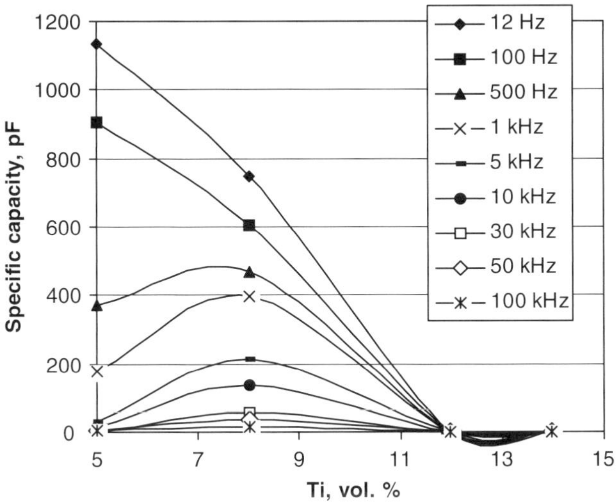

Fig. 20 Measured capacity for Ti/PPX nanocomposites in vacuum at room temperature as a function of the Ti content

This fact is important for the precise in-situ measurement of the percolation-threshold concentration during nanocomposite synthesis.

3.9 Electrochemical Characterization

Figure 21 shows the cell potential as a function of the specific capacity for pure PPX, 10 vol% TiO_2/PPX matrix (sample 10) and 14 vol% TiO_2/PPX matrix (sample 12) electrodes. From this figure, the performance of the PPX and 10 vol% TiO_2/PPX matrix electrodes is the same. This indicates that the PPX polymer has some reversible capacity and that at 10 vol% TiO_2 the active component is still the PPX polymer matrix. By comparing these potential curves with the potential curve of the 14 vol% TiO_2/PPX film, the curve differs in that the potential for the 14 vol% TiO_2 is higher on discharge and lower on charge. Since the reduction and oxidation potential of TiO_2 lies at a flat potential of approximately 1.8 V, TiO_2 seems to be active in the 14 vol% TiO_2 film. This observation is seen despite the fact that this sample is thicker than the other films and the current at which this cell was tested was greater than that of the 10 vol% TiO_2 film. As a note, the currents are directly comparable, since the electrode areas for all cells were the same.

Since the reduction and oxidation potential of TiO_2 is flat vs. Li metal at 1.8 V, the 14 vol% TiO_2 film was cycled at lower currents in order to observe the intercalation of TiO_2. In Fig. 22 this potential is shown in the flat part of the potential curve. Although the flat potential is slightly off of what has been reported, the cell resistance may account for this difference. The reversibility of the 14 vol% TiO_2

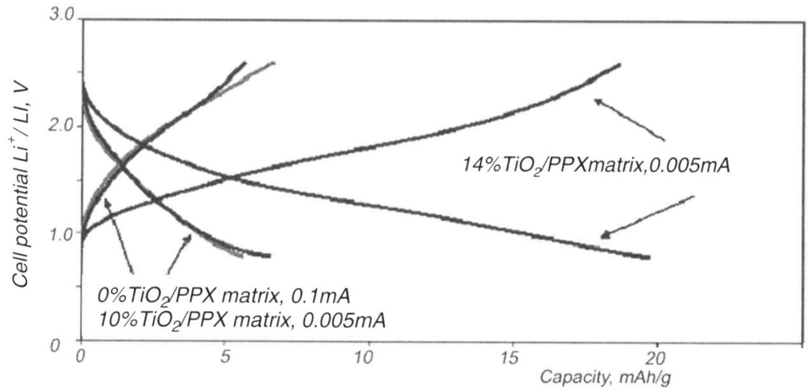

Fig. 21 Percolation of TiO_2 vol.% in PPX matrix

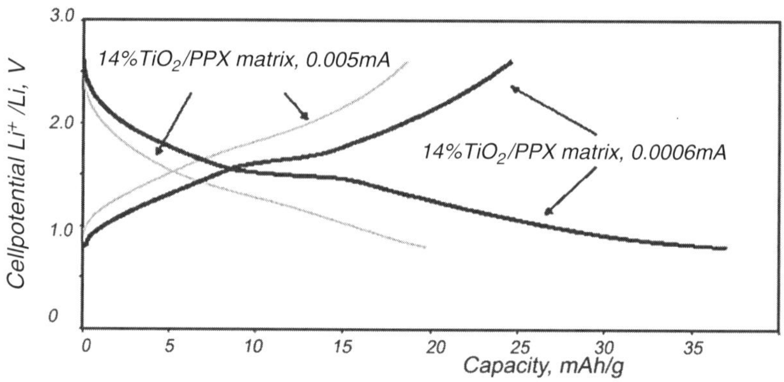

Fig. 22 Reversibility of TiO_2/PPX nanocomposites at different current densities

film seems to be good even though the capacity is low for current rates up to 0.28 μA cm^{-2}. However, for the very low rate of 0.03 μA cm^{-2}, the efficiency decreases as shown in Fig. 23.

4 Concluding Remarks

Thin-film metal (metal oxide)/polymer nanocomposites with different inorganic phase contents were obtained by using the cold-wall vacuum co-deposition technique. A range of metals was shown to be applicable to form nanocomposite thin films with PPX, i. e., Al, Ti, Pd, and Sn. AFM studies show the metal nanoparticles to have a size of 7–50 nm. Within the composite the polymer forms more or less

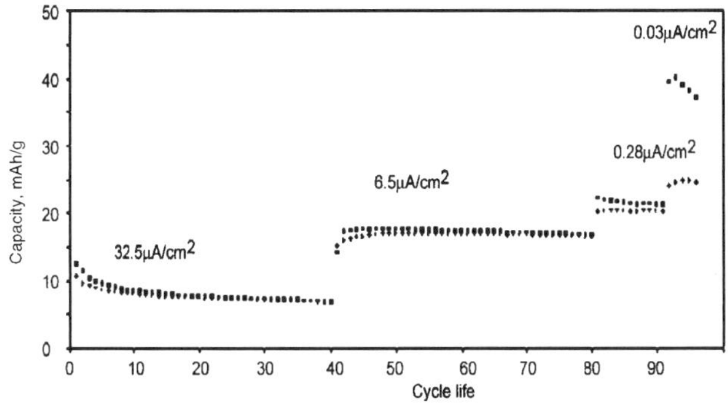

Fig. 23 Specific capacity of sample 12 (14% TiO_2/PPX) vs. cycle life

spherical globules with a maximum size of about 200 nm. The interfacial regions between neighbouring polymeric spherulites contain nanoparticles of the inorganic filler.

The chemical composition, surface morphology, electrical conductivity, optical absorption and Li-ion conductivity have been investigated. It is found that in a relatively narrow concentration range of the inorganic phase, i. e., from 7 till 14 vol% within the polymeric matrix, the properties of these thin-film nanocomposites critically depend on content. Below the percolation threshold, the inorganic phase consists of nanoparticles of the metal oxide (Al_2O_3, TiO_2, SnO_2) and beyond this threshold, of nanoparticles of the metal (Al, Ti, Sn). We have found experimentally that the composites with a metal content beyond the percolation threshold show a positive temperature-coefficient, indicating a conductivity determined by a continuous network of metal-like contacts between the nanoparticles. In contrast, when the metal concentration is below the percolation threshold, the temperature coefficient is negative, indicating a conductivity determined by a network of semiconductor-like contacts. The percolation threshold can be determined by the variation of the temperature dependence of the electrical resistance, which is found to be about 10 vol% of metal. The synthesised nanocomposites demonstrate two types of electrical conductivity, i. e., the electrical conductivity in vacuum is limited by (i) the non-conducting polymer layer and (ii) the conductivity of the metal nanoparticles.

It has been shown that the nanocomposite structure determines the oxidation behaviour of Al, Sn and Ti nanoparticles within the polymeric matrix under air exposure. Measurements of the electrical resistance as a function of temperature in vacuum reveal metal-like conductivity of samples above the percolation threshold, whereas samples below the threshold have semiconductor-like conductivity. For the metal content below the percolation threshold the metal particles became non-conducting within several seconds, whereas for the samples above the threshold the observed resistance growth is several percent and full oxidation takes several hours.

Adhesion strength of the thin-film pure PPX samples is not sensitive to sample thickness below 6 μm, whereas for films with a thickness of $h > 6\,\mu m$. adhesion strength is remarkably increasing.

The admittance of thin-film TiO_2/PPX nanocomposites was modelled using an equivalent RC-p-chain. Comparison of the modelled frequency dependences of admittance with the experimental data proves the hopping mechanism of the charge transfer for the low frequency region only (below 500 Hz). For the high-frequency region (500 Hz–100 kHz), the sufficient deviations from the hopping mechanism are observed. Capacity dependencies on the AC frequency were found to be a function of the Ti content within nanocomposites in vacuum. It is shown that when the Ti content within Ti/PPX nanocomposite is increasing above the percolation threshold, the capacity is decreasing up to zero, which is important for the precise in-situ measurement of the percolation threshold concentration during nanocomposite synthesis.

The synthesized thin films of nanostructured TiO_2/poly-*para*-xylylene composites exhibit a high specific capacity and good cycling performance as anode in Li-ion batteries.

Acknowledgements We are grateful to Dr. Radmir Gaynutdinov (Shubnikov Institute of Crystallography, Russian Academy of Science) for the atomic force microscopy characterization, to Dr. Dan Simon (Delft University of Technology, The Netherlands) for electrochemical characterization and to The Netherlands Organization for Scientific Research for financial support of the project (NWO, grant #047.011.2003.004).

References

1. M.C. Roco, R.S. Williams, and A.P. Alivisatos, *Nanotechnology Research Directions: IWGH Workshop Report. Vision for Nanotechnology R&D in the Next Decade*, Dordrecht/Boston/London, Kluwer Academic Publishers, 2000.
2. EL. Nagaev, Small metal particles, *Adv. Phys. Sci. (in Russian),* 1992 **162**(9) 49–124.
3. B. O'Regan and M. Grätzel, *Nature*, 1991 **353** 737.
4. J. Schoonman, Nanostructured materials in solid state ionics, *Solid State Ionics*, 2000 **135** 5–19.
5. M. Szwarc, *Polym. Eng. Sci.,* 1976 **16** 473.
6. J.J. Senkevich, S.B. Desu, and V. Simkovic, Temperature studies of optical birefringence and X-ray diffraction with poly(*p*-xylylene), poly(chloro-*p*-xylylene) and poly(tetrafluoro-*p*-xylylene) CVD thin films. *Polymer*, 2000 **41** 2379–2390.
7. S. Zavyalov, A. Timofeev, A. Pivkina, and J. Schoonman, Metal–polymer nanocomposites: Formation and properties near the percolation threshold, in *Nanostructured Materials: Selected Synthesis Methods, Properties and Applications*, P. Knauth and J. Schoonman(editors), Boston/Dordrecht/London, Kluwer Academic Publishers, 2002, pp92–117.
8. S. Zavyalov, A. Pivkina, and J.Schoonman, Formation and characterization of metal-polymer nanostructured composites, *Solid State Ionic*, 2002 **147** 415–419.
9. A.T. Fromhold and E.L. Cook, Kinetics of oxide films growth on metal crystals: Electron tunneling and ionic diffusion, *Phys. Rev.,* 1967 **158** 610–612.
10. N. Cabrera and N.F. Mott, Theory of oxidation of metals, *Rep. Prog. Phys.*, 1948–1949 **12** 163–184.

11. A.A. Volinsky, N.R. Moody, and W.W. Gerberich, Interfacial toughness measurements for thin films on substrates, *Acta Mater.*, 2002 **50** 441–466.
12. D. Kramer, H. Huang, M. Kriese, J. Robach, J. Nelson, A. Wright, D. Bahr, W.W. Gerberich, *Acta Mater.*, 1999 **47** 333.
13. Y. Wei and J.W. Hutchinson, *J. Mech. Phys. Solids.*, 1997 **45** 1137
14. W.F. Beach, Macromolecules, 1978 **11** 72–87
15. M. Pollak, T.H. Geballe, *Phys. Rev.*, 1961 **122** 1742–1753

Structure and Mechanical Properties of Nanocomposites with Rod- and Plate-Shaped Nanoparticles

S.J. Picken[1], D.P.N. Vlasveld[2], H.E.N. Bersee[3], C. Özdilek[4], and E. Mendes[1]

1 Introduction

Particle-reinforced polymer composites or compounds have been used for decades to increase the stiffness and strength of polymers and to reduce thermal expansion. Polymer nanocomposites based on exfoliated layered silicates have been developed more recently [11, 23, 24] for improved mechanical properties, barrier properties, and reduced flammability. Polymer nanocomposites are polymers filled with finely dispersed particles that have at least one dimension in the nanometer range. Compared to composites containing larger dispersed particles, nanocomposites have the advantage of achieving the optimal properties at relatively low filler content, resulting in a lower density and better surface smoothness and transparency. This is due to the large aspect ratio and high stiffness of the particles, resulting from the exfoliation of the layered silicate particles. This can be especially favorable for moisture-sensitive polymers like polyamides, which loose a lot of their stiffness under moist conditions [10, 12].

The types of polymer nanocomposite that will be discussed here are the plate-shaped layered silicate nanocomposites, and polymer nanocomposites containing rod-shaped high aspect ratio boehmite needles. In both cases we will consider polyamide-6 (PA6) as the matrix material. For the layered silicate polymer nanocomposites, the increase in stiffness depends strongly on the degree of exfoliation of the silicate layers, see Fig. 1.

[1]Nanostructured Materials, Delft University of Technology, Julianalaan 136, 2826 BL Delft, The Netherland

[2]Promolding BV, Laan van Ypenburg 100, 2497GB Den Haag, The Netherlands

[3]Design and Production of Composite Structures, Faculty of Aerospace Engineering, Delft University of Technology, Kluyverweg 3, 2629 HS Delft, The Netherlands

[4]Fundamentals of Advanced Materials, Faculty of Aerospace Engineering, Delft University of Technology, Kluyverweg 3, 2629 HS Delft, The Netherlands

P. Knauth and J. Schoonman (eds.), *Nanocomposites: Ionic Conducting Materials and Structural Spectroscopies.*

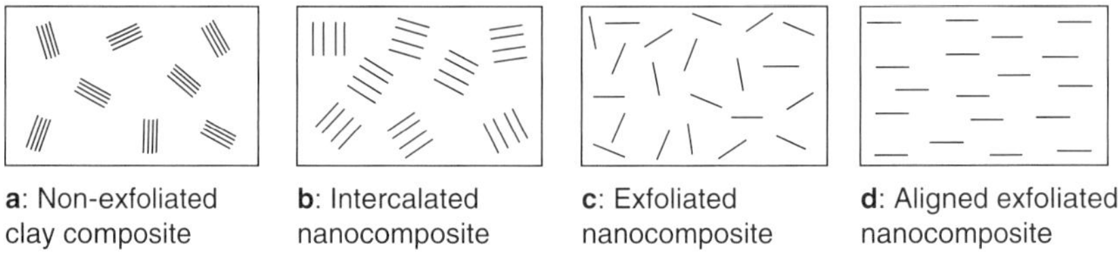

Fig. 1 Different levels of dispersion in layered silicate nanocomposites

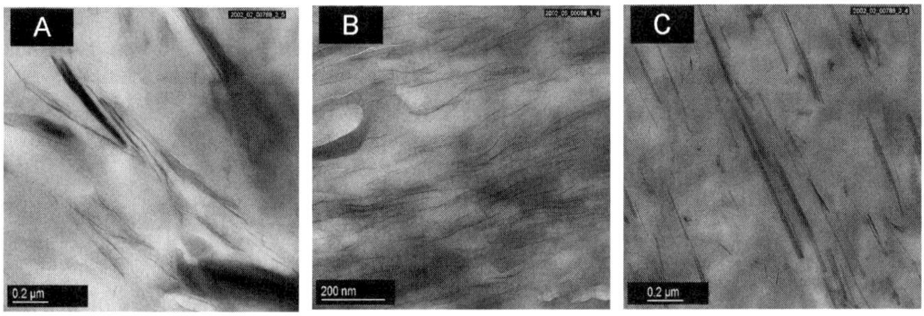

Fig. 2 TEM-images of PA6 nanocomposite with 5 wt% nanoparticles. **a**, unmodified silicate (ME-100); **b**, modified silicate (MEE); and **c**, Sepiolite

This depends on both the mixing method and the interaction between the polymer and the silicate platelets [21]. For most polymers, the interaction and exfoliation can be improved by using organic surfactants on the hydrophilic silicate surface as compatibilizer. Since most layered silicate platelets are negatively charged, cationic surfactants such as quaternary ammonium ion based surfactants are often used to optimize the interaction between the platelets and the polymer. Figure 2 shows some TEM images of PA6 nanocomposites containing various types of nanoparticles with and without surface modification.

Also, with the rod like boehmite needles, it is useful to examine the effect of organic modification, as this may influence the interaction with the semicrystalline PA6 matrix. In addition, for the boehmite system we will discuss the potential for self-organization of the filler nanoparticles to form colloidal nematic liquid crystalline structures in the polymer nanocomposite. It is found that such self-organized structures of the boehmite nanoparticles can act as a template for the PA6 crystallization.

It is worth noting that the advantages in mechanical properties of polymer nanocomposites, such as the higher modulus and yield stress, are usually accompanied by an increase in melt viscosity and a change in rheological behavior. The higher melt viscosity is a disadvantage for processing techniques such as extrusion and injection molding, although it can be beneficial for film extrusion. Both the modulus and the viscosity are influenced by the same particle-related parameters, such as the particle shape and concentration. Therefore, it is necessary to optimize the polymer nanocomposite both in terms of the achieved mechanical properties as well as their processability.

Finally, we will also discuss the appropriate rules of mixing for establishing the mechanical stiffness (modulus) of polymer nanocomposites. It appears that the modulus can be described using the Halpin–Tsai model, which gives a means to determine the effective aspect ratio of the filler particles. It is found that these effective aspect ratios are in good agreement with other techniques using vapor (water) absorption rate and from direct imaging. The result of this analysis explains why in general it is preferable to use semicrystalline polymers for the matrix, as it appears that a parallel additive contribution of the filler particles is retained above the glass-transition due to the higher level of the rubber plateau. This phenomenon gives rise to a dramatic improvement of the so-called Heat Distortion Temperature, in some cases by more than 100°C, at relatively low nanoparticle content. The result of this is that polymer nanocomposite technology provides a relatively easy means to upgrade standard engineering polymers to much higher performance levels that may compete with more expensive and harder to process high-temperature polymers.

2 Experimental Methods

Considering the influence of sample preparation methods on the polymer nanocomposite properties here we give a detailed summary of the experimental techniques used to make and characterize the samples. For readers that are mainly interested in the obtained properties this section can be skipped.

2.1 Materials

2.1.1 Polyamide 6 Grade [36]

Polyamide 6: low molecular weight (LMW PA6) Akulon K222D, injection-molding grade PA6 from DSM, The Netherlands; M_n 16,000, M_w 32,000 g mol^{-1}, T_m 220°C.

2.1.2 Commercial Polyamide 6 Nanocomposites [36]

Two commercial PA6 nanocomposites are used (MW unknown). From Ube, Japan, with 2.5% silicate and from Unitika, Japan, with 4.6% silicate. These nanocomposites are made by in-situ hydrolytic polymerization of ε-caprolactam in the presence of swollen organically modified silicates. The organic surfactant is the initiator for polymerization and so the polymer chains are bound to the surfactant via covalent bonds. Since the surfactants have an ionic bond with the silicate layers, the polymer is ionically bound to the silicate layers, unlike in the melt-processed nanocomposites.

2.1.3 Organically Modified Layered Silicate [36]

Cloisite® 30 B (montmorillonite) from Southern Clay Products, USA.
Surfactant: methyl bis-2-hydroxyethyl tallow quaternary ammonium (32 wt%).
Nanomer® I 30 TC (montmorillonite) from Nanocor, USA.
Surfactant: tri-methyl tallow quaternary ammonium (33 wt%).
Somasif® MAE (synthetic fluorine mica) from Co-op Chemicals, Japan.
Surfactant: di-methyl di-tallow quaternary ammonium (40 wt%).
Somasif® MEE (synthetic fluorine mica) from Co-op Chemicals, Japan.
Surfactant: methyl bis-2-hydroxyethyl coco quaternary ammonium (28.5 wt%).

The manufacturer has added the organic surfactant on the silicate platelets, and the amount of organic surfactant was determined with thermogravimetric analysis (TGA) in a Perkin–Elmer TGA-7 Thermal Gravimetric Analyzer at 800°C for 1 h in air.

2.1.4 Unmodified Layered Silicate [36]

Somasif ME-100 (Synthetic Mica) from Co-op Chemicals, Japan. This is water-swellable synthetic fluorine mica, which does not contain any organic surfactant. The inorganic part is identical to Somasif MEE and MAE.

2.2 *Synthesis*

2.2.1 Boehmite Rods

Colloidal boehmite rods were synthesized [17] according to the method of Buining et al. [2]. The synthesis of particles and their characterization by TEM were done in collaboration with David van der Beek, in the group of Prof. H.N.W. Lekkerkerker (Colloid Chemistry Group,Utrecht University). Boehmite was synthesized by using aluminium isopropoxide (Janssen) and aluminium tri-sec butoxide (Fluka). The aluminium precursors were dissolved in demineralized water that was acidified with 37% hydrochloric acid. The resulting solution was stirred for 1-week to obtain complete hydrolysis of alkoxides, from which polymeric aluminium hydroxides were formed. The latter species were hydrothermally crystallized into boehmite upon autoclaving at 150°C for 22 h. To remove alcoholic by-products, the colloidal dispersions were dialyzed against demineralized water for 1 week. At the end of the synthesis, well-defined boehmite rods dispersed in aqueous medium were obtained. TEM characterization revealed average particle dimensions as 280 nm in length and 20 nm in width, see Fig. 3. In general, boehmite rods exhibit a high extent of polydispersity. A similar work performed with the same synthesis method using identical amounts of aluminium precursors reports the extent of polydispersity in boehmite rods as 30% in length and width [26].

The final dispersions in water were very stable and no aggregation with time was observed. After dialysis, pH was determined to be 5.5. These dispersions exhibited flow birefringence starting from 0.8% w/w particle concentration and turned permanently birefringent above 1% w/w, which is the onset of nematic phase formation by these particles with the given dimensions. Usually, for the in situ polymerization, a boehmite concentration of 1% (w/w) in water was used.

2.2.2 Titanate-Modified Boehmite [18]

We have used a titanate-type coupling agent for the surface modification of boehmite: titanium IV, tris[2-[(2-aminoethyl)amino]ethanolato-O],2-propanolato (obtained from Kenrich Chemicals with the commercial name KR-44) [15]. First, *n*-propanol (Acros organics) was added drop wise into the boehmite (aq) dispersion under ultrasonication. This was followed by azeotropic distillation with further addition of *n*-propanol to maintain a constant volume. In the end, a stable dispersion of boehmite in propanol was obtained. The change of solvent was required as the first step, because of insolubility of the Ti-coupling agent in

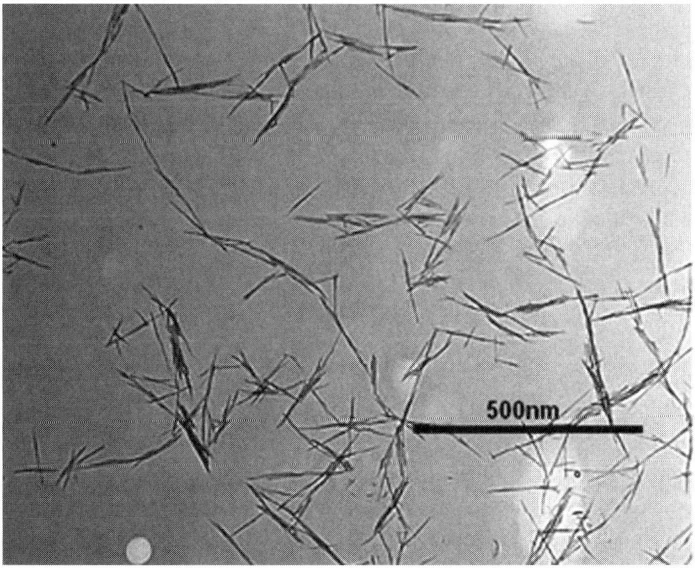

Fig. 3 TEM image of the Boehmite particles

water. Finally, the coupling agent in propanol was introduced in a drop wise manner into the dispersion, accompanied by ultrasonication.

2.2.3 In Situ PA6 Synthesis

For polymerization of ε-caprolactam, Fig. 4, the hydrolytic route was chosen [20]. Although anionic polymerization would be less time-consuming, it appears to be less suitable for nanocomposite formation due to the positive charge of the Boehmite surface. In the hydrolytic route, water initiates the reaction via opening the caprolactam ring and generating aminocaproic acid. The polymer then grows by reaction of the generated amino acid with the cyclic monomer. Polymerization reactions were performed in a 500 ml glass reactor equipped with an automatic stirrer and a temperature controller. The ε-caprolactam obtained from Fluka was used without further purification. About 40 g ε-caprolactam, 10 g aminocaproic acid, 0.25 g adipic acid, and 10 ml water were mixed. The mixture was heated at 140–150°C for 2 h to remove excess water, followed by heating at 230°C for 4 h. The polymeric product could be poured out while still in the melt. Removal of unreacted monomer and cyclic/linear oligomers was carried out by Soxhlet extraction (for 12 h) using methanol.

2.2.4 Preparation of PA6-Boehmite and Titanate-Modified Boehmite Nanocomposite [18, 19]

Synthesis of the composites was carried out according to the method described for PA6. The only difference was the use of aqueous boehmite dispersions instead of water. The final boehmite concentrations in PA6, which is obtained by varying the amount of boehmite dispersion used in the polymerization, range up to about 10 wt%.

In the presence of titanate-modified boehmite needles, the procedure was slightly modified. The dispersion of particles in *n*-propanol was combined with ε-caprolactam to form a homogeneous mixture. Propanol was expelled completely in the rotary evaporator, leaving a solidified mixture of monomer containing the titanate-modified boehmite particles, which were then introduced to the melt polymerization set up. As a result of this treatment, PA6 nanocomposites containing about 1–15 wt% titanate-modified boehmite were obtained.

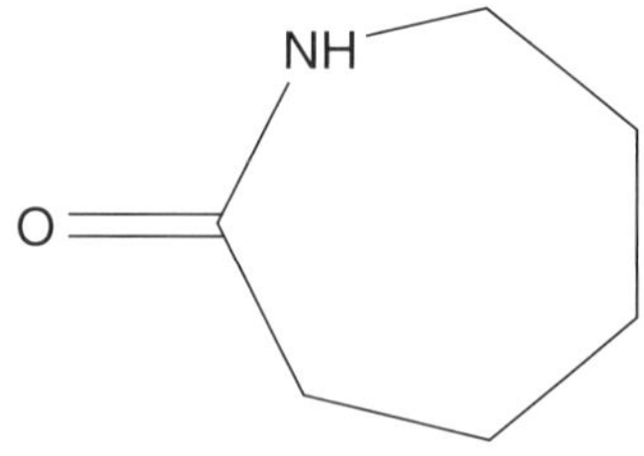

Fig. 4 ε-Caprolactam

2.3 *Sample Preparation*

2.3.1 Montmorillonite and Synthetic Mica – PA6 Nanocomposites [36]

The nanocomposites with montmorillonite and synthetic mica were prepared by mixing in PA6 in a Werner and Pfleiderer ZDS-K28 corotating twin-screw extruder. The screw layout was designed to produce high shear stresses, achieved by incorporating several kneading blocks followed by small backflow elements. The modified layered silicate powder was mixed with the polymer granules and fed into the extruder at a constant rate via a Plasticolor 2500 feeding unit. The extruder was operated at a screw speed of 200 rpm and a feeding rate of approximately 3 kg h^{-1}. The temperature in the feeding zone was 150°C; for PA6 all the other zones were heated to 230°C. Cooling was applied to keep the temperature constant since the high shear forces in the melt can produce too much heat. First, a master batch with a high concentration (12 wt% based on the inorganic content of the filler) was made. Other concentrations were made by diluting the master batch with unfilled polymer in a second extrusion step.

2.3.2 Somasif ME-100-PA6 Polymer Nanocomposites [36]

The nanocomposites with Somasif ME-100 were made by feeding a mixture of cryogenically milled PA6 and ME-100 powder in a Werner and Pfleiderer ZSK 30/44 D corotating twin-screw extruder. To enhance the exfoliation of the water-swellable ME-100 silicate, water was injected into the extruder at a rate of 25 ml min^{-1}, and removed by venting at the end of the extruder. The extruder was operated at a temperature of 240°C at a rotation speed of 200 rpm and a feeding rate of approximately 10 kg h^{-1}.

2.3.3 Injection Molding [36]

Dumbbell-shaped samples according to ISO 527 standards were injection molded on an Arburg Allrounder 221–55–250 injection-molding machine. The feeding zone was heated to 150°C, the melting and mixing zones heated to 240°C, and the nozzle was heated to 270°C.

2.3.4 Sample Preparation for DMA

All the studied PNC samples were pressed into thin films by applying a force of approximate 180 kN in a hydraulic press at 250°C. The final thickness of the films was typically about 0.3 mm. The films were extensively dried for several weeks in a vacuum oven at 80°C before DMA characterization was carried out.

2.4 *Experimental Analysis Techniques*

2.4.1 Transmission Electron Microscopy

Transmission electron microscopy (TEM) was performed using a Philips CM30T electron microscope with a LaB_6 filament operated at 300 kV. Ultramicrotomed slices of our samples were placed on Quantifoil carbon polymer supported on a copper grid.

2.4.2 Differential Scanning Calorimetry (DSC)

The measurements were performed on a Perkin–Elmer DSC7 differential scanning calorimeter. Samples were heated from 25 to 250°C at a rate of 10°C/min and were held for 1 min at the maximum temperature. They were cooled back to 25°C at a rate of 10°C/min, which was followed by a second heating run identical to the first heating run.

2.4.3 Dynamic Mechanical Analysis (DMA)

Storage moduli of the samples were measured in the extension mode at a frequency of 1 Hz by using a Perkin–Elmer DMA 7e dynamical mechanical analyzer. The measurements were taken at a heating/cooling rate of 5°C min^{-1}. Each one of the pressed samples was cut into a small rectangle of 9–7 mm long and 2.5 mm wide. The thickness of the samples after pressing typically was 0.3 mm.

2.4.4 Thermogravimetric Analysis (TGA)

The exact amount of nanoparticles in composite samples was determined by using TGA. The samples were heated from 25 to 800°C at a rate of 50°C min^{-1} and were kept at this temperature for 30 min. Since PA6 degrades completely without leaving any residue, the remaining part gives us the w/w concentration of the filler. In the case of boehmite, the amount of weight loss as observed by TGA on the pure boehmite, which occurs via 2 AlOOH => $Al_2O_3 + H_2O$, was taken into account.

2.4.5 Optical Polarization Microscopy [19]

Nanocomposite films that were confined between glass microscopy slides were placed in a Mettler Toledo FP82HT hot stage to apply heating–cooling cycles between 180 and 240°C at a rate of 5°C min^{-1}. The samples could be followed real-time during the temperature cycles by using a Nikon Eclipse E600POL optical polarization microscope. Identical temperature intervals and heating rates were used in WAXS and OPM for a direct comparison of the results. The sign of the birefringence was determined using a lambda plate by examining if the birefringence in the samples was additive or subtractive. Comparing the two perpendicular optical axes in the sample indicated whether the overall retardation (or birefringence) is positive or negative.

2.4.6 Wide Angle X-ray Scattering (WAXS) [19]

The amount of orientation in the nanocomposites was analyzed with an X-ray D8-Discover diffractometer from Bruker-Nonius. The sample holder was a home-built heating unit, which made it possible to confine the polymer films between polyimide X-ray windows in a vertical position. The temperature was controlled using a thermocouple and a fast response Eurotherm PID controller driving a Delta Electronica DC power supply (maximum heating rate, 300°C min^{-1}), which allowed a temperature range of 25–350°C at less than 0.1°C variation. The measurements were carried out using 0.154 nm Cu K_{α} incident radiation. The scattering data was recorded on a 2D Hi-Star detector (1,024 × 1,024), the sample to detector distance was 6 cm. Samples were subjected to heating–cooling cycles at a rate of 5°C min^{-1} in the 180–240°C interval and each cycle was paused at 30, 180, and 240°C for a 1,800 s data acquisition time.

2.4.7 Moisture Diffusion [34]

The diffusion coefficients were determined from rectangular samples of 27 × 20 × 4 mm, cut from the wide part of ISO 527 test bars. The increase in water content of the nanocomposites and unfilled PA6 was measured at 70°C, conditioning both at 65% and 92% RH. Dry as molded samples were dried in a vacuum oven at 80°C for at least 48 h before testing. Moisture conditioning was performed at 70°C to increase the speed of moisture transport. The sample weight was measured after drying and during conditioning on a Mettler AE-240 to determine the moisture level. The moisture content was calculated based on the amount of matrix material (PA6 and surfactant) and so the weight of the silicate phase was not included.

Samples ultimately containing 3% water were prepared by conditioning in a Heraeus climate chamber at 70°C and 65% relative humidity and samples ultimately containing 6% water were prepared by conditioning at 70°C and 92% relative humidity. The samples with equilibrium moisture content were stored until a constant weight was obtained. This took different times for different silicate contents, and could take up to 10 weeks for the highest silicate content.

3 Results

3.1 Moduli of PA6 Silicate Nanocomposites

The first aspect to be discussed concerns the effect of nanoparticle concentration on the modulus of PA6. Figure 5 shows some representative data for a variety of dry and humidified PA6 samples. In all cases the stiffness increase is comparable; however, the humid samples start at a much lower initial modulus value. The fact that the matrix modulus can be modified via control of the humidity is an important experimental tool to verify the validity of the mechanical Halpin–Tsai model (see Sect. 4 for details). This allows the matrix modulus contribution to be varied systematically while maintaining the overall sample morphology. If the mechanical model makes sense then the calculated aspect ratio of the nanoparticles should be independent of the value of the matrix modulus. This case is demonstrated in Fig. 6, which shows a systematic decrease of the effective particle aspect ratio with increasing weight fraction of silicate nanoparticles.

The most important result from these aspect ratio calculations is the fact that the data for three different moisture contents overlap in Fig. 6. This shows that the modulus of a nanocomposite can be explained by the combination of a very stiff filler particle with a high aspect ratio and the normal matrix modulus, which in this case is varied by changing the moisture content. Apparently, the Halpin–Tsai composite theory can be used to describe the modulus of nanocomposites for a wide variety of matrix moduli, both below and above T_g of the matrix polymer, with the

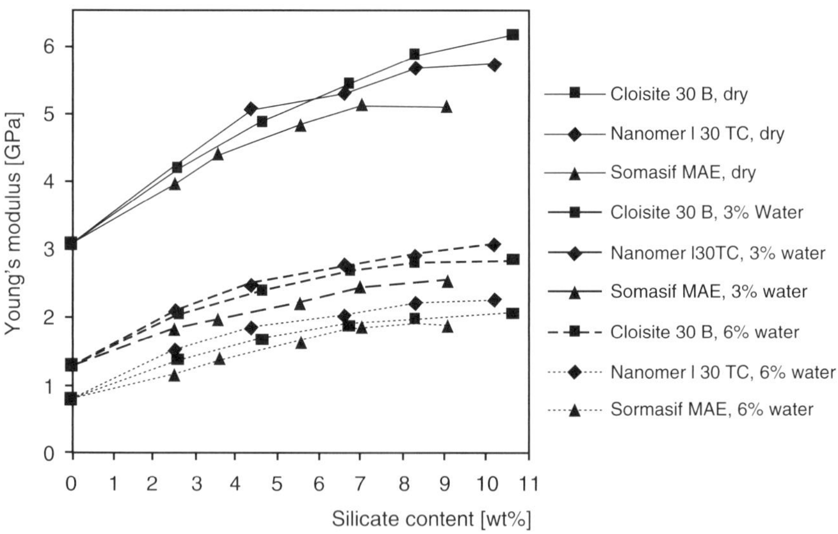

Fig. 5 Modulus as a function of silicate content for three different moisture contents

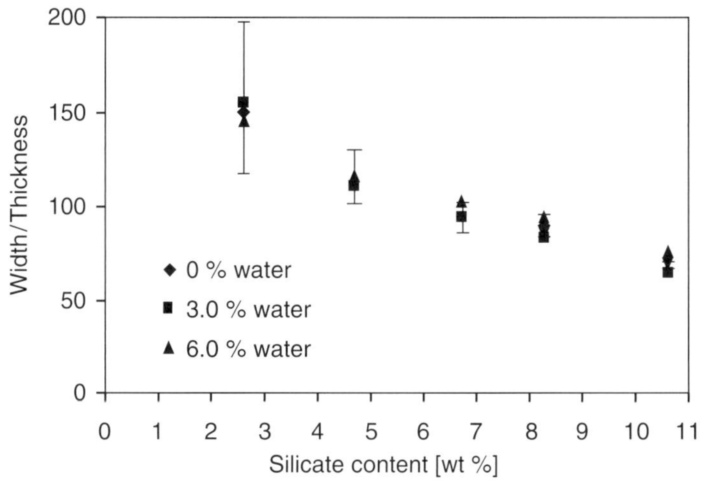

Fig. 6 Aspect ratios calculated from the moduli as a function of silicate content for three different moisture contents (Cloisite 30B nanocomposites)

same aspect ratio for all matrix moduli. The fact that the modulus of nanocomposites can be explained by the Halpin–Tsai theory, independent of the modulus of the matrix, proves that the reinforcing mechanism in nanocomposites is similar to traditional composites. No additional stiffening of the matrix due to confinement of the polymer has to be assumed to explain the high modulus of nanocomposites.

To show the influence of errors in the modulus measurements, error bars are displayed for the dry series in Fig. 6 (for clarity the other series do not show the error bars). The error bars correspond to a constant error of ± 0.1 GPa in the modulus measurement. It can be seen that at low silicate concentrations, a small error in the modulus measurement has large effects on the calculated aspect ratio; however, for higher concentrations the error is much smaller and the calculated aspect ratios are more reliable. On the basis of these calculations it has to be concluded that the effective reinforcing effect of the particles seems to decrease with increasing silicate loading. Similar results follow from moisture absorption measurements, shown in the next section. The effective aspect ratios from the model are realistic values: the aspect ratios are between 70 and 150, which is in the range of what can be observed in the TEM image (see Fig. 2 as an example). With increasing concentration the exfoliation becomes less perfect, resulting in an effective aspect ratio that is half of the highest value. This means that when perfect exfoliation is assumed in the low concentrations, an average stack size of 2 platelets is present at the highest concentrations. This is also in reasonable agreement with TEM analysis. It should be noted that analyzing the effective particle aspect ratios using mechanical modeling provides a much faster method to asses the level of exfoliation than by TEM imaging, which always suffers from potential artefacts, e. g., from sample preparation and limited statistics.

3.2 DMA Results on PA6 Silicate Nanocomposites

To further explore the mechanical properties of PA6 nanocomposites, it is useful to determine the dynamic moduli. In Sect. 4 we will discuss in more detail how such dynamic results can be compared to the dynamic version of the Halpin–Tsai model, and this shows that some of the effects on the T_g may be purely due to the composite model itself without requiring any change in the matrix polymer dynamics. Figures 7 and 8 show the DMA results for MEE composites versus temperature at various particle loading levels. When unmodified silicate particles are used, the moduli are lower, as can be seen in Fig. 9.

The moduli of the nanocomposites with organically modified (MEE) particles are higher than of those with unmodified (ME-100) particles, because the surface modification improves the exfoliation, leading to more individual particles with higher aspect ratios. We will return in more detail to the DMA results in Sect. 4, where we discuss the Halpin–Tsai model more extensively.

3.3 Moisture Diffusion Measurements

The effective aspect ratios of the nanoparticles can also be assessed from moisture diffusion measurements. In Sect. 5, we give a brief outline how such an analysis is

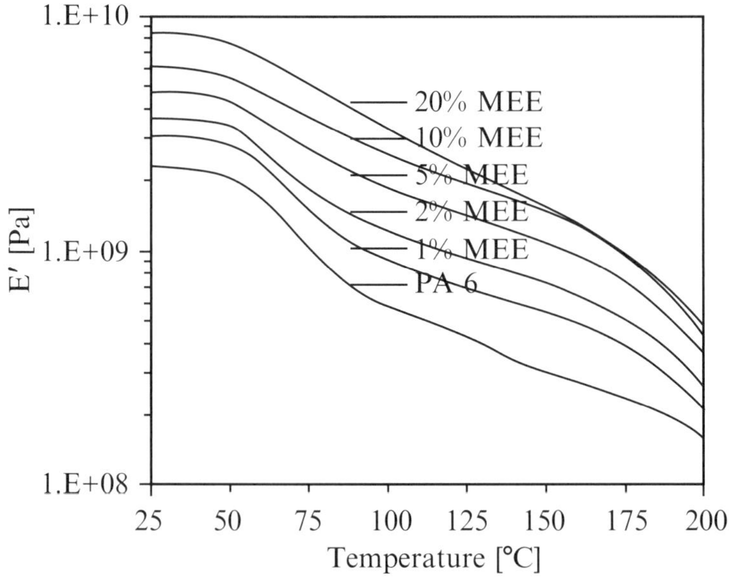

Fig. 7 Measured dynamic storage moduli for exfoliated layered silicate nanocomposites with a filler content up to 20 wt%

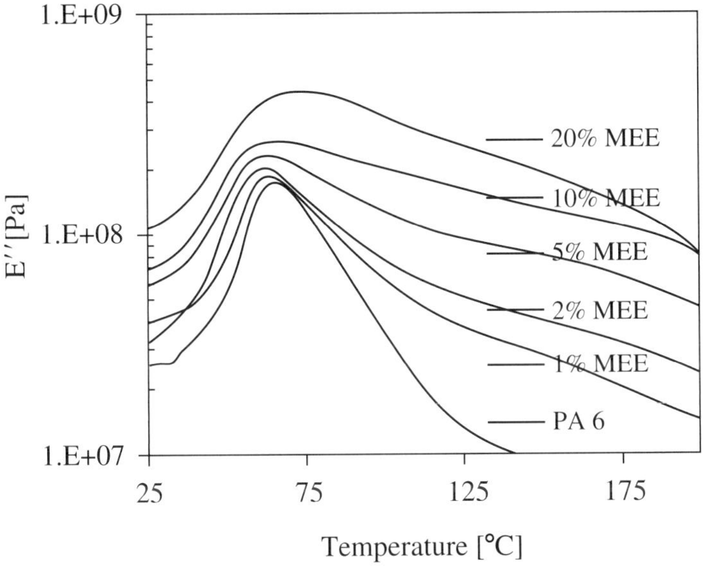

Fig. 8 Measured dynamic loss moduli for exfoliated layered silicate nanocomposites with a filler content up to 20 wt%

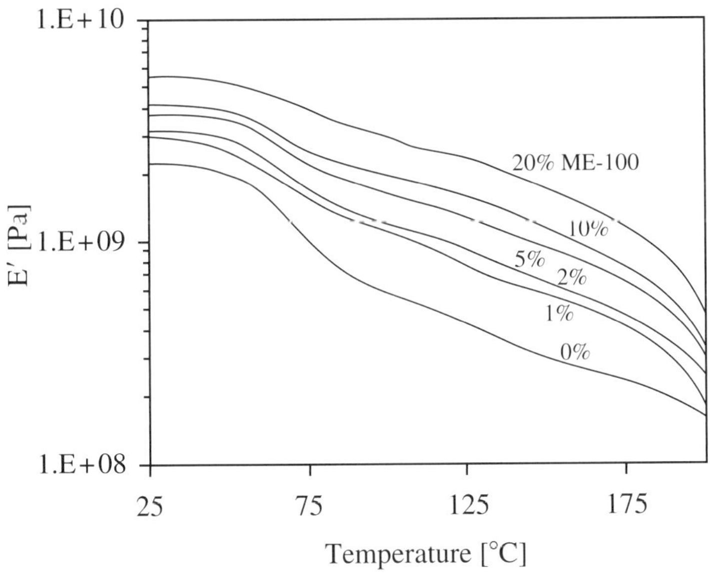

Fig. 9 Measured dynamic storage moduli for unmodified layered silicate nanocomposites with a filler content up to 20 wt%

performed, based on the idea that the nanoparticles form a labyrinth that reduces the effective mobility for the diffusion of water. Figure 10 shows the moisture uptake kinetics at various particle concentrations, which can be converted to diffusion coefficient data as given in Fig. 11. From these results using the tortuous path model it is possible to determine the effective particle aspect ratios as shown in Fig. 12.

Figure 12 shows the aspect ratios calculated from the ratio of diffusion coefficients from the matrix and the nanocomposite. The general trend of decreasing aspect ratio for increasing silicate content and the values are similar to the aspect ratios derived from the mechanical model, shown in Fig. 6. The lower aspect ratios for MAE nanocomposites can be explained by the lower compatibility with PA6 with this very hydrophobic clay.

3.4 Determination of the Yield Stress in the Melt

The complex modulus of unfilled polymers and composites without a yield stress converges to zero at zero shear frequency. However, nanocomposites with high exfoliated silicate content can form a structure in the melt, which leads to a yield stress. This effect is present in the MEE nanocomposites, as can be seen in Fig. 13.

For unfilled PA6 and low MEE volume fractions, the plot in Fig. 13 shows a straight line converging to zero for $\omega \rightarrow 0$, while the modulus for higher concentrations levels off to a finite value. An apparent yield stress can be determined from this value, and this is shown in Fig. 14.

The unfilled matrix has zero yield stress, while the nanocomposites containing unmodified particles have very low yield stresses, even up to 20 wt% filler. The MEE nanocomposites show a steep increase in the yield stress above 5 wt%. At 20% MEE silicate, τ_{yield} is more than 3 decades higher than at 20 wt% ME-100. A high yield stress can be a disadvantage for applications where flow occurs at low speed, such as in the impregnation of (macro-) fiber composites.

Unmodified silicate nanocomposites obviously have a lower modulus, as was shown in the previous section in Fig. 9.

A comparison between the modulus increase and the viscosity increase should now be made to find the particles that provide the best compromise for specific applications, such as injection molding, extrusion, or fiber composite matrix. The modulus increase can be compared with the viscosity increase by plotting the relative modulus ($E_{\text{composite}}/E_{\text{matrix}}$) at room temperature versus the relative viscosity in the melt. This is shown in Fig. 15 at two shear frequencies: at 0.46 rad s^{-1} (Fig. 15a) and at 100 rad s^{-1} (Fig. 15b).

From Fig. 15a it can be seen that for low shear rate processes, such as fiber impregnation and wetting, nanocomposites based on unmodified silicate have the advantage of a lower viscosity compared to those based on modified silicate for equivalent modulus enhancement, especially at loadings above 5%. The advantage becomes even clearer when also the yield stress is considered (Fig. 14). However, when processes with high shear rates (Fig. 15b) are applied, such as injection molding, both types of nanocomposite show a similar increase in viscosity. At high shear

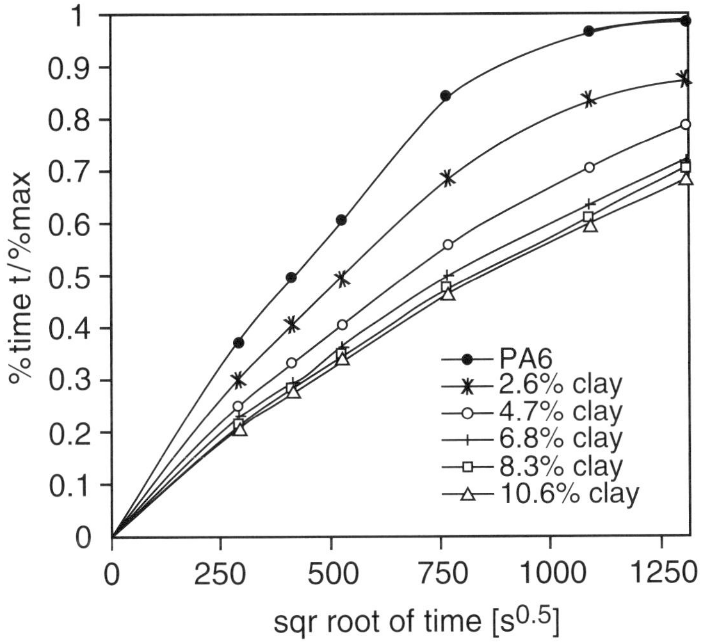

Fig. 10 Moisture uptake as fraction of maximum content versus the square root of time: the initial slope is used to derive the diffusion coefficient (Cloisite 30B nanocomposite)

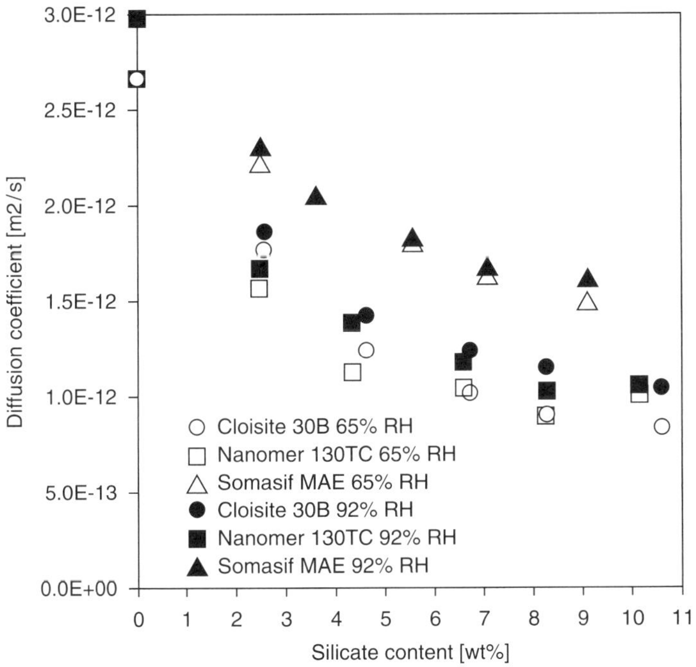

Fig. 11 The diffusion coefficients as a function of silicate content for two different moisture conditions and three different clays. Clearly one clay (MAE) has a higher diffusion coefficient, which is related to a lower degree of exfoliation. This is also apparent from the lower modulus of these samples

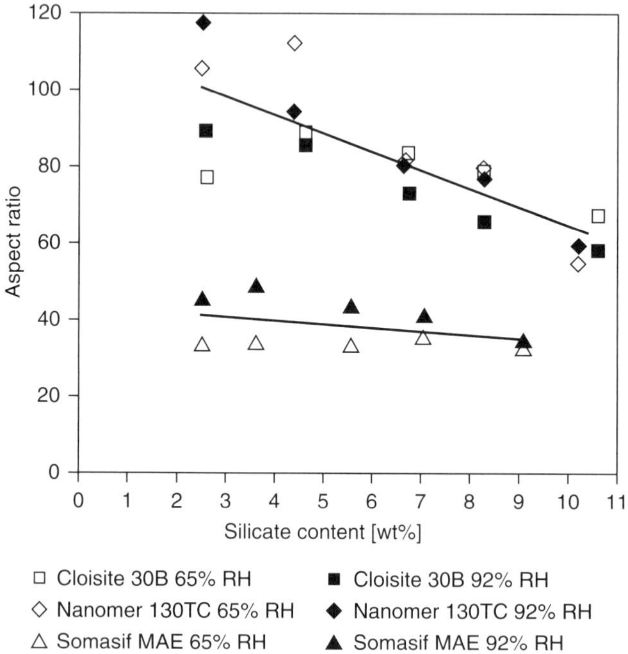

Fig. 12 Aspect ratios derived from the ratio of diffusion coefficients as a function of silicate content for two different moisture conditions and three different clays

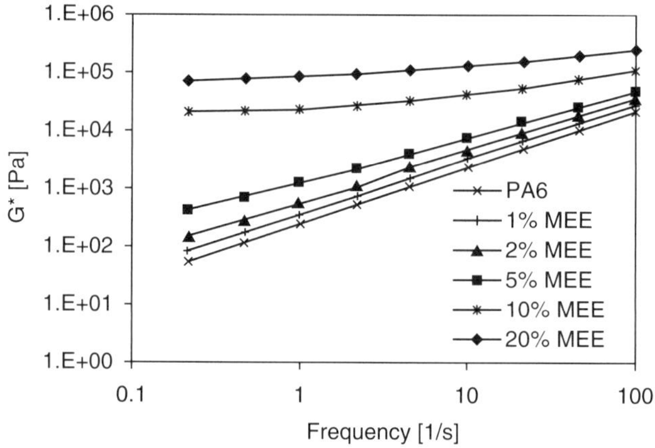

Fig. 13 Complex modulus MEE nanocomposites

rates the penalty in the viscosity for better exfoliation is reduced, probably because the platelets become aligned. A fiber-shaped nanofiller (sepiolite) causes a larger increase of the viscosity than the platelet filler at a similar modulus increase at high shear rates. This is supported by theories, which predict a quadratic dependence of the viscosity on the aspect ratio for fibers influenced by Brownian motion, and a

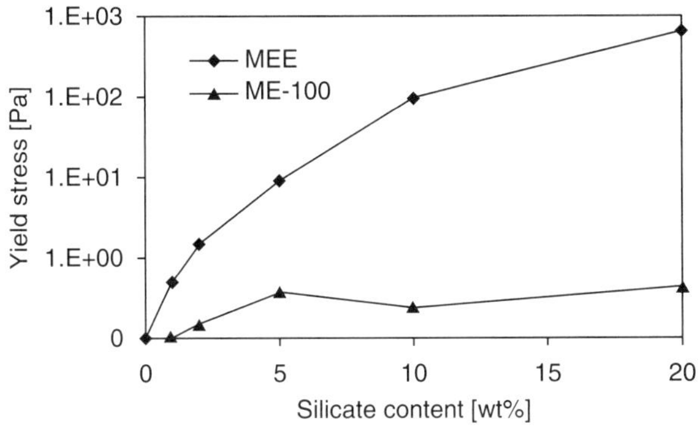

Fig. 14 Melt yield stress for modified and unmodified layered silicate

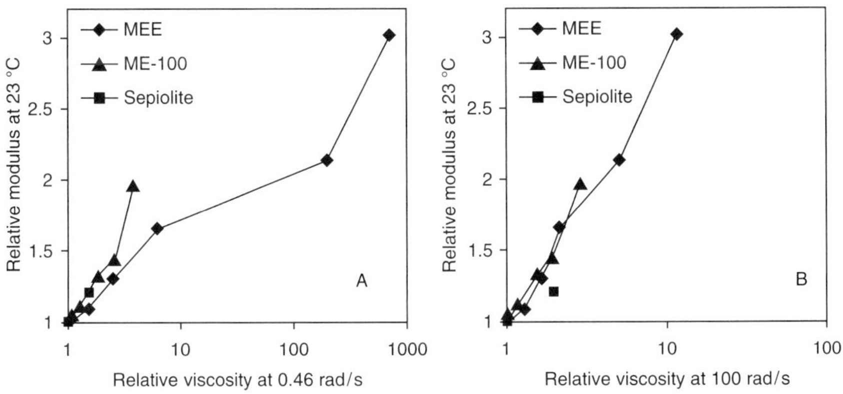

Fig. 15 Relative modulus compared with relative viscosity (uncorrected for yield stress): (**a**) 0.46 rad s^{-1} and (**b**) 100 rad s^{-1}

linear dependence for platelets [32]. Note that the solid-state modulus predicted by the Halpin–Tsai model shows a linear dependence for all particle shapes.

3.5 Anisotropy and Templating in Liquid Crystalline PA6–Boehmite

Figure 16 shows the dynamic storage moduli of PA6–Boehmite nanocomposites in the temperature interval 37–160°C. Initially, there is no mechanical reinforcement observed with 1% (wt) Boehmite. The reinforcement becomes effective at 1.3% and increases systematically up to 3.74 GPa at 9% Boehmite content. Regarding the similarity between the 7.5% and 9% curves, one can argue that the storage moduli reach a saturation point after a certain filler concentration and they do not increase further beyond this point. In Fig. 17, dynamic storage moduli

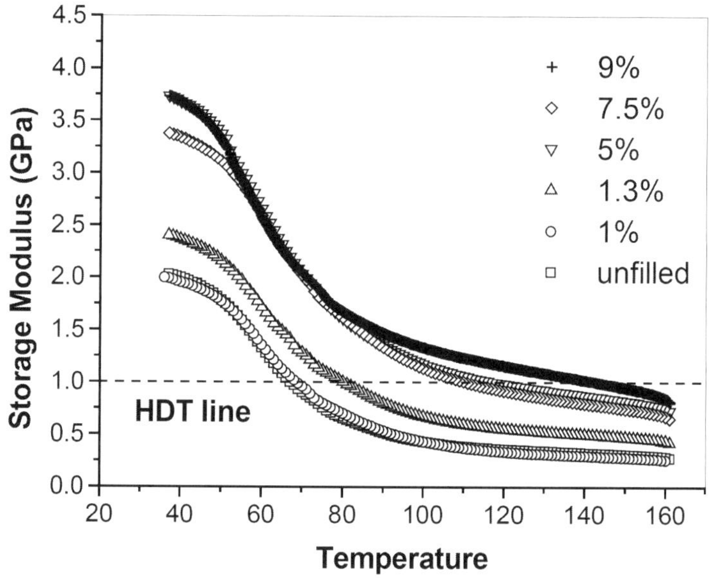

Fig. 16 Dynamic storage moduli of PA6–Boehmite nanocomposites

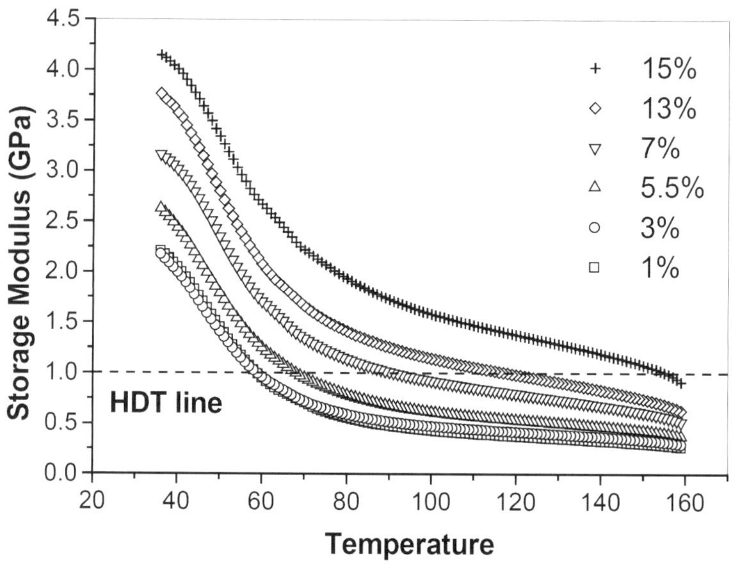

Fig. 17 Dynamic storage moduli of PA6–Ti–Boehmite nanocomposites

of the PA6–Ti–Boehmite nanocomposites in the temperature interval 36–160°C are shown. The first two samples with 1% and 3% (wt) Ti–Boehmite do not have a significant difference. Starting from the 5.5% sample, storage moduli systematically increase and reach 4.23 GPa at 15% Ti–Boehmite content. Contrary to the

PA6–Boehmite samples, the increase in storage modulus of the PA6–Ti–Boehmite samples is continuous.

Another information that can be derived from storage modulus–temperature curves is the Heat Distortion Temperature (HDT). HDT of a polymeric material is defined as an index of its heat resistance under applied constant stress. Storage modulus of a polymer decreases with increasing temperature and when it reaches a critical modulus value, the corresponding temperature is recorded as the HDT of the sample. Apart from the standard testing method (ASTM D648), modulus–temperature curves can also be used to obtain the same kind of information by reading off the temperature where the modulus reaches approximately 1 GPa.

Among various polymer–clay nanocomposites reported in the literature, PA6–clay systems show the most dramatic improvement in their HDT's. With these nanocomposites, an 80°C increase in HDT has been reported at 4.2 wt% clay loading [11]. In the case of PA6–Boehmite nanocomposites, the HDT values are determined from the storage modulus curves in Figs. 16 and 17. In accordance with the ISO-A standard, the temperature at which the modulus curve intersects the constant 1 GPa line is taken as the HDT of the sample. The HDT's of Boehmite and Ti–Boehmite nanocomposites are presented in Table 1.

In the PA6–Boehmite series, the HDT's increase gradually from 67°C of the unfilled polymer to 143°C of the nanocomposite with 9% Boehmite (Table 1). With PA6–Ti–Boehmite, the 1% and 3% samples have even lower HDT values than that of the unfilled polymer and an improvement starts to occur at 7% Ti–Boehmite content. This initial decrease in HDT can be attributed to the plasticizing effect of the surface modifier (Ti-agent) contained in these samples. In this series, the maximum HDT is obtained as 155°C at 15% Ti–Boehmite content.

The HDT increments of the Boehmite nanocomposites at 5–5.5% filler contents are much lower when compared to that of the PA6–clay nanocomposites at 4.2% filler content [11]. This is in agreement with the argument that Boehmite rods give a smaller modulus increase as compared to the clay platelets, because rods reinforce in one dimension instead of two for plates, and the aspect ratio of Boehmite is lower (see also Fig. 21). Related to this effect, one may expect lower barrier properties and less remarkable improvements in HDT. At the highest Boehmite contents, the increments obtained in HDT of PA6–Boehmite and PA6–Ti–Boehmite samples are 76°C and 88°C, respectively (Table 1). These are significant improvements, but they occur at much higher filler concentrations with respect to PA6-layered silicate nanocomposites.

Table 1 Heat distortion temperatures of PA6–Boehmite and PA6–Ti–Boehmite samples

Sample	HDT at 1 GPa (°C)	Sample	HDT at 1 GPa (°C)
PA6-unfilled	67	PA6-Ti-Boeh 1%	58
PA6-Boeh 1%	68	PA6-Ti-Boeh 3%	58
PA6-Boeh 1.3%	80	PA6-Ti-Boeh 5.5%	68
PA6-Boeh 5%	117	PA6-Ti-Boeh 7%	91
PA6-Boeh 7.5%	110	PA6-Ti-Boeh 13%	117
PA6-Boeh 9%	143	PA6-Ti-Boeh 15%	155

3.6 Comparison with Mechanical Models

The results obtained with the PA6–Boehmite and PA6–Ti–Boehmite systems have been compared to several composite models, like the parallel model and the Halpin–Tsai model. As seen in Fig. 18, the Halpin–Tsai model provides a good fit for the experimental data. In these calculations, the matrix modulus is taken as 2.03 GPa, which is the modulus measured for the unfilled in-situ polymerized PA6. The filler modulus is taken as 253 GPa. To convert weight percentages to volume fractions, densities of PA6 and Boehmite are used as 1.13 g cm^{-3} and 3.01 g cm^{-3}, respectively.

The mechanical properties of the PA6–Boehmite and PA6–Ti–Boehmite nanocomposites can be summarized as follows: a twofold improvement in storage modulus of the polymer matrix is observed and the experimental data can be fitted by the Halpin–Tsai model. Although both modulus curves follow a similar trend, use of Ti-modified Boehmite allows higher Boehmite contents in the polymer, which obviously leads to higher modulus values. In addition, heat distortion temperatures of PA6–Boehmite and PA6–Ti–Boehmite systems have increased from 67°C of the unfilled polymer to 143°C and 155°C, respectively.

3.7 Nematic Phase Behavior of Ti–Boehmite in PA6

The nematic phase behavior of Ti–Boehmite rods in PA6 is demonstrated by optical polarization microscopy and WAXS. In OPM studies, the polymer crystallites

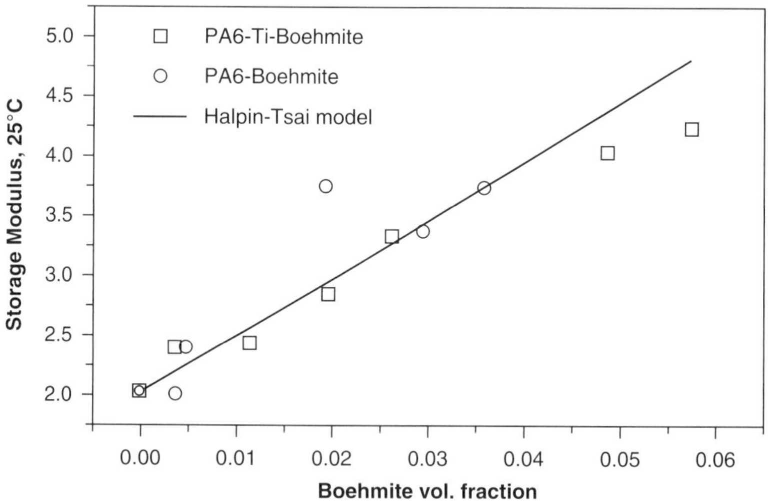

Fig. 18 Storage moduli of the PA6–Boehmite and PA6–Ti–Boehmite nanocomposites in comparison to the Halpin–Tsai model

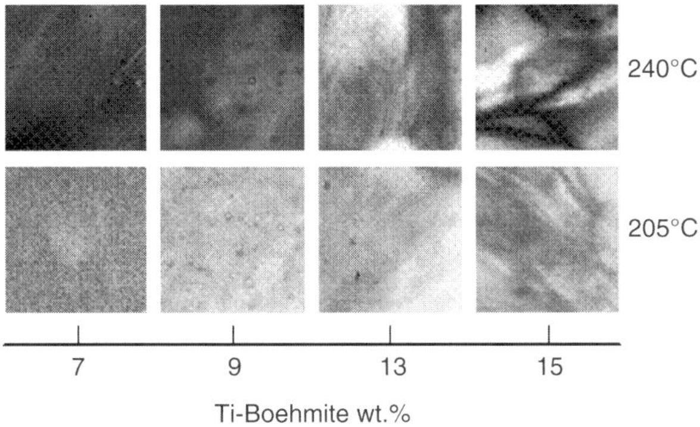

Fig. 19 OPM pictures of PA6–Ti–Boehmite nanocomposite samples

have significant contribution to the birefringence and therefore the observations are made above the melting point of PA6. Below the melting point, all samples have a certain amount of birefringence. As the polymer starts to melt, samples with 5% and lower Ti–Boehmite contents lose their birefringence and become completely isotropic, whereas the samples with 7% and higher Ti–Boehmite contents remain birefringent also above the melting point. The OPM images of the nematic samples are shown in Fig. 19. The images illustrate the state of the samples at temperatures below (205°C) and above (240°C) the melting point of PA6.

As seen in Fig. 19, all samples are birefringent below the melting point. The 13% and 15% samples show the highest amount of birefringence below and above the melting point, which clearly indicates their nematic behavior. In comparison, the 7% and 9% samples have a much lower birefringence. As a result of this analysis, it is concluded that the samples with Ti–Boehmite contents of 5% and lower are isotropic; the 13 and 15% samples are nematic and the 7% and 9% samples may be in the biphasic region.

The orientation in the nematic samples is also evident from WAXS analysis. In Fig. 20, the WAXS images of 7% and 15% Ti–Boehmite– PA6 nanocomposites at 30°C are compared to their OPM images. The inner- and outermost scattering peaks are the Boehmite peaks and the other peaks belong to PA6. The general trend in the WAXS spectra is that the shape of Boehmite scattering peaks change from isotropic to anisotropic as the weight percentage of Ti–Boehmite is increased and the anisotropy is particularly strong at 13% and 15%. This observation is in good correlation with the OPM analysis. In addition, the increase in anisotropy of the PA6 peaks is proportional to that of the Boehmite peaks, meaning that the filler particles form a template for polymer crystallization and induce further orientation of the polymer chains.

Although one might anticipate that this templated crystallization would give rise to an enhanced modulus and anisotropy of the PA6–Boehmite samples, this was not found in practice due to the still rather low level of orientation of the samples. This can be assessed by a more detailed analysis of the XRD results as is discussed in more detail in one of our publications [19].

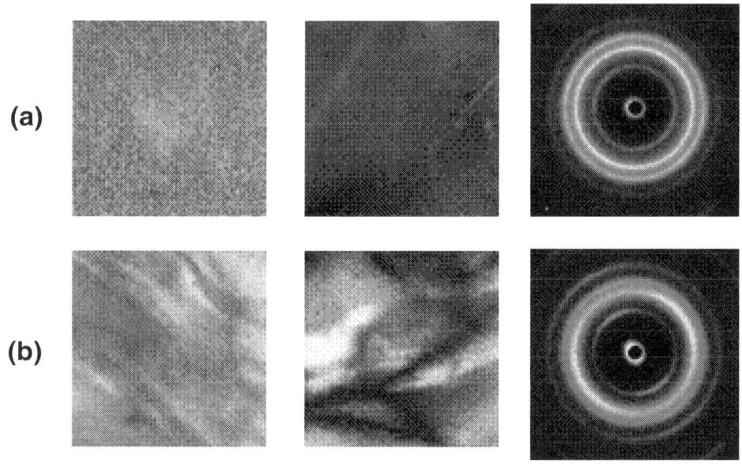

Fig. 20 Left to right: OPM images (at 205°C and 240°C, respectively) and the room temperature WAXS spectra of PA6–Ti–Boehmite nanocomposites at (**a**) 7% and (**b**) 15% filler contents

4. Mechanical Modeling

There are a variety of possible approaches to start assessing the mechanical properties of polymer nanocomposites. One of the aspects that is still unclear at present is to what extent the presence of nanoparticles gives rise to a change in the local polymer dynamics. For instance if there is a strong polymer–particle interaction one may anticipate a local decrease in dynamics and a corresponding increase of the local glass transition temperature. Similarly a poor interaction may give rise to a faster polymer mobility and local T_g decrease. At present we prefer not to address these issues directly and primarily deal with what should happen to the mechanical properties purely on the basis of the particle aspect ratio and concentration. So the question is what should happen without considering any change in the matrix polymer dynamics? It will become clear that already in this case interesting phenomena may occur especially at higher particle aspect ratios.

To understand the mechanical properties of our nanocomposites have chosen to use the Halpin–Tsai model (1), which was originally developed to describe the mechanical properties of semicrystalline polymers [6, 7], using the shape factors derived by Van Es [27].

$$\frac{E_c}{E_m} = \frac{1+\varsigma\eta\phi_f}{1-\eta\phi_f} \text{ in which } \eta = \frac{(E_f/E_m)-1}{(E_f/E_m)+\varsigma} \tag{1}$$

Here E_c composite Young's modulus; E_f filler modulus; E_m matrix modulus; ζ shape factor – depending on geometry, aspect ratio, and orientation; ϕ_f filler volume fraction. Similar expressions can be derived for the other moduli like G.

It should be noted that in the literature often the Halpin–Tsai equations are described as a semiempirical approach. This however does not do justice to their original derivation as they are the result of closely examining the work of Kerner et al. [6] that analyzed composite stiffness in terms of self-consistent mechanical models, where the particles, of certain aspect ratio, are embedded in a polymer matrix and this is embedded in a continuum (consisting of polymer and particles). The self-consistency constraint then gives rise to equations for the various moduli that under certain assumptions can be reduced to the form proposed by Halpin and Tsai. This means that the Halpin–Tsai equations should be considered as a mechanical mean-field model and not as an arbitrary choice.

With the appropriate shape factors for different particles shapes and orientations, this model can successfully describe Young's and shear moduli. The specific shape factors can be determined by comparing the model with experimental results or with more fundamental theories, i. e., the Eshelby theory, the Mori–Tanaka theory, and 3D finite element modeling [22, 27, 28].

The shape factors for the tensile moduli of platelet-reinforced composites (*a* width; *b* thickness) are [27]

$$E_{11} \text{ or } E_{22}\ \varsigma = \frac{2}{2}\left(\frac{a}{b}\right) \text{(in the radial direction of the platelets)}$$

$$E_{33}\ \varsigma = 2 \text{ (perpendicular to the platelets).}$$

The shape factors for fibers (*a* length; *b* width) are [27]

$$E_{11} \text{ or } E_{22}\zeta = 2 \text{ (Perpendicular to the fibre direction)}$$

$$E_{33}\ \varsigma = 2\left(\frac{a}{b}\right) \text{(in the fibre direction).}$$

When the stiffness of the composite, matrix, and filler are known, the Halpin–Tsai model can also be used to back-calculate the aspect ratio of the reinforcing particles.

This will be an effective aspect ratio, because the particles can have different shapes, sizes, and thickness, as has been pointed out in various articles [4, 14, 37, 22]. Instead of using image analysis of TEM images to estimate the aspect ratio distribution [4, 14, 22], we simply use the effective aspect ratio that the model gives based on the experimental data. This effective aspect ratio is at least a reasonable estimate of the average aspect ratio [33, 34] and provides a useful parameter to compare different nanocomposite compositions. The Halpin–Tsai equations predict that particles are hardly effective at very low aspect ratios (<10). The maximum stiffening effect is reached only at aspect ratios above 1,000 for platelets and 100 for fibers [28] (Fig. 21).

When (1) is rewritten in one equation as is shown in (2), the physical nature of the Halpin–Tsai model becomes more transparent:

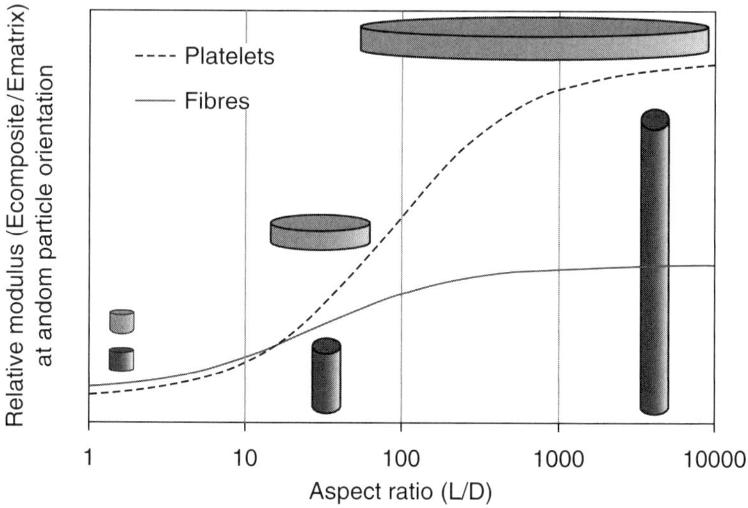

Fig. 21 Influence of the aspect ratio on the modulus for platelets and fibers [28]

$$E_c = \zeta E_m \frac{(\phi_f + 1/\zeta)E_f + (1-\phi_f)E_m}{(1-\phi_f)E_f + (\phi_f + \zeta)E_m}. \quad (2)$$

When the aspect ratios are much smaller than the ratio of filler and matrix modulus ($\zeta\, E_f / E_m$), the Halpin–Tsai model gives results close to a series model:

$$E_c = \frac{E_m E_f}{E_f(1-\phi_f) + E_m\phi_f} = \left(\frac{\phi_f}{E_f} + \frac{1-\phi_f}{E_m}\right)^{-1}. \quad (3)$$

When the aspect ratios are much larger than the ratio of filler and matrix modulus ($\zeta\, E_f / E_m$), the Halpin–Tsai model gives results close to a parallel model:

$$E_c = E_f\phi_f + E_m(1-\phi_f). \quad (4)$$

For nanocomposites, the series model underestimates the modulus. Because the volume fraction of filler is low, the first term in (3) dominates giving values close to the matrix modulus. Similarly the parallel model (4) overestimates the composite modulus at low matrix moduli, since it assumes a continuous reinforcing phase.

The Halpin–Tsai model (2) leads to results that are in between these two extremes when the aspect ratio is of the same order of magnitude as the ratio of moduli ($\zeta \approx E_f/E_m$). The aspect ratio of the particles has a strong influence on the results and, therefore, determines how close we are to either the series or parallel model. The calculated effect of the aspect ratio on the relative modulus ($E_{composite}/E_{matrix}$) in the Halpin–Tsai model for a platelet filled composite

(E_{matrix} = 3.1 GPa, E_{filler} = 172 GPa, ff = 0.05) is shown in Fig. 22, which shows that the particles are hardly effective below w/t = 10, and only when w/t is above 1,000 the maximum reinforcement effect is reached.

Further insight in the Halpin–Tsai model can be obtained if (2) is approximated for large aspect ratios and low filler fractions as shown in (5):

$$E_c = \frac{E_m \zeta}{E_f + E_m \zeta}\left(E_f \phi_f + E_m (1 - \phi)\right). \tag{5}$$

Here we observe that if the matrix modulus remains sufficiently high, the prefactor will reduce to a value close to 1 and the parallel model is obtained. When the matrix modulus times the ζ factor get below E_f the prefactor will start forcing the composite modulus to lower values and ultimately the soft matrix will dominate the behavior of the material. This explains why a nanocomposite with large aspect ratio particles can perform so well at elevated temperatures using a semicrystalline polymer matrix. Above the T_g, the Young's modulus of a semicrystalline polymer is normally too low for many applications. But in nanocomposites, the physical network from the crystalline regions allows the large aspect ratio nanoparticles to contribute to the modulus roughly in accordance with the parallel model. This means that the filler contribution is additive to the modulus of the soft semicrystalline matrix and a useful stiffness results. Note that this does not work for an

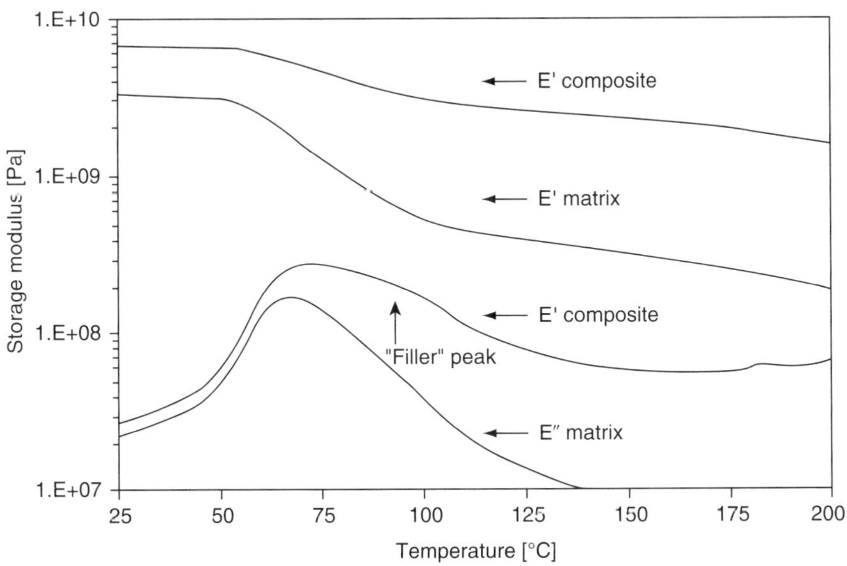

Fig. 22 Calculated storage and loss moduli, using (1) or (2), based on measured data for the polymer matrix for the composite (E_c' and E_c") and the polymer matrix (E_m' and E_m") using ζ = 500 and ϕ_f = 0.03

amorphous polymer, as the modulus in the rubber plateau of amorphous polymers is far too low for the prefactor to be useful.

Concerning the possible change in polymer dynamics already mentioned, it should be stressed that the Halpin–Tsai model itself already gives rise to an additional relaxation peak. This is best seen by making use of the analytic continuation theorem, which implies that (1) or (2) may also be used in their complex form to compute the storage and loss moduli of the composite. The result of this is shown in Fig. 22 and indeed at higher aspect ratio an additional loss peak is observed. The low T loss peak results from the changing polymer matrix dynamics at T_g, and the second high T peak results from the transition from "parallel" to "series"-like behavior and can be understood by examining the prefactor in (5). It will be clear that the prefactor will cause the stiffness of the composite to drop from roughly $E_f\phi_f$ (at not too low E_m) to 0 (or $\phi_f E_m\zeta$) when the value of $E_m\zeta$ decays to values below E_f. This means that there is a second stepwise change in the storage modulus and as a result we get a second peak in the loss modulus at the point where $E_m\zeta = E_f$. This peak is the inevitable consequence of the Kramers–Kronig relations.

This second "filler-peak" only turns up at rather high aspect ratios ($\zeta > 500$) and in most experiments should not be visible, especially considering the particle polydispersity. However at practical aspect ratios, this effect will cause an *apparent* rise of the T_g of the composite due to partial overlap of the polymer peak and the "filler-peak." Therefore extreme caution is indicated when claiming that adding nanoparticles causes a change in polymer dynamics or T_g as the underlying cause may actually be due to the shape dependent "filler-peak."

The model can be used to investigate the effect of variation of the filler content and the filler shape, shown in Figs. 23 and 24.

Figures 23 and 24 show that both an increase in aspect ratio of the particles (~ shape factor) and an increasing volume fraction of the filler cause a larger rise in E'' above

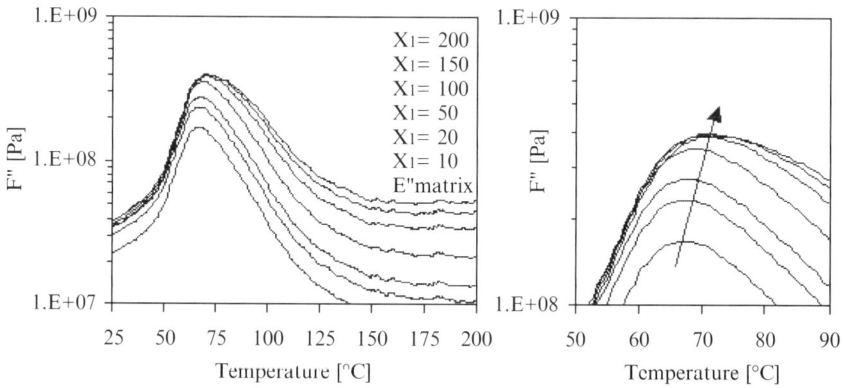

Fig. 23 Modeled loss modulus curves as a function of shape factor, based on a measured E'' from unfilled PA6 matrix and a constant filler volume fraction of 0.04. On the right a part of the figure around T_g is shown in more detail, showing the virtual T_g shift caused by the rise in loss modulus above T_g

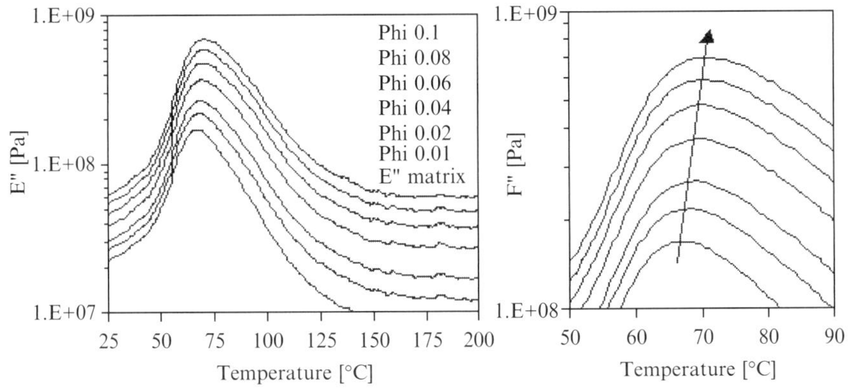

Fig. 24 Modeled loss modulus curves as a function of filler volume fraction, based on a measured E'' from unfilled PA6 matrix and a constant shape factor of 70. On the right a part of the figure around T_g is shown in more detail, showing the virtual T_g shift caused by the rise in loss modulus above T_g

T_g than below. In both cases, the model shows that this leads to a apparent rise in T_g, which is clearly visible in the enlarged sections on the right of Figs. 23 and 24.

At least qualitatively the model produces some features that are also seen in the experimental data in Sect. 3, such as an extra plateau or even a "filler peak" at temperatures above T_g, a remarkably high E' and E'' above T_g, and an apparent shift in the T_g peak in the loss modulus to higher temperatures.

It is worth stressing that one should *not* use the tanδ peak to assess any changes in T_g. This should already be clear from the definition of tanδ (= E"/E') where changes in E' especially above T_g may give rise to tanδ peak shifts. In fact this is already the case for unfilled semicrystalline or amorphous polymers (or networks) where entanglement levels, crosslinking, or varying crystallinity gives rise to shifts in the tanδ peak while analyzing the E'' does not indicate any change in T_g. Similar, or in fact worse, problems will arise here as E' changes both above and below T_g depending on the particle loading and aspect ratio, which may cause erratic changes in the location of the tanδ maxima.

5 Modeling of Moisture Diffusion

The moisture uptake diffusion coefficients D can be used to compare the diffusion speed in the unfilled polymer and nanocomposite and to estimate how much time total moisture saturation of a sample takes, depending on the dimensions. In addition, it is possible to calculate the effective aspect ratio of the silicate sheets in the nanocomposites from the ratio of diffusion coefficients of the nanocomposite and the matrix. Nielsen [16] has derived a simple theory, which relates the diffusion coefficient to the increased path that molecules have to travel when diffusing around impenetrable platelets. Many simplifications have been made in this theory, which reduce the accuracy of the results. An important difference between Nielsen's model and real nanocomposites is the fact that Nielsen's theory is a two-dimensional theory in which the third dimension of the plates is considered infinitely long. In addition, it assumes perfectly overlapping ribbons, while real particles are randomly oriented, at least in the long dimensions of the platelets.

Therefore, this theory underestimates the real diffusion coefficient. Van Es [28] has derived a more accurate equation for diffusion in nanocomposites filled with aligned platelets, which is based on a combination of models by Brydges et al. [3] and Hatta et al. [8]:

$$\frac{D_{\text{matrix polymer}}}{D_{\text{nanocomposite}}} = \left(\phi_{\text{f}}\beta + 1 - \phi_{\text{f}}\right)\left(\phi_{\text{f}}\beta\gamma(1-\gamma)+1\right),$$

where $\beta = \frac{8}{5\pi}\frac{w}{t}$ and w/t = width/thickness = aspect ratio platelets, and γ= overlap factor. The overlap factor γ= 0.5 for perfect overlap of the platelets, 0 for no overlap, and 0.25 for platelets that are randomly placed in two-dimensions.

This equation is valid for diffusion with a wide variety of particle concentrations, and the validity was confirmed by comparison with three-dimensional finite element modeling. This equation is based on diffusion through a material around impenetrable platelets that are all aligned parallel to each other. This is a reasonable assumption in film and injection-molded samples [13, 29, 4], especially on the outside of the sample where the first diffusion is measured. WAXS measurements on these nanocomposites have indeed shown highly oriented silicate layers [33]. However, within the plane of the platelets, there is no specific ordering and so the random placement of the sheets leads to an overlap factor γ of 0.25 in this model.

6 Summary and Outlook

In this chapter, we have discussed the properties of polymer nanocomposites and have compared them with the Halpin–Tsai composite model. It is found that a wide range of effects observed in the experiments, like the modulus increase, the level of exfoliation, the apparent shift of the T_g from DMA, etc. can be successfully explained purely based on the particle aspect ratios and the nanoparticle concentration. Moreover the aspect ratios obtained from the mechanical modeling via Halpin–Tsai are in rather good agreement with those found from measuring moisture diffusion. We believe that it is possible to use this analysis for optimization of polymer nanocomposite properties for various applications. For instance in some cases it is necessary to find an optimum in processability and stiffness, e. g., when using polymer nanocomposites for glass- or carbon-fiber-reinforced composites where a very high viscosity leads to incomplete wetting of the fiber reinforcement. The use of such models also has implications for the analysis of the polymer dynamics in nanocomposites. Any reports in the literature on increased T_g levels should be closely examined before attributing this to an actual change of the polymer chain mobility by confinement or interaction with the nanoparticles. We have shown that such T_g changes may be purely attributable to the response of the dynamic Halpin–Tsai model.

Also we have shown that the presence of rod-shaped nanoparticles may give rise to a change in the crystal morphology of the polymer matrix when the nanoparticles form a colloidal nematic phase inside the polymer matrix. Unfortunately, so far we have not observed any benefits of this phenomenon in terms of a further enhanced stiffness (and anisotropy) of the nanocomposites, which is probably due to a too low level of alignment of the colloidal nematic phase.

Finally it is worth noting that adding a relatively low level of nanoparticles can substantially enhance the maximum use temperature of PA6, as is evident from the considerable rise of the heat-distortion temperature. This combined with the ability to process polymer nanocomposites at the normal processing temperature of PA6 is an aspect that deserves further attention as it would make relatively cheap and easy to process PA6 a suitable material for high-end applications. As these materials also show reduced flammability and have self-extinguishing properties, this would make PA6 polymer nanocomposites a very interesting option, e. g., aerospace applications. This is an aspect that we are exploring further at present.

Acknowledgements Dr. P.J. Kooyman of the National Center for High Resolution Electron Microscopy, TU Delft, is acknowledged for performing electron microscopy.

Ben Norder of the NanoStructured Materials section, TU Delft, is highly acknowledged for the DMA and DSC measurements and for his help in the interpretation of the DSC data. Gerard de Vos of the NanoStructured Materials section, TU Delft, is acknowledged for assistance with the extrusion, injection molding, SEM investigation, and sample preparation.

Dr. David van der Beek, from the group of prof. H.N.W. Lekkerkerker (Colloid Chemistry Group,Utrecht University) is acknowledged for assistance with the Boehmite synthesis and characterization.

DSM is acknowledged for kindly providing us with ME-100 composites.

The work of Daniel Vlasveld and Ceren Özdilek formed part of the research programme of the Dutch Polymer Institute.

References and Notes

1. Akkapeddi MK. Glass fiber reinforced polyamide-6 nanocomposites. Polym Compos 2000;21(4):576–85.
2. Buining PA, Pathmamanoharan C, Jansen JBH, Lekkerkerker HNW. Preparation of colloidal boehmite needles by hydrothermal treatment of aluminium alkoxide precursors. J Am Ceram Soc 1991;74:1303.
3. Brydges WT, Gulati ST, Baum G. Permeability of glass-ribbon reinforced composites. J Mater Sci 1975;10(12):2044–49.
4. Fornes TD, Paul DR. Modelling properties of nylon6/clay nanocomposites using composite theories. Polymer 2003;44(17):4993–5013.
5. Giannelis EP. Polymer layered silicate nanocomposites. Adv Mater 1996;8(1):29.
6. Halpin JC, Kardos JL. Halpin-Tsai Equations – Review. Polym Eng Sci 1976;16(5):344–52.

7. Halpin JC. Stiffness and expansion estimates for oriented short fiber composites. J Compos Mater 1969;3:732–34.
8. Hatta H, Taya M, Kulacki FA, Harder JF. Thermal diffusivities of composites with various types of filler. J Compos Mater 1992;26(5):612–25.
9. Kelly A, Tyson WR. Tensile properties of fibre-reinforced metals: copper/tungsten and copper/molybdenum. J Mech Phys Solids 1969;13:329–50.
10. Kohan MI. Nylon plastics handbook. Munich: Carl Hanser; 1995.
11. Kojima Y, Usuki A, Kawasumi M, Okada A, Fukushima Y, Kurauchi T, Kamigaito O. Mechanical-properties of nylon 6–clay hybrid. J Mater Res 1993;8(5):1185–89.
12. Kojima Y, Usuki A, Kawasumi M, Okada A, Kurauchi T, Kamigaito O. Sorption of water in nylon-6 clay hybrid. J Appl Polym Sci 1993;49(7):1259–64.
13. Kojima Y, Usuki A, Kawasumi M, Okada A, Kurauchi T, Kamigaito O, Kaji K. Novel preferred orientation in injection-molded nylon-6-clay hybrid. J Polym Sci B Polym Phys 1995;33(7):1039–45.
14. Kuelpmann A, Osman MA, Kocher L, Suter UW. Influence of platelet aspect ratio and orientation on the storage and loss moduli of HDPE-mica composites. Polymer 2005;46(2): 523–30.
15. Monte SJ. Neoalkoxy titanate and zirconate coupling agent additives in thermoplastics. Polym Polym Compos 2002;10(1):1–52.
16. Nielsen LE. J Macromol Sci A 1967;5(1):929–42.
17. Özdilek C, Kazimierczak K, Van der Beek D, Picken SJ. Preparation and properties of polyamide-6-boehmite nanocomposites. Polymer 2004;45:5207–14.
18. Özdilek C, Kazimierczak K, Picken SJ. Preparation and characterization of titanate-modified boehmite–polyamide-6 nanocomposites. Polymer 2005;46:6025–34.
19. Özdilek C, Mendes E, Picken SJ. Nematic phase formation of boehmite in polyamide-6 nanocomposites. Polymer 2006;47(6):2189–97.
20. Pielichowski J, Puszynski A. Polymer preparation methods. Krakow: Technical University of Krakow; 1978.
21. Reichert P, Nitz H, Klinke S, Brandsch R, Thomann R, Mulhaupt R. Poly(propylene)/organoclay nanocomposite formation: Influence of compatibilizer functionality and organoclay modification. Macromol Mater Eng 2000;275(2):8–17.
22. Sheng N, Boyce MC, Parks DM, Rutledge GC, Abes JI, Cohen RE. Multiscale micromechanical modeling of polymer/clay nanocomposites and the effective clay particle. Polymer 2004;45(2):487–506.
23. Usuki A, Kawasumi M, Kojima Y, Okada A, Kurauchi T, Kamigaito O. Swelling behaviour of montmorillonite cation exchanged for omega-amino acids by epsilon-caprolactam. J Mater Res 1993;8(5):1174–8.
24. Usuki A, Kojima Y, Kawasumi M, Okada A, Fukushima Y, Kurauchi T, Kamigaito O. Synthesis of nylon 6–clay hybrid. J Mater Res 1993;8(5):1179–84.
25. Usuki A, Koiwai A, Kojima Y, Kawasumi M, Okada A, Kurauchi T, Kamigaito O. Interaction of nylon-6 clay surface and mechanical properties of nylon-6 clay hybrid. J Appl Polym Sci 1995;55(1):119–23.
26. Van Bruggen MPB, Donker M, Lekkerkerker HNW, Hughes TL. Anomalous stability of aqueous boehmite dispersions induced by hydrolyzed aluminium poly-cations. Colloids Surf A 1999;150:115–28.
27. Van Es M, Xiqiao F, Van Turnhout J, Van der Giessen E. Comparing polymer–clay nanocomposites with conventional composites using composite modeling. Specialty polymer additives. Oxford: Blackwell Science; 2001.
28. Van Es M. Polymer clay nanocomposites. The importance of particle dimensions. Thesis, Delft University of Technology; 2001.
29. Varlot K, Reynaud E, Kloppfer MH, Vigier G, Varlet J. Clay-reinforced polyamide: Preferential orientation of the montmorillonite sheets and the polyamide crystalline lamellae. J Polym Sci B Polym Phys 2001;39(12):1360–70.
30. Vlasveld DPN, de Jong M, Bersee HEN, Gotsis AD, Picken SJ. The relation between rheological and mechanical properties of PA6 nano- and micro-composites. Polymer 2005;46(23):10279–89.

31. Vlasveld DPN, Groenewold J, Bersee HEN, Mendes E, Picken SJ. Analysis of the modulus of polyamide-6 silicate nanocomposites using moisture controlled variation of the matrix properties. Polymer 2005;46:6102–13.
32. Vlasveld DPN, Groenewold J, Bersee HEN, Picken SJ. Moisture absorption in polyamide-6 silicate nanocomposites and its influence on the mechanical properties. Polymer 2005;46(26):12567–76.
33. Vlasveld DPN, Vaidya SG, Bersee HEN, Picken SJ. A comparison of the temperature dependence of the modulus, yield stress and ductility of nanocomposites based on high and low MW PA6 and PA66. Polymer 2005;46(10):3452–61.
34. Wu YP, Jia QX, Yu DS, Zhang LQ. Modeling Young's modulus of rubber-clay nanocomposites using composite theories. Polym Test 2004;23(8):903–09.

Gaining Insight into the Structure and Dynamics of Clay–Polymer Nanocomposite Systems Through Computer Simulation

Pascal Boulet[1], H. Christopher Greenwell[2], Rebecca M. Jarvis[2], William Jones[3], Peter V. Coveney[4], and Stephen Stackhouse[5]

1 Introduction

Clay minerals belong to a wider class of solids known as layered materials, which may be defined as 'crystalline materials wherein the atoms in the layers are cross-linked by chemical bonds, while the atoms of adjacent layers interact by physical forces' [1]. Both clay sheets and interlayer space have one dimension in the nanometre range. Cationic clays are the predominant naturally occurring minerals with alumino-silicate sheets carrying a negative charge. Therefore, the interlayer guest species are positively charged to compensate the layer charge [2]. In anionic clays, also known as layered double hydroxides (LDHs), the interlayer guest species carry a negative charge and the inorganic mixed metal hydroxide sheets are positively charged. In recent times, there has been a growing interest in anionic clays, although initial attention was focussed almost exclusively on the cationic clay materials. Reviews have appeared that often emphasise interesting properties and the use of experimental techniques to determine or at least infer the local structure of the clay sheet or intercalated material [3–5]. However, clays are polycrystalline materials and precise experimental location of interlayer species is extremely difficult.

Only rarely can sufficiently large crystals for full structural determination by conventional single-crystal X-ray diffraction be obtained. Powder X-ray diffraction (PXRD) gives some indication of the bulk structure of the material (e. g., the spacing between layers), but, in general, clay nanocomposites are characterized by the absence of significant long-range order. Interlayer arrangements (e. g. guest orientation)

[1]UMR 6121 CNRS, Université de Provence - Aix Marseille I Centre Saint Jérôme, 13397 MARSEILLE Cedex 20, France

[2]Centre for Applied Marine Sciences, School of Ocean Sciences, University of Wales, Bangor, Menai Bridge Anglesey LL59 5Ab, UK

[3]Department of Chemistry, University of Cambridge, Lensfield Road, Cambridge, CB2 1WE, UK

[4]Centre for Computational Science and Department of Chemistry, University College of London, 20 Gordon Street, London WC1H 0AJ, UK

[5]Department of Earth Sciences, University College of London, Gower Street, London WC1E 6BT, UK

P. Knauth and J. Schoonman (eds.), *Nanocomposites: Ionic Conducting Materials and Structural Spectroscopies.*

may, however, sometimes be postulated from the interlayer spacing determined from PXRD measurements, but are frequently little more than educated guesses based solely on the assumed molecular dimensions of the guest [6]. The distinction between mono-layer and bi-layer arrangements, and the orientation of anisotropic interlayer guests may be inferred from the interlayer spacing determined by PXRD. The lateral arrangement of interlayer guest molecules, however, cannot and this is an area in which computer simulation can be usefully applied. Moreover, interlayer arrangements depend strongly on the interlayer water content of the clay as well as the nature of the intercalate, although this aspect is frequently overlooked [7]. Neutron diffraction can be used to determine the interlayer spacing in clays. It can also be used to determine the positions and self-diffusion of interlayer species that have been isotopically labelled [8, 9]. Other techniques such as Fourier transform infrared (FTIR) spectroscopy, near edge X-ray adsorption fine structure (NEXAFS) analysis and Mössbauer spectroscopy have been used to characterize clays, but these give only limited information about the molecular structure and arrangement of guests within the interlayer region [10].

Because of these limitations, the use of computational methods for studying these layered solids has increased in recent years [11, 12]. This review summarizes developments in computational methods for studying clay–polymer nanocomposite systems. Section 2 provides an outline of computer simulation techniques to introduce the reader, unfamiliar with these methods, to terminology and algorithms used in the field of quantum and classical molecular simulation. Sections 3, 4, 5 and 6 then review the application of these computational techniques in the study of clay–polymer nanocomposites. Conclusions and the outlook for future simulation of clay–polymer nanocomposite materials are given in Sect. 7.

2 Computer Simulation Techniques

The application and development of simulation techniques represent a huge area, with a diverse range of continually improving methods and techniques. The topic is dealt with in various textbooks and monographs [13]. To familiarize the reader with some of the terminology necessary for the remainder of this article, a brief summary is provided here. Techniques for the simulation of atomistic systems may, in general, be separated according to the accuracy with which they calculate interatomic interactions and type of structural and statistical data that they provide.

2.1 *Definition of the Potential Energy Surface*

The potential energy surface of a system describes the way in which the energy of the system changes with configuration, and plays an important role in simulation techniques. The potential energy surface is a complex function of inter-electronic, electron–nuclei and nuclei–nuclei interactions, and can rarely be calculated analytically, except in a few cases such as the hydrogen-like atoms and simple molecules (i. e., containing a single electron).

In quantum chemistry, calculations on large systems make use of the Born–Oppenheimer approximation [14] that assumes that, because of the higher weight of the nuclei with respect to that of the electrons, the motion of the two types of particles can be decoupled. As a consequence the total wavefunction can be written as a product of two wavefunctions, one for the electrons and the other one for the nuclei. Therefore, the positions of the nuclei are parameters in the electronic wavefunction and the nuclei wavefunction is a solution of a Schrödinger-like equation that describes the potential energy surface (PES). The PES is the so-called Born–Oppenheimer (BO) potential energy surface. This approximation is broadly used in all quantum methods (ab initio, Density-Functional Theory (DFT) and semi-empirical techniques) when the coupling between electrons and nuclei can be neglected (which is always assumed). Strictly speaking, this approximation prohibits the study, for instance, of vibronic spectra or PES crossings, although some techniques exist to overcome this difficulty (see for instance Sato et al. [15]). At the classical level, since electron motion is not considered, the PES is simply calculated from a functional form used to describe the interatomic interactions (vide infra).

2.1.1 Quantum Mechanics

Quantum mechanical simulations attempt to solve, to a good approximation, the fundamental equations of quantum mechanics, in order to model the interactions between all the particles (electrons and nuclei) of a system of atoms. The advantage of quantum mechanical simulations is that they allow the modelling of electron dynamics in a process, such as bond making–bond breaking, as they have an explicit representation of electrons. In addition, the only input data necessary is the atomic number and initial configuration of the nuclei and total number of electrons. The major disadvantage of the method is the associated huge computational cost – at present electronic structure calculations are limited to the study of hundreds of atoms, even when using large parallel machines.

In the context of materials chemistry, two families of quantum mechanical techniques are commonly used. The first one relies on solving the Schrödinger equation to find the wavefunction for the system of interest. In the theoretical chemistry community, these techniques are called the ab initio methods. The starting point of these methods is the Hartree–Fock method [16, 17] for which, the wavefunction is formally written as a Slater determinant [18] in order to account for the Pauli exclusion principle [19]. As mentioned above, the BO approximation is also used. The Hartree–Fock method is a self-consistent field (SCF) theory since each electron is moving in a mean field created by all the other electrons. The molecular orbitals that are solutions of the Hartree–Fock (HF) equations appear within the HF operator itself, and consequently, the equations have to be solved iteratively. Because of the Pauli exclusion principle, a repulsion energy term appears in the HF equations between electrons of same spin. This is the so-called exchange energy. This energy is represented by the exchange operator in the HF Hamiltonian, and accounts for about 10–20% of the total energy. By nature this operator is non-local, which means that, the result of this operator when applied to an electron depends on the application of this operator to all the other electrons. The advantage of the HF methods is to evaluate exactly the exchange energy of the electrons. The downside is that this operator is computationally expensive to calculate. In addition, the HF method does not account for the correlation interactions between electrons bearing opposite spins. Although the numerical importance of the correlation energy is marginal compared to the other contributions (it only represents about 1–5% of the exchange energy), its implications can be non-negligible in some chemical and physical processes (in particular in the search for transition states of reactions). The calculation of the correlation energy can be achieved using post-HF methods, the first of which is based on the many body Rayleigh perturbation theory (especially the Møller-Plesset series of methods [20]: MP2, MP3, MP4, where the numbers correspond to the order of the perturbation development). Another set of efficient methods is based on the linear combination of a reference wavefunction for the system with several wavefunctions where the electrons are promoted into virtual orbitals. The reference wavefunction is usually the HF one. Single, double or higher excitations can be taken into account and the corresponding methods are called CIS (for Configuration Interactions Single excitations), CISD (for single and double excitations) or even Full-CI, if all the combinations between electrons and virtual orbitals and excitation types are considered. Usually a limited number of electrons, relevant to the process of interest, are promoted into virtual orbitals, and constitutes the active space. If, within this active space, all the possible excitations are calculated, the method is called the CAS-MCSCF method (Complete Active Space Multi Configurational Self Consistent Field). Note, however, that in the case of the CI methods only the coefficients of the wavefunctions are optimized during the SCF process, whereas both the wavefunction coefficients and those of molecular orbitals are optimized during a CAS-MCSCF calculation. Another, similar set of methods is the coupled cluster methods (CCSD, CCSDT and CCSD[T]) that share the same goal as the CI and MCSCF methods. Recently, because of the development of high performance supercomputer facilities, new methods called MRCI, for Multi-Reference Configuration Interaction, have been used. The idea is to use a multi-reference wavefunction (such as an optimized CAS-MCSCF wavefunction) as a starting point to generate a set of single and double excitations, the goal being to optimize this new wavefunction to obtain a very good description of the system of interest. Unfortunately, all these methods are extremely computationally expensive and cannot be used for our purposes in the study of clay–polymer materials.

More recently, a second family of quantum methods appeared to be a good alternative for studying large systems with reasonable accuracy, namely, which includes both exchange and correlation energies. These methods belong to the DFT [21] and are based on the resolution of a set of Schrödinger-like equations, the so-called Kohn-Sham equations [22]. The central entity of this theory is not the wavefunction of the system but its electron density as it has been proved by Hohenberg and Kohn

[23] that knowledge of the electron density of a system implies knowledge of the external potential (i. e., the potential created by the nuclei and external fields, if any) applied to this system, and vice versa. One can immediately see a great computational advantage of using the electron density instead of the wavefunction, as the former function depends only on the three coordinates of position, whereas the latter depends on three coordinates of N positions, those of the N electrons. Another advantage is that interaction terms can be easily added to the Kohn-Sham Hamiltonian. As a consequence, the exchange and correlation potential energies are straightforwardly included within the equations by means of simple analytical expressions. An important point in the theory of Hohenberg and Kohn is the fact that the expression of the exchange-correlation functional does not depend on the external potential but uniquely on the electron density of the system. In other words, the rule for calculating the exchange-correlation energy is universal, though the theoretical framework provides no indication of what the functional form of this is. Approximate postulated expressions are instead used, which lead to a series of functional forms belonging to the local density approximation (LDA – where the functional expression depends on the density only) and to the generalized gradient approximation (GGA – where the gradient of the electron density is also accounted for). DFT is an appealing method as it provides a very good compromise between the effectiveness to generally yield quantitatively accurate results and speed of performance. For routine calculations, such as that of the potential energy, geometry minimization or frequency analysis, GGA results are better than the HF ones at a similar cost. Systems with up to hundreds of atoms can be studied. Unfortunately, as for every quantum methods, there is a price to pay for effectiveness. For instance, it is also well-known that van der Waals interactions are poorly reproduced with DFT. However, alternatives now exist to overcome the latter difficulties [24–27].

2.1.2 Classical Molecular Mechanics

Simulation methods based on classical mechanics consider atoms as a single unit and the forces between them are modelled by potential functions based on classical physics. There is no description of interactions at the sub-atomic scale, i. e. interactions involving electrons. In addition to initial atomic positions, one must also provide a set of suitable parameters for the interaction potential functions, known as a force-field (FF). Parameters for FFs are derived from experimental data (in particular from infrared spectra) and/or quantum mechanical calculations (mainly at the HF or DFT levels of approximations), in general on a set of finite systems (molecules and embedded clusters). Therefore the question arises as to how well a FF is able to model the properties of systems dissimilar to those from which it was derived.

As a consequence, for more than a decade numerous FFs have been developed to describe all sorts of chemical systems: the water molecule [28], organic compounds [29], inorganic materials (zeolites [30], clays [31], carbon nanotubes [32]), metal oxides [33–35] and biological systems [36–38]. Recently, tentative attempts have been made to derive FFs able to account for bond making–bond breaking [39, 40].

When bonds between atoms are not important for the description of the process of interest (for instance, proteins folding or physisorption within constrained media) groups of atoms, in particular methyl CH_3 and methylene CH_2 groups, can be represented by a single atom bearing a proper set of FF parameters. This is the so-called united atom technique that helps to speed up computations without loss of accuracy [41].

In principle, since bond-breaking/bond-forming processes are not amenable to classical mechanics calculation, only non-bond interactions and relevant FF parameters, i. e., the van der Waals and electrostatic interactions, should be considered in these calculations. Usually, two forms of pair potentials are used to calculate the van der Waals interactions, namely the Lennard-Jones one

$$U(\boldsymbol{r}) = 4 \in \left[\left(\frac{\sigma}{\boldsymbol{r}} \right)^{12} - \left(\frac{\sigma}{\boldsymbol{r}} \right)^{6} \right] \tag{1}$$

or the Buckingham one

$$U(\boldsymbol{r}) = A\mathrm{e}^{-\frac{r}{\sigma}} - C\left(\frac{\sigma}{\boldsymbol{r}} \right)^{6}, \tag{2}$$

where for both expressions, ε, σ, A and C are parameters pertaining to a pair of atoms, and $\boldsymbol{r}$ is the distance separating these atoms. In these expressions, the first term corresponds to the short-distance repulsion and the second term is the long-distance attraction. In practice, the computer calculation of the van der Waals interaction energy is easily performed and a cut-off radius is applied (typically 12–15 Å) beyond which, the interaction pairs are neglected. The calculation of the electrostatic contribution is more difficult, especially in the case of periodic systems like clays. The electrostatic interaction is a double sum over all particles of the Coulombic term that reads, in vacuum

$$E_{\mathrm{elec}} = \sum_{i=1}^{N} \sum_{j=1}^{M} \frac{q_i q_j}{4\pi \in_0} \frac{1}{\left| r_i - r_j \right|}, \tag{3}$$

where q and r are the charge and position of atoms i and j, and ε_0 is the permitivity of vacuum.

Owing to its long range the sum is very slow to converge and scales as $O(N^2)$ where N is the number of atoms. An efficient way to evaluate the Coulomb energy is to use a distribution of charges (e. g., a Gaussian distribution) in the central cell and to divide the sum into two contributions evaluated in the direct and reciprocal spaces, respectively, and an additional term for cancelling the self-interaction of the charge distribution with the corresponding point charge located at the centre of the distribution. This is the Ewald summation [42]. In practice, for computer programs that take advantage of fast Fourier transforms, the Ewald summation algorithm scales as $O(N^{3/2})$. A detailed explanation of the physical aspects of the Ewald summation can be found in Allen and Tildesley's monograph [13b] and equations are given by Frenkel and Smit in [13c]. More recently, new techniques have been elaborated to calculate the electrostatic interaction. The original formulation was described by Hockney and Heastwood [43] and is called particle–particle/particle–mesh technique (PPPM). Though this is very similar to Ewald summation, it uses a mesh in reciprocal space in order to discretize the Fourier transform. This dramatically speeds up the calculation that scales the algorithm as $O(N \ln N)$.

Other algorithms have been developed to accelerate calculations. A detailed description of these can be found in Plimton's work [44] that includes atom-, force- and spatial-decomposition algorithms, and neigbour-list and link-cell methods.

Usually more complex FFs, which comprise bond, angle, dihedral or more interaction terms in addition to the dispersive and electrostatic terms, are built. FFs are therefore divided into two classes, namely those that comprise only simple interaction terms (i. e., bond, angle, dihedral, out-of-plane and improper interactions) called Class I FFs, and those having more complex terms where bond–bond, bond–angle, angle–angle etc. interactions are also included and known as Class II FFs. Class II FFs are especially well suited for the study of complex systems such as clay–polymer nanocomposites [31b–c, 45].

In conclusion, classical simulations are well suited for modelling phenomena that do not involve bond breaking/making. The use of simple inter-atomic potentials means that it is possible to handle up to millions of atoms and therefore model much larger and realistic systems.

2.2 *Structural and Statistical Data*

Having calculated the PES, there are various means by which it can be traversed and searched, of which we shall concern ourselves with three broad methods: geometry optimization, molecular dynamics and Monte Carlo.

2.2.1 Geometry Optimization

Finding the equilibrium geometry of a structure, that is the atomic configuration that represents the minimum energy for the structure, is the starting point to predict physical and chemical properties of materials. Indeed, it is also one of the most difficult tasks to achieve in quantum chemistry and classical modelling. Over recent decades, various algorithms have been developed to find minima on a potential energy (hyper)surface, from the most simple one (the steepest descent algorithm) to more sophisticated ones based on, for instance, the calculation of the Hessian matrix that corresponds to the second derivative of the energy with respect to the atomic coordinates. Simple algorithms such as the steepest descent one are best suited when the starting structure is far from its minimum. As the structure is converging towards the bottom of the potential well, the algorithm usually starts to oscillate between two or more geometries. It is now well established that in the vicinity of the minima, the quasi-Newton method [46] is better suited techniques for reaching convergence. These methods rely on the estimate of the hessian matrix, which is however a time-consuming task. The Hessian estimate is periodically updated during the process of optimization using the BFGS (Broyden-Fletcher-Goldfarb-Shanno) scheme [46]. Since the Hessian matrix elements are stored in memory, this procedure can be prohibitive for very large systems. Another technique used to find structures is the GDIIS approach [47] (geometry optimization using the Direct Inversion in the Iterative Subspace, formally used for calculating the wavefunction during the self-consistent field procedure). This technique uses linear extrapolations of the available structures for minimizing the magnitude of the error vector. The new geometry is then calculated by combining this error vector and the estimated Hessian matrix. Recent improvements of this method have been shown to be powerful, especially in the case where potential energy hypersurfaces are flat [48] (which is typical of very large molecules) or when other techniques converge towards transition states rather than minima.

We have seen that the energy of a configuration of atoms may be minimized with respect to geometry by iteratively varying bond parameters, in a systematic but non-trivial way, to follow the gradient of a potential energy well until a minimum is reached. In theory, this should correspond to the expected ‘real life’ atomic structure. The methods, however, can never assure that a global minimum has been located and since these techniques neglect thermal motion, only local minima on the PES may be searched. In addition, these techniques are impracticable in the case when systems, such as clay–polymer nanocomposites, comprising millions of atoms have to be simulated (see below). Alternatively, techniques such as molecular dynamics, Monte Carlo, genetic algorithms and simulated annealing have been developed to overcome as much as possible these difficulties. In this review, we will focus on techniques based on statistical ensembles, namely molecular dynamics and Monte Carlo simulations.

2.2.2 Molecular Dynamics (MD)

This technique is similar to performing an experiment: a system of interest is prepared in predefined manner and left evolving in time until equilibrium is reached. Measurements are then carried out on the equilibrated system. Similarly, molecular dynamics simulations aim at describing the equilibrium state of the system of interest by integrating Newton’s law of motion. Put into practice, the procedure is as follows: (1) one gives the positions and velocities of the particles; (2) the forces on the particles are computed. These are calculated from the expression

$$f(\boldsymbol{r}) = -\frac{\partial U(\boldsymbol{r})}{\partial \boldsymbol{r}}, \tag{4}$$

where $U(\boldsymbol{r})$ is the energy expression given above (equation 1 or 2); (3) New positions are calculated from these forces and the positions obtained at the previous step. Steps 2 and 3 are repeated several thousand of times until equilibration is reached. The main advantage of the method is that the dynamical evolution of a system, with time, may be followed, which allows comparison with additional experimental techniques such as NMR and quasi-elastic neutron scattering. It still remains a challenge, however, to follow the evolution of a system beyond the timescale of 1–10 ns, even when using classical mechanics simulations.

The most time-consuming part of a molecular dynamics simulation is the computation of forces. For an N particle system interacting by means of pairwise forces, a brute force algorithm scales as $O(N^2)$. Some algorithms have been designed to speed up the force evaluation. For instance, these rely on the construction of a neighbour list, i. e., a list of atoms bonded to a particular atom, for which the force must be calculated. Instead of double summing over all particles, the forces are only calculated using the atoms in the neighbour list. This list is updated periodically during the simulation. Other such algorithms have been developed (see Frenkel and Smit [13c] for more details).

Finally, because the system evolves with time towards equilibrium state during molecular dynamics simulations, time-averaged quantities, such as the diffusion coefficient, the thermal and electrical conductivities can be easily evaluated from MD simulations [49] and compared to experimental results.

2.2.3 Monte Carlo (MC)

Monte Carlo simulations involve searching the PES of a system by sampling many different configurations, generated by imposing random changes to a system according to a set of predefined rules. In this respect, Monte Carlo and molecular dynamics simulations aim at reaching the same goal, which is obtaining a description of the equilibrium state of the system of interest. The main difference in the Monte Carlo procedure is that the motion of the particles is not driven by the Newton equations but rather by a sequence of random steps. Ultimately, a set of configurations is selected onto which statistical averages are computed to obtain physical properties of the system. It is important to note that, for a system in an equilibrium state, statistical averages performed on configuration ensembles (as in Monte Carlo) is equivalent to a time average over a dynamic profile of a system as obtained in molecular dynamics. This is called the ergodic hypothesis, which is simply assumed to hold, though it depends on specific dynamical properties that cannot be established for most real systems. More details on the special features of the Monte Carlo method can be found in Frenkel and Smit's monograph [13c].

Of the latter two methods, both have advantages depending upon the information desired from the simulation. MD simulation offers the advantage that the dynamical evolution of a system with time may be followed allowing comparison with time resolved experimental techniques such as nuclear magnetic resonance (NMR) or fast Fourier transform Infrared (FTIR) spectroscopies. However, current algorithms and computer power hardly permit the simulation of systems over a microsecond, which is about the time for spins to flip in an NMR experiment. MC simulation, by contrast, is very efficient for calculating thermodynamic averages for a system and can rapidly search a set of low energy configurations and find the global energy minimum but allows no deterministic pathway to be followed across the PES. A question therefore arises as to which system would be simulated better by using either MC or MD. In practice, it is usually observed that MC is better suited for simulating phase equilibria and adsorption equilibria, whereas for complex systems such as polymers and biological systems (*e.* g. proteins) MD is preferred, although in this case new MC algorithms (parallel tempering [13c]) may also be useful.

MC simulations can still produce averaged sets of system low energy configurations that can be compared with quasi-elastic neutron scattering (QENS) or XRD data from experiment. In general, MC simulations are carried out using rigid clay sheets and fixed interlayer spacing (except in a few cases, see Refson et al. [50]) to allow the rapid calculation of the arrangement and loading of interlayer species, whereas MD simulations are carried out, often with flexible clay sheets, and a variable interlayer spacing to determine the effects that interlayer species have on the interlayer separation.

2.3 *Statistical Ensembles*

The methods of traversing the PES described above may be carried out with various conditions imposed upon the ensemble of microstates that collectively define the system in a statistical mechanical sense. For the most widely used ensembles these conditions include constant number of atoms (N), volume (V) and temperature (the canonical or NVT ensemble); constant number of atoms (N), pressure (P) and temperature (T) (the isobaric-isothermal NPT ensemble) and the constant chemical potential (μ), volume (V) and temperature (T) (the grand-canonical μVT ensemble). During energy minimization an NVT ensemble is generally employed to remove unphysical interactions within the initial structures and then the unit cell parameters systematically varied between cycles of energy minimization to attain a lowest energy configuration. The NPT ensemble most closely represents laboratory conditions in that conditions of constant external pressure and temperature are maintained. To simulate processes such as swelling the system must be able to alter its volume, hence ruling out constant volume ensembles. However, in certain scenarios such as when the clay interlayer molecular loading and arrangement is being calculated for a known *d*-spacing from X-ray diffraction studies constant volume ensembles, e. g. μVT, may be employed and the model clay sheets kept locked at the experimentally observed separation. The amount of intercalants can then be straightforwardly compared to that obtained experimentally.

More recently, this list of ensembles has been completed by the devising of new methods referred to as Gibbs ensemble techniques [51]. This ensemble is mainly used for the study of first-order phase transitions such as the liquid–liquid and liquid–vapour equilibria.

2.4 Modelling Periodic Systems

To model the bulk structure of materials (greater than 10^{23} atoms) using relatively small models (generally less than 10^6 atoms), two approximations are often employed. These are (a) the use of super-cells, where the original unit-cell, usually derived from a crystal structure, is replicated several times and then redefined as one larger simulation cell and (b) imposition of periodic boundary conditions on the simulation cell, where the super-cell is considered to be replicated infinitely in all three orthogonal space directions. The use of periodic boundary conditions is a good way of replicating a crystalline, extended system. In addition, on a technical viewpoint, it provides a good framework to calculate long-range electrostatic interactions (vide supra). However, the use of periodic boundaries imposes artefacts unique to itself, due to the influence of these images on the system being simulated. In particular, when the original unit cell is too small, the flexibility of a clay layer cannot be observed during a MD simulation because of the too rigid boundary limits. In effect, Suter et al. [52] and Thyveetil et al. [53] have studied mechanical properties of the double hydroxide and montmorillonite clays, respectively, using large-scale molecular dynamics simulations, namely, utilizing models in the range 10^5–10^7 atoms. They have shown that for such large models undulations of the clay layers can be observed (see Sect. 5 of this review).

2.5 Data Analysis

Computer simulations provide large amounts of information on the electronic, atomic and molecular structure. Depending on the simulation method employed, further information can be extracted such as self-diffusion coefficients, radial distribution functions (RDF's), which show the co-ordination environment of atoms, as well as mechanical properties such as elastic constants. Much of this computed data can be compared more or less directly with data from experimental measurements. However, when data cannot be obtained from experiment, simulations are of great interest to experimentalists.

3 Interlayer Structure and Dynamics in Clay–Polymer Nanocomposites

The starting point for running simulations is to obtain a good structure for the model. In the case of polymer–clay nanocomposites, one has to know the atomic positions of the polymer and clay atoms. Generally speaking, organo-clay models are built on the basis of data gleaned from two experimental analytical techniques – X-ray diffraction (XRD) and thermogravimetric analysis (TGA). XRD measurements provide an averaged spacing between successive clay sheets (the *d*-spacing), while

TGA gives the amount of organic material and water present in the sample. It should be noted that XRD cannot give accurate information on the relative proportion of exfoliated and non-exfoliated sample, or the conformational organization of the clay sheets, though lack of registry between successive layers is often inferred from the shape of the peaks in the XRD pattern. TGA provides only the sample-averaged data and cannot always distinguish between surface adsorbed and intercalated organic material. However, it provides crucial information on the interlayer content of organic and water molecules. The simulation super-cell is then set up these experimentally observed *d*-spacing and organic and water contents.

3.1 Cation Dynamics in Clay–Polymer Composites: Lithium Ion Conduction

Understanding the dynamics of exchangeable cations between the clay sheets in the presence of intercalated polymers is important as it can lead to the development of new materials with interesting electrical properties. A further application is in the field of the petrochemical industry, since the swelling properties of clays lead to borewhole instabilities (vide infra) [54–58]. Intercalated or non-exfoliated clay–polymer nanocomposites comprise polymer chains confined between clay mineral layers, stacked together in a regular arrangement. The study of the dynamics of interlayer cations has been prompted by the postulated use of Li^+-clay polymer systems as composite polymer electrolytes [59]. Additionally, the dynamics of interlayer cations is accessible through both solid-state NMR experiment and simulation, thereby providing validation for simulation methods.

An initial simulation study of the dynamics of Li^+ in a poly(ethylene oxide) (PEO)–montmorillonite system was reported by Yang and Zax who used a model Hamiltonian to elucidate the line shape of the ^7Li NMR spectra [60]. The study concluded that the main limitation on diffusion in the clay–polymer system was the weak co-ordination of the Li^+ cations to the PEO backbone oxygen atoms, which resulted in the Li^+ moving along the clay sheets in short jumps or 'hops'. This study was followed by two studies by Giannelis and co-workers, who performed Monte Carlo (MC) and molecular dynamics (MD) simulations to examine the dynamics of Li^+ and PEO intercalated between layers of the clay mineral montmorillonite, as a function of varying hydration and layer charge [61, 62]. The simulations showed that the polymer chains form a bilayer structure, but were relatively disordered in the plane parallel to the interlayer surface. The authors also examined the effect of the presence of a small amount of water within the nanocomposite. In the absence of water it was observed that the cations bind to the interlayer surface, but that when water is present two types of cation environment are discernible. The first was for those cations bound to the interlayer surface, as before, and the second for cations hydrated by water in the interlayer region. In both cases the extent of co-ordination of cations by PEO was minimal. In addition, the calculated diffusion coefficients showed that cations bound to the interlayer surface diffused much more slowly than those that were hydrated in the interlayer. By repeating the calculations with varying charge (and hence cation content), the authors show that the degree of intercalation of polymer was lower in systems with high cation content as these restricted co-ordination of the PEO to the clay surface, which is driven by replacement of weakly bound water.

Kuppa and Manias investigated the dynamics of Li^+ in both bulk polymer and when confined in the nanometre range galleries of montmorillonite using MD simulation [63]. The authors found that the motion of the lithium cations in the nanocomposites was temperature independent and followed a hopping mechanism, whereby cations move from one ditrigonal cavity to another, in agreement with Yang and Zax assumptions mentioned above. This is in contrast to the behaviour of lithium cations in bulk PEO for which, a hopping motion at low temperatures and random Brownian-like diffusion at higher temperatures were observed [63]. In contrast to the previous studies [61, 62], the authors found in their simulations that the Li^+ dynamics was driven by competitive adsorption of the Li^+ between the PEO chains and the clay surface, with correlation between the Li^+ dynamics and the PEO segment dynamics [63]. This study was backed up by further experimental studies of ^7Li and ^{23}Na NMR by Reinholdt et al., which confirmed that the dynamics of Li^+ described a hopping motion, and also suggested that the Li^+ hydration state is unaffected by intercalation of PEO polymers [64].

Further studies [65] have addressed the interlayer structure, and hence cation arrangements, in clay–polymer systems albeit without setting out to study cation dynamics. In the studies reported in this section, it has been assumed that the clay sheets are rigid planar structures. This is not necessarily the case, and permitting the layer atoms to move freely has recently demonstrated the emergence of undulatory phenomena in clay sheets as described in Sect. 7 [53].

3.2 Interlayer Arrangement in Organo-Modified Clay-Composites

In many clay–polymer application areas, exfoliated (delaminated) nanocomposites are preferred. These are comprised of disordered and dispersed clay mineral layers in a continuous polymer matrix. In the synthesis of such clay–polymer nanocomposites it is often necessary to pre-treat the clay mineral by exchange of the natural cations with long-chain alkylammonium ions. This renders the originally hydrophilic clay mineral layers organophilic, thereby increasing favourable enthalpic interactions with the intercalating polymer and hence promoting exfoliation. The selection of a suitable alkylammonium ion, which will provide a high interfacial strength between the clay mineral layers and polymer matrix, is however not always straightforward. Regarding the interlayer arrangement in nanocomposites, the bulk of simulations of clay–polymer systems address organo-modified clays and the literature in this area may be separated into two groups: (a) those papers addressing the structure of the organo-modified clay only and (b) those papers considering the interaction of the organo-modified clay with the polymer.

3.2.1 Structure of Organo-Modified Clays

Using classical MD, Pospísil et al. compared the interlayer structure of organo-ammonium surfactants with cetyl pyridinium and cetyltrimethyl ammonium headgroups [66]. These organic chains are of the same length, and form disordered liquid-like arrangements. The results of the simulations suggest that the cetyltrimethyl ammonium species has far stronger electrostatic interactions with the montmorillonite clay sheets. Using MD simulations Zeng et al. examined the interlayer structure and dynamics of a set of increasingly large alkylammonium surfactants (from tretramethyl- to octadecyldihydroxylethylmethyl-ammonium) in the interlayer of montmorillonite [67]. It was found that the ammonium headgroup of the surfactants tended to align adjacent to the clay surface and depending on the surfactant size, the aliphatic tails formed mono-layers, bi-layers, tri-layers and even pseudo-tetralayers between the clay sheets. It was also shown for two intermediate (in size) surfactants that they form a paraffin-like structure. In follow-up work, the structure of the dioctadecyldimethylammonium surfactants in the quadrilayer was further investigated [68]. The difference in diffusion properties of the alkyl chains was monitored; it was found that the diffusion increased towards the tail of the molecules, suggesting more liquid and disordered behaviour. Again, the atomic coordinates in the clay sheets were fixed during MD in these early studies.

In direct comparison with experimental work, Heinz et al. used MD simulations to gain insight into the phase transitions observed in alkylammonium mica, obtaining good quantitative agreement [69]. The authors investigated systems at less than 100% cation exchange capacity, and proposed a geometric parameter λ for the clay surface saturation by the alkyl chains. λ reflects the percentage of clay surface that is needed so that the alkyl chains can lie perpendicular to the surfaces. λ is related to the tilt angle θ formed by the alkyl chain and the clay surface by the relation $\lambda = \cos\theta$. The authors predicted that the alkylammonium species tilt by 37° and 66° for dioctadecyldimethylammonium and octadecyltrimethylammonium ions, respectively, when adsorbed within the sheets of mica. This study necessitated some FF development, particularly with respect to the charges assigned to the system [70]. In effect, the choice of the atom charges is critical and since charges are not observables, schemes to evaluate them from quantum calculations yield a plethora of different results. In this study, the atomic charges were derived from a Born method [71, 72] (based on thermodynamic cycles) and modified by the authors so as to account for the covalent and ionic nature of bonds.

To determine the suitability of surfactant-modified clays for preparing fully exfoliated nanocomposites, Pospísil et al. attempted MD simulations to calculate the sublimation energy, interaction energy and the exfoliation energy of simulated octadecylammonium–montmorillonite systems further intercalated with alkyl amine species [73]. Results suggested that the presence of Na^+, and shorter chain length alkyl amines both have a detrimental effect on the exfoliation energy. In a combined experimental/MD simulation study, Paul et al. determined that a linear relationship existed between *d*-spacing increase and the mass ratio between organic and clay

[74]. Where surfactants had hydroxyl-ethyl units, increased packing density occurred, which was attributed to increased hydrogen bonding between these units and the clay sheets. The authors also identified that the structure of the interlayer in surfactant-modified clays is significantly more disordered than many models with parallel alkyl chains, or precisely inclined monomers, might suggest. A FF-based energy optimization route was chosen to determine the variation in total energy versus the basal spacing for a selection of organo-modified montmorillonite, with the resulting curves being fitted to Morse potentials [75]. Agreement with experimental *d*-spacings was found with three of the surfactants, dimethyl, 2-ethylhexylocatdecyl ammonium, bis(2-hydroxyethyl) octadecyl ammonium, benzyl dimethyloctadecyl ammonium, but a poor fit was found for the dimethyl dioctadecyl ammonium. This study used rigid layers, and did not include thermal effects (namely, the authors performed molecular mechanics simulations) and so the discrepancy may be due to the lack of molecular motion resulting in the surfactant molecules becoming trapped in local energy minima [76].

3.2.2 Interactions of Polymer with Organo-Modified Clays

Tanaka et al. performed energy minimization and molecular dynamics simulations to investigate the interaction of nylon 6,6 with an isolated layer of the clay mineral montmorillonite, treated with alkylammonium ions. These showed that as the molecular volume of the alkylammonium ions increases the binding energy between nylon 6,6 and the clay mineral decreases, while that between nylon 6,6 and the alkylammonium ions increases. In addition, the authors indicated that the presence of polar functional groups, such as –OH and –COOH, in the alkylammonium chain, increases binding energies with nylon 6,6 [77]. Also investigating quaternary ammonium-modified montmorillonite nylon 6 composites with atomistic MD simulations, Fermeglia et al. showed that the binding energy of the polymer matrix with the clay sheets was decreasing, as the volume fraction of surfactant was increased. However, the binding between the chemical species, i. e. polymer–ammonium salt and ammonium salt–clay increased with increasing ammonium salt content. Again, the presence of polar groups increased the interaction energy between the ammonium salts and the polymer [78].

Subsequent studies investigated other types of polymers. Polypropylene organoclay systems have been investigated using MD simulations by Toth et al. and Minisini and Tsobnang [79, 80]. In the former work, similar trends to those identified in the study on nylon 6 – organoclay, by Fermeglia et al., [78] were found. Minisini and Tsobang examined the interactions between ocatdecyldimethyl 2-ethylhexyl ammonium montmorillonite and polypropylene using classical MD and showed that the presence of maleic anhydride as an additive improved the interactions between polypropylene and the organoclay [80].

The structure and energetics of the biodegradable caprolactone organoclay composites have been simulated by Katti and co-workers [81–83], and by Garbedien et al. [84]. Garbedien and co-workers systematically varied the amount of poly (ε-caprolactone) and observed that the interlayer adopted a quadrilayer, with a greater portion of extended chain structures relative to a bulk polymer. A detailed analysis of the energetics was performed by splitting the total energy into intermolecular contributions, namely the montmorillonite–surfactant, montmorillonite–poly (ε-caprolactone) and surfactant–poly(ε-caprolactone) interaction energies, where the surfactant was the dimethyl 2-ethylhexyl *n*-octadecylammonium. Examination of the energetics by the authors showed that both polar and non-polar interactions played a significant role in nanocomposite formation [84]. In particular, the binding between poly(ε-caprolactone) and surfactant is mostly maintained by van der Waals, hydrophobic interactions. The clay–surfactant interactions is mainly electrostatic in nature and finally, clay–poly(ε-caprolactone) interaction energy is equally divided into electrostatic and van der Waals contributions. Similar results were noted for the energetics of caprolactam in 12-aminolauric acid-modified montmorillonite by Sikdar et al. [81]. The organic modifier was found to lie parallel to the clay sheet, and strong non-bonded interactions occurred between the modifier and the polymer. The authors also quantified the relative surface coverage of modifier and intercalated polymer on the clay surface, calculating 54%:46% respectively [81]. In this study constraints were again placed on the boundaries of the clay sheet, though the *d*-spacing was allowed to vary. In a follow-up study, Katti et al. used high-resolution photoacoustic FTIR to gain insight into bonding modes within the caprolactam organoclay composite, which was then compared with the atomic level detail available from the MD-simulations [82]. Sikdar et al. undertook a comprehensive MD simulation examination of the interaction energetics in these systems [83]. It was found that energetically the polymer organoclay composite system yields more favourable situation than a mixture of the polymer and organo-modifier only and, as previously discussed, non-polar interactions of polymer backbone atoms played a significant role in nanocomposite structure.

Other composite systems have also been investigated such as the ethylene–vinyl alcohol co-polymer organoclay nanocomposite. Aleperstein et al. used the Universal FF to energy minimize the montmorillonite structure and the COMPASS FF for the molecular dynamics simulation of the intercalated polymers. The simulations used fixed atomic co-ordinates in the clay sheets and constrained interlayer separations to examine the co-polymer organo-modifier interactions, giving insight to aid interpretation of experimental data, including viscosity measurements [85]. These authors showed that ethylene–vinyl alcohol is unable to intercalate into the untreated clay interlayer since the loss of conformational entropy of the polymer cannot be compensated by more favorable energetic interactions. By contrast, when the clay is organo-modified with octadecylamine as a surfactant, the polymer enters the gallery, implying an increase of the interlayer space. This theoretical result is experimentally correlated to an increase of the viscosity of the materials. In addition, the polymer chains mainly bind to the surfactant by van der Waals interactions, in full agreement with Garbedien's observations [84] regarding the poly(ε-Caprolactone)/(dimethyl 2-ethylhexyl *n*-octadecylammonium–montmorillonite) nanocomposite.

3.3 Simulation of Non-Organo-Modified Clay–Polymer Systems

Following on from their earlier work, which addressed the dynamics of Li^+ cations in the clay interlayer, Kuppa et al. went on to use MD simulations to examine the structure and dynamics of polymer segments when confined in the clay-interlayer. An initial study compared the behaviour of PEO in Li^+–montmorillonite with that of polystyrene in an organo-modified octadecyl-ammonium fluorohectorite model; in both cases physi-sorption of polymer segments to the clay surface resulted in slower segment dynamics [86]. In a further study, the authors examined the effect of the geometric confinement on polymer structure and dynamics, through varying the cation exchange content, and hence number of Li^+ ions in the interlayer, and by comparing with the bulk PEO–Li^+ system. Results showed that increasing cation content had considerable effect owing to the strong interactions between PEO backbone oxygen atoms and the Li^+ [87]. The increase in the Li^+ content lead to a more disordered arrangement of the intercalated PEO monomers. This is a consequence of the PEO chain distortions to chelate the ions. The chelation of the Li^+ cation can be observed in Fig. 1: the polymer chains wrap the cations in order to facilitate interactions with terminal hydroxyl groups.

Though long chain alkyl ammonium species can be used to render the clay interlayer, more organophilic for intercalating polymers, in instances where low molecular weight primary amines are utilized, the only interlayer spacing observed correspond

Fig. 1 Snapshot taken after 1 ns of MD simulation, showing interactions between poly(ethylene glycol) and Li^+ cations in the PEG/Li^+–montmorillonite nanocomposites. The polymer wraps the cation to promote favourable electrostatic interactions. The color scheme is Purple, Li^+ cation; Red, oxygen; Grey, carbon; White, hydrogen; Yellow, silicon and Green, aluminium

to a monolayer arrangement of organic material. FTIR analysis of these materials indicated that increased hydrogen bonding was occurring within the interlayer region relative to analogous primary ammonium intercalated clay, similar to systems where a mixture of ammonium and amine species were co-intercalated [88].

Simulation studies using large-scale MD methods, with non-rigid clay sheets, showed that at the experimental organic loadings a monolayer of the poly(propylene oxide) diamine monomer forms [89]. If the amine groups are protonated, i. e. to form ammonium groups, a conformational change in the monomers occurs, whereby the ammonium cation strongly coordinate with the surface oxygen atoms of the tetrahedral clay sheet and a slight increase in basal spacing occur. In models in which only some of the amine groups are protonated to form the ammonium species, both intra- and inter-molecular H-bonds form between amine N atoms and ammonium H atoms, accounting for the increased H-bonding observed in the FTIR spectra and indicating that a mixture of ammonium and amine species were present in the interlayer of the experimental system, as suggested by the experimental evidence [89].

In some instances where a high clay fraction is required, such as the stabilization of oilfield well-bores [90], it is desirable to produce clay–polymer nanocomposites that are able to prevent borehole instabilities. The principal cause of shale instability is the macroscopic swelling of sodium-saturated smectites such as Na^+-montmorillonite [91,92]. The cost to oil industries due to wellbore instabilities is estimated to amount to at least 600 million dollars per year. Although the substitution of Na^+ cations by K^+ within the interlayer space of montmorillonite prevents the clay from swelling, the toxicity of KCl prohibits wide-spread utilizations. It is then desirable to find new, active additives such as bio-degradable polymers. The study of clay swelling inhibitors was initiated by Bain et al. in 2001 [93]. They simulated, using Monte Carlo simulations in the grand canonical ensemble and molecular dynamics, the behaviour of small chains of polyethylene glycol intercaled into Na^+–montmorillonite. They showed that the small polymers act in two ways to inhibit swelling: first, hydrophobic interactions between the polymer tails prevents the uptake of water and second, the hydrophilic head of the polymers encapsulates the Na^+ cations and favours their binding to the clay layer therefore prohibiting the intercalation of water. More recently, Zhang and co-workers investigated by molecular dynamics simulations the intercalation of polyacrylamide polymers into the K^+-otay clay [94]. As expected, in presence of polymer the water and cations diffusion into the interlayer space is decreased from 2.3156×10^{-5} cm^2 s^{-1} to 2.065×10^{-5} cm^2 s^{-1} and from 0.8377×10^{-5} cm^2 s^{-1} to 0.6116×10^{-5} cm^2 s^{-1}, respectively. In addition, when the polyacrylamide polymers are intercalated, the hydration of the K^+ cations decreases, therefore preventing the clay from swelling.

Recently, in situ polymerization of small monomer molecules within the clay galleries has been studied as a new synthetic route for clay–polymer nanocomposites with high clay fraction by Coveney and co-workers [95]. This synthesis process is known as self-catalysed in situ intercalative polymerization, in which two monomers are intercalated between the clay mineral layers, which then spontaneously copolymerize [65, 88]. In a series of combined experimental and theoretical studies, MC, energy minimization and MD simulations were performed to model the sorption of pairs of organic monomers into the interlayer of the clay mineral montmorillonite. Again, a non-rigid clay model was used. These showed that small organic molecules are easily sorbed into the interlayer. In addition, the observed arrangement of molecules within the interlayer indicated the clay mineral may play the role of an active template that aligns the molecules in a favourable orientation for in situ polymerization. Some evidence for cross-linking between molecular bi-layers was also seen. To pursue this work, Boulet et al. implemented the clay–organic FF, developed by Teppen [31b,c], within the highly scalable Large Atomistic/Molecularly Massively Parallel Simulator (LAMMPS) MD code, developing potential parameters for the Li^+ cation [45].

To rationalize reactivity in these systems, it is necessary to understand the interlayer arrangement of the reactive centres, where polymerization or cross-linking occurs. FF-based simulations have been used to examine issues such as how the nature of the monomer backbone, monomer head-group and identity of interlayer cations affect the arrangement of intercalated monomers [65, 88]. For simulated poly(ethylene glycol) (PEG) nanocomposites no evidence was observed for hydrogen-bond interactions between the protons of the PEG alcohol groups and the tetrahedral oxygen atoms of the clay surface [65]. It seems therefore that, in the presence of water and cations, poly(ethylene glycol) is unlikely to form strong H-bonds to the clay surface. When functionalized with terminal acrylate or alcohol groups, the poly(ethylene oxide) chains tend to orientate with the O atoms towards the mid-plane for the Na^+ and Li^+ clays, away from the cations, which reside at the clay sheet face.

The choice of monomer was also found to affect the cation distribution across the composite interlayer. In the poly(ethylene glycol) composites, hydroxyl groups retained some of the cations and associated hydrations shells within the mid-plane of the interlayer region. The magnitude of this effect was dependent upon the cation present in the simulated clay composite, with the high surface charge density Li^+ more susceptible than Na^+, while the majority of the K^+ ions migrated to the face of the clay sheet. Snapshots of the PEO/Li^+–montmorillonite and PEO/K^+–montmorillonite systems after 1 ns of MD simulation are shown in Figs. 2 and 3. Since the cations are retained in the interlayer region they are also more closely associated with the monomer backbone O atoms (see Fig. 4 for the case of the PEO/Na^+–montmorillonite nanocomposite). Therefore, in the radial distribution functions, the order of interaction for both the poly(ethylene glycol) hydroxyl O atoms and the backbone O atoms with the cations is $Li^+ > Na^+ > K^+$ [65]. Conversely, the results showed that the poly(ethylene oxide) diacrylate monomer, having no hydroxyl groups, does not retain the cations in the interlayer region, resulting in the vast majority of the Li^+ and Na^+ cations migrating into vacancies on the tetrahedral layer of the clay sheet, with the K^+ cations migrating to the face of the clay sheets. This prevents the Li^+ cations, effectively charge-shielded by the O atoms at the clay surface and associated water molecules, from interacting with the monomer oxygen atoms. Comparison of the interaction between the different cations and the poly(ethylene oxide) diacrylate backbone and endgroup O atoms confirms this, showing preferential interaction with the low surface charge density cations, i. e. in the order $K^+ > Na^+ > Li^+$ [65].

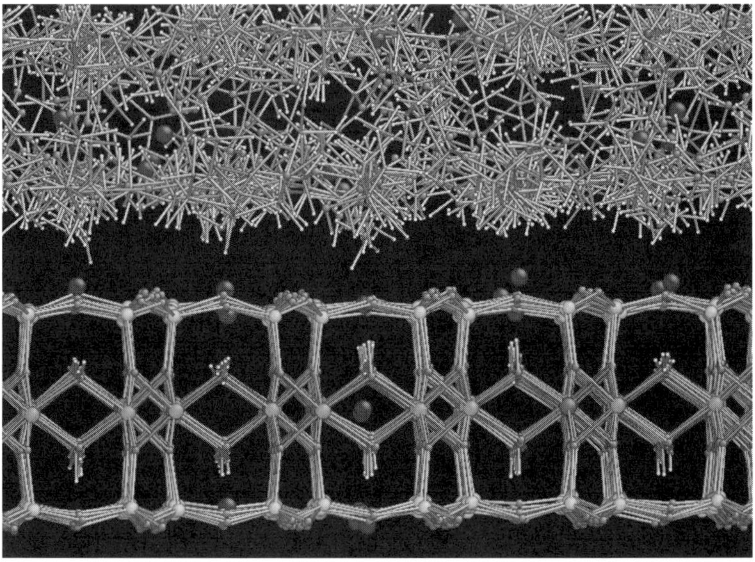

Fig. 2 Snapshots taken after 1 ns of MD simulation showing the interlayer arrangement in poly(ethylene glycol)–Li^+–montmorillonite nanocomposites. The Li^+ cations are chelated both by the clay sheets and the polymer chains. As a consequence Li^+ can migrate towards the centre of the interlayer space. Interestingly, some cations are also observed to migrate into the octahedral layer of the clay sheets. The color scheme is Blue, Li^+; Red, oxygen; Grey, carbon; White, hydrogen; Yellow, silicon and Green, aluminium

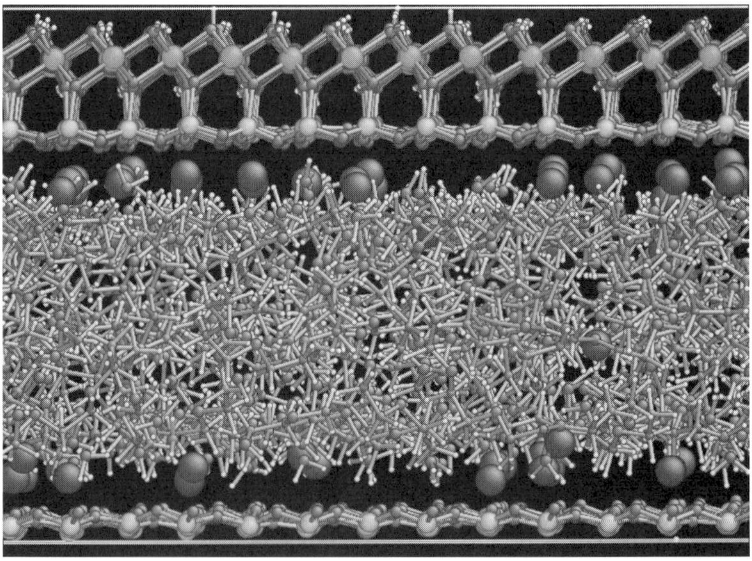

Fig. 3 Snapshots taken after 1 ns of MD simulation showing the interlayer arrangement in poly(ethylene glycol)–K^+–montmorillonite nanocomposites. The K^+ cations are not easily chelated by the polymers and are therefore adsorbed on the clay sheets. The color scheme is Purple, K^+; Red, oxygen; Grey, carbon; White, hydrogen; Yellow, silicon and Green: aluminium

Fig. 4 Snapshots taken after 1 ns of MD simulation showing the interactions in poly(ethylene glycol)–Na^+–montmorillonite nanocomposites. The polymer wraps the cations to promote interactions with the backbone oxygen atoms. The color scheme is Brown, Na^+; Red, oxygen; Grey, carbon; White, hydrogen; Yellow, silicon and Green, aluminium

Caprolactone-intercalated clays are of interest since caprolactone is a biodegradable polymer, and composites containing it have potential application in coatings and packaging. In a classical MD study, Gaudel-Siri et al. examined the interlayer arrangement in Na^+–montmorillonite caprolactone composites. In this study, the clay atoms were constrained. The caprolactone oxygen atoms were observed to replace water in the co-ordination spheres of Na^+ in the dry clay systems [96].

There have been studies on clay minerals other than the layered silicates, such as montmorillonite and hectorite. In an unusual study, which looks at organo-clay interactions of bulky molecular species, Fois et al. examined why the pigments in Maya blue paint are extremely stable. They showed that the indigo dye molecules were incorporated and able to diffuse within the channelled palygorskite clay used within the pigment. However, after a period of time, the dye molecules reached sites where they became locked into the clay structure and diffusion ceased [97]. Toth et al. used classical MD simulations in a study of LDH–polymer composites. LDH are anionic clays where the charge of the clay sheets is positive and the interlayer counter ions are negative. The study examined the structure and energetics of composite systems comprising seven nonsteroidal anti-inflammatory drugs (NSAIDs), two biocompatible polymers and an MgAl LDH [98]. The study showed that the more polar, smaller NSAIDs were readily effected by the presence of water in the interlayer. LDHs have a hydroxylated surface, and it was observed that the water lessened the clay–NSAID interactions. The presence of polymers was found to increase the interaction energy between the NSAIDs and the LDH host. It should be noted that the structure of the LDH system during the MD runs was constrained in this work.

4 Electronic Structure Studies of Reactivity in Clay–Polymer Nanocomposites

Most of the electronic structure calculations reported so far in the literature, either using the DFT or ab initio methods, have been devoted to the study of acid–base, adsorption or other reactivity properties. Recent reviews have been published on these topics [99, 100].

Interest in the catalytic properties of clay minerals has also arisen out of a desire to better understand and control new synthetic routes to clay–polymer nanocomposites. Experimental work by Coveney et al. indicated that when the natural and unmodified clay mineral montmorillonite is treated with a solution of methanal and ethylenediamine under mild conditions, the monomers spontaneously copolymerize to form an intercalated clay–polymer nanocomposite material with desirable properties [90, 95]. In this context Stackhouse et al. performed DFT calculations on a periodic montmorillonite model to investigate the catalytic role played by the clay mineral in the reaction [101]. A variety of possible Brønsted and Lewis acid sites were investigated to understand their role in increasing the susceptibility of the methanal C=O carbonyl towards nucleophilic attack. Initial simulations indicated that methanal could only undergo nucleophilic attack by ethylenediamine when suitably activated either by protonation or co-ordination to a suitable Lewis acid.

These original studies considered only the interlayer species of the natural clay, various cations and water molecules, and showed that the interlayer cation, when modelled in vacuo with the two organic species, could feasibly be sufficiently activating to promote the reaction [101]. Having noted that the interlayer cations and water may have a limited role in clay catalysis, the effect of the structure of the clay sheet was considered. Stackhouse et al. also investigated the effects of isomorphous substitution (Al^{3+} by Mg^{2+} or Si^{4+} by Al^{3+}) upon Brønsted acidity of hydroxyl groups located in the octahedral layer, the tetrahedral layer and at edge sites. Protonation of the methanal molecule was not observed in any of these scenarios, suggesting that the initial step in the in situ polymerization reaction was unlikely to be Brønsted acid catalysed. The Lewis acidity of exposed Al atoms at edge sites on the clay sheets was therefore considered. The octahedral and tetrahedral sites were shown to exhibit a catalytic effect, the magnitude of which was found to be strongly dependent upon the degree of substitution of Al^{3+} by Mg^{2+} in the octahedral layer of the clay sheets.

There have been other studies where clay–organic interactions have been investigated using electronic structure calculations, but these have not explicitly been aimed at clay–polymer composites. Aquino and co-workers have looked at the role of cationic clays in the adsorption of organic matter, and also clay catalyzed peptide formation [102, 103]. The former study was motivated by environmental chemistry considerations, while the latter was driven by prebiotic chemistry. Greenwell et al. used DFT and periodic cells to examine the reactivity of *tert*-butoxide intercalated LDHs in *trans*-esterification reactions [104].

5 Accessing Material Properties and Phase-Diagrams of Clay–Polymer Nanocomposites Using Simulation Methods

Computer simulations can give remarkable insight into the structure of clay–polymer systems. Classical MC and MD simulations can be analysed to give atom density profiles across the interlayer, the distribution of atoms relative to each other and other atoms, power spectra comparable to infra red spectra, diffusion coefficients and visual images of the clay interlayer. All of this data is of much interest to those studying the chemistry and interactions occurring in the clay–polymer composite system; however, to design clay–polymer systems with desirable properties in silico, it is necessary to be able to calculate the materials properties of the clay–polymer system.

One method by which the elastic properties of clay–polymer systems can be ascertained is by subjecting the cell to stress and measuring the forces. This technique was employed by Manevitch and Rutledge, who used MD to obtain the elastic properties of a single infinite two-dimensional montmorillonite sheet [105]. Katti et al. used MD simulations to investigate the mechanical properties of amino acid-modified montmorillonite to ascertain their potential for use in clay polymer composites. The simulations showed that the system is in the order of three times stiffer when under tension, compared to when placed under compression [106].

In a study of polypropylene amine/ammonium intercalated systems, Greenwell et al. reported that undulations were visible in very large, fully dynamical, molecular dynamics simulations (see Fig. 5) [89]. Suter et al. subsequently utilized distributed high performance multiprocessor machines located within Europe and the USA (exploiting grid computing techniques) to systematically vary super-cell sizes

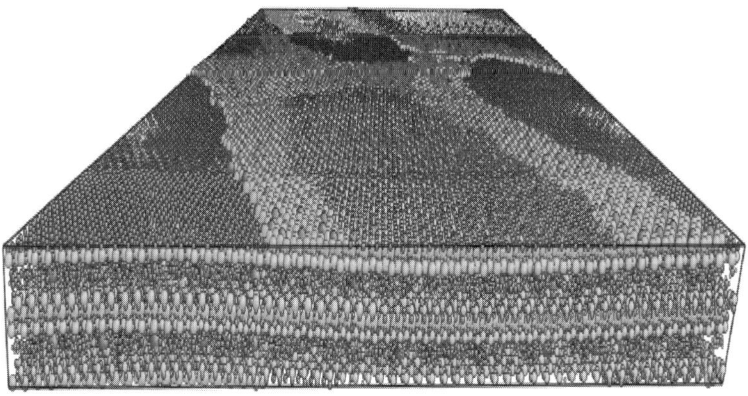

Fig. 5 Snapshot of 350,840 atom super-cell ($14 \times 14 \times 2$ replication of unitcell) of poly(propylene oxide) diammonium Na^+–montmorillonite after 0.5 ns of MD simulation showing a perspective view of the rectilinear super-cell, the clay sheets exhibiting gentle undulations. The colour scheme is Gray, C. white, H; Red, O; Blue, N; Orange, Si; Green, Al; Magenta, Mg and Brown, Na. Periodic boundary conditions were imposed in all 3 orthogonal space directions

up to some 10 million atoms, to investigate these effects in considerable detail within montmorillonite clays [52]. The authors reported that thermal fluctuations only become apparent in the clay-systems above a certain critical system size, i. e., finite size effects limit the observation of emergent properties. Direct analysis of the undulations (using methods related to those previously employed for lipid bi-layer systems [107]), and coupled stress–strain calculations, allowed the determination of mechanical properties of the clay systems, giving a bending modulus of 1.6×10^{-17} J, which corresponds to an in-plane Young's modulus of ca. 230 GPa. Using a similar approach, Thyveetil et al. also calculated the previously undetermined materials properties of anionic clays (chloride containing LDHs) [53]. The layered hydroxides, having substantially thinner clay sheets, consisting of only mono-octahedral layers, were found to have a bending modulus of ca. 1.0×10^{-19} J, which corresponds to an in-plane Young's modulus of 90 GPa for the clay sheets, or 63 GPa for the hydrated system. These developments are important given the current difficulty in determining such materials properties of clay platelets by experimental means.

Recently, interest in the simulation across multiple time and length scales (so called multi-scale modelling) have grown within the community of the computational chemistry. Although multi-scale modelling – i. e., simulating several length and time scales at once – is not yet feasible, authors have used several techniques devoted to explore various time and length scales (quantum mechanics, molecular mechanics and continuum methods) to study the materials properties of clay–polymer, and other, composites. Sheng et al. used micromechanical, continuum methods to investigate the behaviour and structure of clays at a composite level and also at the level of clay–polymer interactions [108]. Zhu and Narh developed a finite element approach to simulate the tensile modulus of clay-based nanocomposites [109]. The authors investigated the effect of the distribution of the clay platelets within the polymer matrix. Results showed that maximum strain occurred in the interlayer near the ends of the clay sheets and increased significantly when the interlayer modulus decreased.

There have been a number of publications reviewing the use of computer modelling at various length and time scales for determining materials properties in nanocomposites in a general sense [110–112]. In a unique study, Ginzburg and Balazs used a density–functional type approach at a coarse particle level to calculate phase diagrams of polymer–platelet systems. The phase diagram was determined to be highly dependent on the shape anisotropy of the filler, the polymer chain length and the strength of the inter-particle interaction [113].

6 Longer Time and Length-Scales: Understanding Formation Mechanisms of Clay–Polymer Nanocomposites Using Coarse-Grain Simulations

As discussed, simulations with an atomistic level of accuracy can access on the order of a nanosecond timescale (1–10 ns) and the size range is limited to the order of tens of nanometres. The molecular dynamics methodology can be extended to include longer times and larger systems by introducing further approximation, i.e., coarse graining the parameter set of the FFs. In coarse-grained simulations, a number of atoms are counted as one bead – the beads are then connected by simple harmonic functions and inter-molecular interactions are based on Lennard-Jones type functions. Using such methods, Smith et al. investigated the matrix-induced interaction between nanoparticles in clay–polymer composites [114]. By assessing the mean force potential of the system, the authors found that for weaker polymer–nanoparticle interactions, the polymer matrix promoted aggregation, which was overcome when the strength of the polymer–clay interaction was increased, resulting in dispersion occurring.

In a series of coarse-grained studies, Farmer and co-workers looked at the behaviour of stacks of clay lamellae in both a polymer melt and in a binary fluid (representing a curing agent and a monomer), thereby simulating polymer melt preparation and in situ polymerization preparation methods, respectively [115,116]. Interestingly, the results from the latter intercalation study showed that completely intercalated structures may be formed by simply adjusting the relative concentrations of the binary fluid, or the pressure experienced by the nanocomposite system, with increased swelling suggested by the authors to be indicative of exfoliation observed in some cases [117]. In the simulated polymer melt system, the interaction parameter between the clay sheet and the polymer was adjusted to represent some polymers strongly interacting with the sheet, others having functionalized strongly interacting head groups but weakly attractive (to the clay) polymer segments, and others having no functionality. The studies showed that the strongly interacting polymers 'pinned' clay sheets together around the anterior, impeding the fraction of intercalated material. Low intercalation density and decreased interaction between clay sheets were observed for the end functionalized polymers. The highest intercalation density was found for simulations containing a blend of end-functionalized and non-functionalized polymers [116]. As in large-scale molecular dynamics studies, significant distorsion of the clay sheets was observed in all these studies.

7 Future Directions for Computer Simulation

In summary, though computer simulation has provided insight into the interlayer structure, dynamics and reactivity of clay–polymer systems, which have been recently reviewed, each of the simulation techniques has an inherent set of approximations and underlying assumptions. Often these approximations and assumptions are not clearly defined for the non-specialist and this article seeks to fulfil such a role. A move away from MD simulations where the atoms in the clay sheet are fixed has resulted in further insight gained from simulation through the dynamical motion of the clay sheets.

Future advances in the computer simulation of clay–polymer nanocomposite systems, and materials in general, are likely to come through the increasing amount of computational power available to researchers. These increases do not just arise through faster processors, but also the development of high-end computing systems with many processors interconnected so that communication is rapid. These high-end facilities may be further expanded through connecting several high-end facilities in a computational Grid, for which novel software now exists to facilitate user uptake [117]. Such approaches allow the length and time-scales of any level of approximation to be significantly extended. Simultaneous with advances in computing technology come advances in theory, allowing different levels of simulation to be 'coupled' together not just in a hierarchical fashion [118], but also dynamically[119], whereby more accurate simulations are 'called' as required. As simulation size increases, analysis becomes more difficult and visualization of the simulation becomes desirable. One of the challenges faced by the scientists will be to develop methods to analyse the large-scale simulations fully. Interactive 'steered' simulations and immersive visualization are likely to be employed.

References

1. Schoonheydt, R. A.; Pinnavaia, T. J.; Lagaly, G.; Gangas, N. *Pure and Applied Chemistry*, 1999, **71**, 2367–2371.
2. Grim, R. E. "*Applied Clay Mineralogy,*" Mcgraw-Hill, New York, 1962.
3. Newman, S. P.; Jones, W. *New J. Chem.*, 1998, **22**, 105–115.
4. Cavani, F.; Trifirò, F.; Vaccari, A. *Catal. Today*, 1991, **11**, 173–301.
5. Carlino, S. *Chem. Br.*, 1997, **33**, 59–62.
6. Lagaly, G.; Beneke, K. *Colloid. Polym. Sci.*, 1991, **269**, 1198–1211.
7. Newman, S. P.; Williams, S. J.; Coveney, P. V.; Jones, W. *J. Phys. Chem. B*, 1998, **102**, 6710–6719.
8. Swenson, J.; Schwartz, G. A.; Bergman, R.; Howells, W. S., *Eur. Phys. J. E.*, 2003, **12**, 179–183.
9. Kagunya, W. W. *J. Phys. Chem.*, 1996, **100**, 327–330.
10. Rives, V. (ed.), "*Layered Double Hydroxides: Present and Future,*" Nova Science, New York, 2001.
11. Greenwell, H. C.; Jones, W.; Coveney, P. V.; Stackhouse, S., *J. Mater. Chem.*, 2006, **16**, 708–723.
12. Boulet, P.; Greenwell, H. C.; Stackhouse, S.; Coveney, P. V, *J. Mol. Struct. THEOCHEM*, 2006, **762**, 33–48.
13. (a) Leach, A. R., "*Molecular Modelling, Principles and Applications*", 2nd Ed., Pearson Education, England, 2001; (b) Allen, M. P.; Tildesley, D. J. "*Computer Simulation of Liquids,*" Clarendon, Oxford, 1987; (c) Frenkel, D.; Smit, B. "*Understanding Molecular Simulation: From Algorithms to Applications,*" 2nd Ed., Academic Press, London, England, 2002.
14. Born, D.; Oppenheimer, J. R. *Ann. Phys. Rev.*, 1927, **84**, 457.
15. Sato, T.; Tokunaka, K.; Tanaka, K. *J. chem. Phys.*, 2006, **124**, 024314.
16. (a) Hartree, D. R. *Proc. Cambridge Philos.*, 1928, **24**, 89. (b) Hartree, D. R. *Proc. Cambridge Philos.*, 1928, **24**, 11, ibid. 426.
17. Fock, V. *Z. Phys.*, 1930, **61**, 126.
18. Slater, J.C. *Phys. Rev.*, 1930, **48**, 35.
19. Pauli, W. *Phys. Rev.*, 1940, **58**, 719.
20. Møller, C.; Plesset, M. S. *Phys. Rev.*, 1934, **46**, 618–622.
21. Parr, R. G.; Yang, W. "*Density-Functional Theory of Atoms and Molecules,*" Oxford Science Publication, Oxford, 1989.

22. Kohn, W.; Sham, L. J. *Phys. Rev. A*, 1965, **140**, 1133.
23. Hohenberg, P.; Kohn, W. *Phys. Rev. A*, 1964, **136**, 864.
24. Mooij, W. T. M.; van Duijneveldt, F. B.; van Duijneveldt-van de Rijdt, J. G. C. M.; van Eijck, B. P. *J. Phys. Chem. A*, 1999, **103**, 9872–9882.
25. Elsner, M.; Hobza, P.; Frauenheim, T.; Suhai, S.; Kaxiras, E. *J. Chem. Phys.*, 2001, **114**, 5149–5155.
26. Cortona, P. *Phys. Rev. B*, 1991, **44**, 8455.
27. (a) Wesolowski, T. A.; Warshel, A. *J. Phys. Chem.*, 1993, **97**, 8050; (b) Wesolowski, T. A.; Weber, J. *Chem. Phys. Lett.*, 1996, **248**, 71.
28. (a) Jorgensen, W. L.; Chandrasekhar, J.; Madura, J. D.; Impey, R. W.; Klein, M. L. *J. Chem. Phys.*, 1983, **79**, 926–935; (b) Mahoney, M. W.; Jorgensen, W. L. *J. Chem. Phys.*, 2000, **112**, 8910–8922; (c) Matsuoka, O.; Clementi, E.; Yoshimine, M. *J. Chem. Phys.*, 1976, **64**, 1351–1361; (d) Teleman, O.; Jonsson, B.; Engstrom, S. *Mol. Phys.*, 1987, **60**, 193–203; (e) Berendsen, H. J. C.; Grigera, J. R.; Straatsma, T. P. *J. Phys. Chem.*, 1987, **91**, 6269–6271.
29. Mayo, S. L.; Olafson, B. D.; Goddard III, W. A. *J. Phys. Chem.*, 1990, **94**, 8897–8909.
30. (a) Jackson, R. A.; Catlow, C. R. A. *Mol. Simul.*, 1988, **1**, 207–224; (b) Faux, D. A.; Smith, W.; Forester, T. R. *J. Phys. Chem. B*, 1997, **101**, 1762–1768; (c) Catlow, C. R. A.; Freeman, M.; Vessal, B.; Tomlinson, S. M.; Leslie, M. *J. Chem. Soc. Faraday Trans.*, 1991, **87**, 1947–1950; (d) de Vos Burchart, E.; Ph.D. Thesis, 1992, *"Studies on Zeolites: Molecular Mechanics, Framework Stability and Crystal Growth,"* Table I, Chap. XII.
31. (a) Cygan, R. T.; Liang, J.-J.; Kalinichev, A. G. *J. Phys. Chem., B*, 2004, **108**, 1255–1266; (b) Teppen, B. J.; Rasmussen, K.; Bertsch, P. M.; Miller, D. M.; Lothar Schäfer, L. *J. Phys. Chem. B*, 1997, **101**, 1579–1587; (c) Teppen, B. J.; Yu, C.-H.; Newton, S. Q.; Miller, D. M.; Schäfer, L. *J. Mol. Struct.*, 1998, **445**, 65–88.
32. Walther, J. H.; Jaffe, R.; Halicioglu, T.; Koumoutsakos, P. *J. Phys. Chem. B*, 2001, **105,** 9980–9987.
33. Goddard III, W. A.; van Duin, A.; Chenoweth, K.; Cheng, M.-J.; Pudar, S.; Oxgaard, J.; Merinov, B.; Jang, Y. H.; Persson, P. *Topics Catal.*, 2006, **38**, 93–103.
34. Kornherr, A.; French, S. A.; Sokol, A. A.; Catlow, C. R. A.; Hansal, S.; Hansal, W. E. G.; Besenhard, J. O.; Kronberger, H.; Nauer, G. E.; Zifferer, G. *Chem. Phys. Lett.*, 2004, **393**, 107–111.
35. (a) Sun, H. *J. Phys. Chem. B*, 1998, **102**, 7338–7364; (b) Sun, H.; Ren, P.; Fried, J. R. *Comput. Theor. Polym. Sci.*, 1998, 229–246.
36. MacKerell Jr., A. D.; Bashford, D.; Bellott, M.; Dunbrack Jr., R.L.; Evanseck, J. D.; Field, M. J.; Fischer, S.; Gao, J.; Guo, H.; Ha, S.; Joseph-McCarthy, D.; Kuchnir, L.; Kuczera, K.; Lau, F. T. K.; Mattos, C.; Michnick, S.; Ngo, T.; Nguyen, D. T.; Prodhom, B.; Reiher III, W. E.; Roux, B.; Schlenkrich, M.; Smith, J. C.; Stote, R.; Straub, J.; Watanabe, M; Wiorkiewicz-Kuczera, J.; Yin, D.; Karplus, M. *J. Phys. Chem. B*, 1998, **102**, 3586–3616.
37. Foloppe, N.; MacKerell Jr., A. D. *J. Comp. Chem.*, 2000, **21**, 86–104.
38. Kaminski, G. A.; Friesner, R. A.; Tirado-Rives, J.; Jorgensen, W. L. *J. Phys. Chem. B*, 2001, **105**, 6474–6487.
39. (a) Han, S. S.; Kang, J. K.; Lee, H. M.; van Duin, A. C. T.; Goddard, W. A. *J. Chem. Phys.*, 2005, **123**, 114703; (b) van Duin, A. C. T.; Strachan, A.; Stewman, S.; Zhang, Q. S.; Xu, X.; Goddard, W. A. *J. Phys. Chem. A*, 2003, **107**, 3803–3811; (c) van Duin, A. C. T.; Dasgupta, S.; Lorant, F.; Goddard, W. A. *J. Phys. Chem. A*, 2001, **105**, 9396–9409.
40. Yin, K. L.; Xia, Q.; Xu, D. J.; Ye, Y. J.; Chen, C. L. *Comput. Chem. Eng.* 2006, **30**, 1346–1353.
41. (a) Martin, M. G.; Siepmann, J. I. *J. Phys. Chem. B*, 1998, **102**, 2569–2577; (b) Martin, M. G.; Siepmann, J. I. *J. Phys. Chem. B*, 1999, **103**, 4508–4517; (c) Wick, C. D.; Martin, M. G.; Siepmann, J. I. *J. Phys. Chem. B*, 2000, **104**, 8008–8016; (d) Mundy, C. J.; Balasubramanian, S.; Bagchi, K.; Siepmann, J. I.; Klein, M. L. *Faraday Discuss.*, 1996, **104**, 17–36.
42. Ewald, P. P. *Ann. Phys.*, 1921, **64**, 253–287.
43. Hokney, R. W.; Eastwood, J. W. *"Computer simulations using particles,"* McGraw-Hill, New York, 1981.

44. Plimpton, S. *J. Comput. Phys.*, 1995, **117**, 1–19.
45. Boulet, P.; Coveney, P. V.; Stackhouse, S. *Chem. Phys. Lett.*, 2004, **389**, 261–267.
46. Schlick, T. *"Optimization Methods in Computational Chemistry"* in Reviews in Computational Chemistry, Lipkowitz, K. B.; Boyd, D. B. (ed.), Wiley, New York, Vol. 3, pp. 1–71.
47. Csaszar, P.; Pulay, P. *J. Mol. Struct.*, 1984, **114**, 31–34.
48. Farkas, O.; Schlegel, H. B. *Phys. Chem. Chem. Phys.*, 2002, **4**, 11–15.
49. Hansen, J.-P.; McDonald, I. R. *"Theory of Simple Liquids,"* 2nd Edn., Academic Press, London, 1986.
50. Skipper, N. T.; Refson, K.; McConnell J. D. C. *J. Chem. Phys.*, 1991, **94**, 7434–7445.
51. Panagiotopoulos, A. Z. *Mol. Phys.*, 1987, **61**, 813–826.
52. Suter, J. L.; Coveney, P. V.; Greenwell, H. C.; Thyveetil, M.-A., *J Phys Chem C*, 2007, **111**, 8248–8259.
53. Thyveetil, M.-A.; Coveney, P. V.; Suter, J. L.; Greenwell, H. C., "*Emergence of undulations and determination of materials properties from large-scale molecular dynamics simulation of layered double hydroxides*" preprint (2007).
54. Tambach, T. J.; Boek, E. S.; Smit, B. *Phys. Chem. Chem. Phys.*, 2006, **8**, 2700–2702.
55. Tambach, T. J.; Bolhuis, P. G.; Hensen, E. J. M.; Smit, B. *Langmuir*, 2006, **22**, 1223–1234.
56. Zhang, L. M.; Chen, D. Q. *Macromol. Mater. Eng.*, 2003, **288**, 252–258.
57. Zhang L. M. *J. Appl. Polym. Sci.*, 2001, **79**, 1416–1422.
58. Boek, E. S.; Coveney, P. V.; Skipper, N. T. *J. Am. Chem. Soc.*, 1995, **117**, 12608–12617.
59. Vaia, R. A.; Vasudevan, S.; Krawiec, W.; Giannelis, E. P., *Adv. Mater.*, 1995, **7**, 154–156.
60. Yang, D. K.; Zax, D. B. *J. Chem. Phys.*, 1999, **110**, 5325–5336.
61. Hackett, E.; Manias, E.; Giannelis, E. P. *Chem. Mater.*, 2000, **12**, 2161–2167.
62. Bujdák, J.; Hackett, E.; Giannelis, E. P. *Chem. Mater.*, 2000, **12**, 2168–2174.
63. Kuppa, V.; Manias, E. *Chem. Mater.*, 2002, **14**, 2171–2175.
64. Reinholdt, M. X.; Kirkpatrick, R. J.; Pinnavaia, T. J. *J. Phys. Chem. B*, 2005, 109, 16296–16303.
65. Greenwell, H. C.; Bowden, A. A.; Chen, B. Q.; Boulet, P.; Evans, J. R. G.; Coveney, P. V.; Whiting, A. *J. Mater. Chem.*, 2006, **16**, 1082–1094.
66. Pospísil, M.; Capková, P.; Merínská, D.; Malác, Z.; Simoník, J. *J. Colloid Interface Sci.*, 2001, **236**, 127–131.
67. Zeng, Q. H.; Yu, A. B.; Lu, G. Q.; Standish, R. K. *Chem. Mater.*, 2003, **15**, 4732–4738.
68. Zeng, Q. H.; Yu, A. B.; Lu, G. Q.; Standish, R. K. *J. Phys. Chem. B*, 2004, **108**, 10025–10033.
69. Heinz, H.; Castelijns, H. J.; Suter, U. W. *J. Am. Chem. Soc.*, 2003, **125**, 9500–9510.
70. Heinz, H.; Suter, U. W. *J. Phys. Chem. B*, 2004, **108**, 18341–18352.
71. Born, M. *Verh. Dtsch. Phys. Ges.*, 1919, **21**, 13–24.
72. Haber, F. *Verh. Dtsch. Phys. Ges.*, 1919, **21**, 750–768.
73. Pospísil, M.; Kalendová, A.; Capková, P.; Simoník, J.; Valásková, M. *J. Colloid Interface Sci.*, 2004, **277**, 154–161.
74. Paul, D. R.; Zeng, Q. H.; Yu, A. B.; Lu, G. Q. *J. Colloid Interface Sci.*, 2005, **292**, 462–468.
75. Minisini, B.; Tsobnang, F. *Composites A*, 2005, **36**, 531–537.
76. Greenwell, H. C.; Jones, W.; Coveney, P. V.; Stackhouse, S. *J. Mater. Chem.*, 2006, **16**, 708–723.
77. Tanaka, G.; Goettler, L. A. *Polymer*, 2002, **43**, 541–553.
78. Fermeglia, M.; Ferrone, M.; Pricl, S. *Fluid Phase Equilib.*, 2003, **212**, 315–329.
79. Toth, R.; Coslanich, A.; Ferrone, M.; Fermeglia, M.; Pricl, S.; Miertus, S.; Chiellini, E. *Polymer*, 2004, **45**, 8075–8083.
80. Minisini, B.; Tsobnang, F. *Composites A*, 2005, **36**, 539–544.
81. Sikdar, D.; Katti, D. R.; Katti, K. S. *Langmuir*, 2006, **22**, 7738–7747.
82. Katti, K. S.; Sikdar, D.; Katti, D. R.; Ghosh, P.; Verma, D. *Polymer*, 2006, **47**, 403–414.
83. Sikdar, D.; Katti, D. R.; Katti, K. S.; Bhowmik, R. *Polymer*, 2006, **47**, 5196–5205.
84. Gardebien, F.; Gaudel-Siri, A.; Bredas, J. L.; Lazzaroni, R. *J. Phys. Chem. B*, 2004, **108**, 10678–10686.

85. Aleperstein, D.; Artzi, N.; Siegmann, A.; Narkis, M. *J. Appl. Polym. Sci.*, 2005, **97**, 2060–2066.
86. Kuppa, V.; Foley, T. M. D.; Manias, E. *Eur. Phys. J. E*, 2003, **12**, 159–165.
87. Kuppa, V.; Manias, E. *J. Polym. Sci. B*, 2005, **43**, 3460–3477.
88. Boulet, P.; Bowden, A. A.; Coveney, P. V.; Whiting, A. *J. Mater. Chem.*, 2003, **13**, 2540–2550.
89. Greenwell, H. C.; Harvey, M. J.; Boulet, P.; Bowden, A. A.; Coveney, P. V.; Whiting, A. *Macromolecules*, 2005, **38**, 6189–6200.
90. Coveney, P. V.; Watkinson, M.; Whiting, A.; Boek, E. S. *Stabilizing Clayey Formations*, US Patent Number 6,787,507.
91. Chenevert, M. E. *J. Petrol. Technol.*, 1970, **11**, 1141.
92. Mody, F. K; Hale, A. H. *J. Petrol. Technol.*, 1993, **45**, 1093.
93. Bains, A. D.; Boek, E. S.; Coveney, P. V.; Williams, S. J.; Akbar, M. V. *Mol. Simul.*, 2001, **26**, 101–145.
94. Zhang, J.; Rivero, M.; Choi, S. K. *J. Phys. B*, 2007, **40**, 545–553.
95. Coveney, P. V.; Griffin, J. L. W., Watkinson, M.; Whiting, A.; Boek, E. *Mol. Simul.*, 2002, **28**, 295.
96. Gaudel-Siri, A.; Brocorens, P.; Siri, D.; Gardebien, F.; Brédas, J.-L.; Lazzaroni, R. *Langmuir*, 2003, **19**, 8287–8291.
97. Fois, E.; Gamba, A.; Tilocca, A. *Microporous Mesoporous Mater.*, 2003, **57**, 263–272.
98. Toth, R.; Ferrone, M.; Miertus, S.; Chiellini, E.; Fermeglia, M.; Pricl, S. *Biomacromolecules*, 2006, **7**, 1714–1719.
99. Boulet, P.; Greenwell, H. C.; Stackhouse, S.; Coveney, P. V. *J. Mol. Struct. THEOCHEM*, 2006, **762**, 33–48.
100. Bougeard, D.; Smirnov, K. S. *Phys. Chem. Chem. Phys.*, 2007, **9**, 226–245.
101. Stackhouse, S.; Coveney, P. V.; Sandré, E. *J. Am. Chem. Soc.*, 2001, **123**, 11764–11774.
102. Aquino, A. J. A.; Tunega, D.; Haberhauer, G.; Gerzabek, M. H.; Lischka, H. *J. Comput. Chem.*, 2003, **24**, 1853–1863.
103. Aquino, A. J. A.; Tunega, D.; Gerzabek, M. H.; Lischka, H. *J. Phys. Chem. B*, 2004, **108**, 10120–10130.
104. Greenwell, H. C.; Stackhouse, S.; Coveney, P. V.; Jones, W. *J. Phys. Chem. B*. 2003, **107**, 3476–3485.
105. Manevitch, O. L.; Rutledge, G. C. *J. Phys. Chem. B*, 2004, **108**, 1428–1435.
106. Katti, D. R.; Ghosh, P.; Schmidt, S.; Katti, K. S. *Biomacromolecules*, 2005, **6**, 3276–3282.
107. Lindahl, E.; Edholm, O. *Biophys J.*, **79**, 426, 2000.
108. Sheng, N.; Boyce, M. C.; Parks, D. M.; Rutledge, G. C.; Abes, J. I.; Cohen, R. E. *Polymer*, 2004, **45**, 487–506.
109. Zhu, L. J.; Narh, K. A. *J. Polym. Sci. B*, 2004, **42**, 2391–2406.
110. Buryachenko, V. A.; Roy, A.; Lafdi, K.; Anderson, K. L.; Chellapilla, S. *Compos. Sci. Technol.*, 2005, **65**, 2435–2465.
111. Borodin, O.; Bedrov, D.; Smith, G. D.; Nairn, J.; Bardenhagen, S. *J. Polym. Sci. B*, 2005, **43**, 1005–1013.
112. Valavala, P. K.; Odegard, G. M. *Rev. Adv. Mater. Sci.*, 2005, **9**, 34–44.
113. Ginzburg, V. V.; Balazs, A. C. *Macromolecules*, 1999, **32**, 5681–5688.
114. Smith, J. S.; Bedrov, D.; Smith, G. D. *Compos. Sci. Technol.* 2005, **63**, 1599–1605.
115. Anderson, K. L.; Sinsawat, A.; Vaia, R. A.; Farmer, B. L. *J. Polym. Sci. B*, 2005, **43**, 1014–1024.
116. Sinsawat, A.; Anderson, K. L.; Vaia, R. A.; Farmer, B. L. *J. Polym. Sci. B.*, 2003, **41**, 3272–3284.
117. Coveney, P. V.; Saksena, R. S.; Zasada, S. J.; McKeown, M.; Pickles, *Comp. Phys. Comm.*, 2007, **176**, 406–418.
118. Broughton, J. Q.; Abraham, F. F.; Bernstein, N.; Kaxiras, E. *Phys. Rev. B*, 1999, **60**, 2391–2402.
119. De Fabritiis, G.; Delgado-Buscalioni, R.; Coveney, P. V., *Physical Review Letters*, 2006, **97**, 134501.

X-ray Absorption Studies of Nanocomposites

Alan V. Chadwick and Shelley L.P. Savin

1 Introduction

Roy and co-workers [1–4] first used the term *nanocomposite* in the early 1980s. The most general definition of a nanocomposite is a multi-phase compound in which one of the phases has a length scale in the nanometer range. In 1992 a more practical definition, the starting point for the present article, was been given by Komarneni [5] as 'composites of more than one Gibbsian solid phase where at least one-dimension is in the nanometer range and typically all solid phases are in the 1–20 nanometer range'. The solid phases can be crystalline or amorphous, or a combination of both. They can be inorganic or organic or a mixture of both. In principle this includes a wide range of biological systems, such as bone [6, 7], teeth [7], shells [8, 9], ivory [10] and the accumulation of inorganic species in plants and animals at the cellular level [11, 12]. Whilst many of these systems are of intense current research activity, particularly in the case of regenerative medical applications [13–15], they are beyond the scope of this chapter. Even with the exclusion of the biological nanocomposites, this still leaves a very wide range of types of systems and a further clarification is required. The focus here will be on man-made nanocomposites, specifically (a) nanoparticles in/on an inorganic matrix, (b) nanoparticles in/on a polymer matrix and (c) nanoparticle–nanoparticle composites. Clearly, except for (c), one of the phases, i. e. the matrix, could be have a dimension beyond 20 nm.

The wide variety of nanocomposite materials leads, in turn, to a very wide range of applications. These range from biomedical applications [13, 14, 16] such as dental materials [16] and bone grafting [13] to use in solar cells [17, 18], photocatalysis [19, 20] and flame retardancy [21, 22]. Areas such as the use of nanocomposites in coatings have been widely investigated, with applications ranging from abrasion resistance [23], corrosion resistance [24], hard coatings [25, 26] to paint [27]. Polymer-based nanocomposites are also being studied for use as conducting polymers [28, 29], while a great deal of work has been carried out in the use of nanocomposites

Functional Materials Group, School of Physical Sciences,
University of Kent, Canterbury, Kent CT2 7NH, UK.

P. Knauth and J. Schoonman (eds.), *Nanocomposites: Ionic Conducting Materials and Structural Spectroscopies.*

in batteries [30–34], fuel cells [35, 36] and other energy storage devices [37]. Nanocomposite catalysts based on oxides [38–40] are of great interest; however, other systems such as the use of a Cu-polyaniline nanocomposite in the Wacker oxidation reaction [41] and modified metal oxides [42] are also being investigated. Equally, a variety of novel nanocomposites are of interest in the field of sensors [43–47]. Examples include carbon nanotube-based nanocomposites such as Prussian blue nanoparticles/carbon nanotubes/chitosan for glucose sensing [43] or carbon nanotube-polypyrrole nanocomposites for gas sensing [44] and metal nanoparticles in/on a polymer matrix [45, 46]. Other applications of nanocomposite being investigated include optical [48, 49], hydrogen storage [50], electronic [51], magnetic [52–54], optical storage [55], ferroelectric materials [56], membranes [57], nanolithography [58] and environmental [12, 52, 59].

Successful technological application of nanocomposites clearly requires a detailed understanding of the underlying physics and chemistry, which in turn relies on a thorough knowledge of the structure at the atomic scale. X-ray absorption spectroscopy, XAS, is one of the many techniques that can provide information on the microstructure of materials and it has some specific advantages in the study of composites. In the next section we will outline the principles of XAS, the types of experiment and the particular advantages of the technique. This will be followed by a survey of the application of XAS to nanocomposites, restricting the discussion to the three classes of man-made nanocomposites listed above. The final section will summarise the conclusions of the studies and possible directions for future investigations.

2 Principles of X-ray Absorption Spectroscopy

In X-ray absorption spectroscopy the absorption coefficient, μ (= log [incident intensity/transmitted intensity]) of a sample is measured as a function of the incident photon energy across the absorption edge for the ejection of a core level (K or L) electron. The typical spectrum for condensed matter is shown in Figure 1. Beyond the absorption edge there are oscillations referred to as fine structure, which can be about 1/10 of the size of the edge step and which decrease in magnitude with increasing X-ray energy. The whole region is referred to as X-ray Absorption Fine Structure (XAFS) and is divided into two regions, the X-ray Absorption Near Edge Structure (XANES) extending about 50 eV beyond the edge, and the Extended X-ray Absorption Fine Structure (EXAFS) extending typically 1,000 eV beyond the edge. In addition, there can be pre-edge features in the spectrum. All three types of feature provide structural information about the target atom, the atom that is emitting the photoelectron. The pre-edge features arise from excitations of the core electron to higher partially occupied states within the atom and can fingerprint the oxidation state of the atom. The pre-edge feature can also be used to determine the first neighbour co-ordination geometry [60–62]. The structural information from the EXAFS and XANES regions of the spectrum will now be considered in detail.

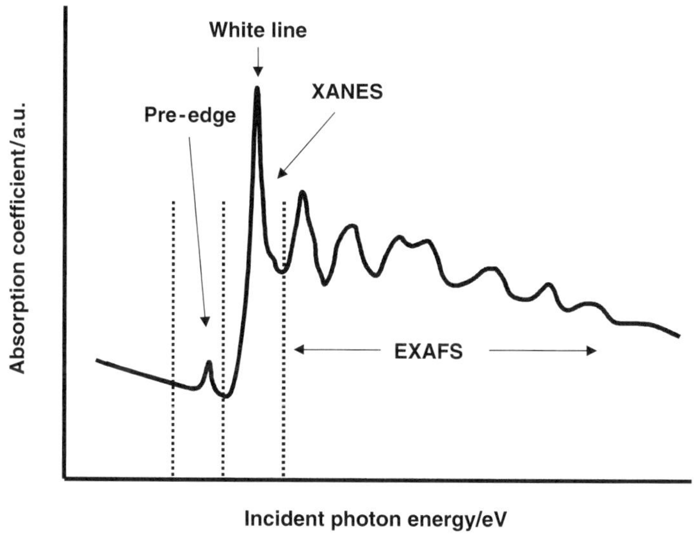

Fig. 1 A typical XAS spectrum

2.1 EXAFS

It is the oscillations in the EXAFS region that can be analysed in detail to provide quantitative information on the local environment. In this region the photoelectron is moving rapidly and is subject predominantly to single scattering events with the surrounding atoms. The theory of EXAFS can be found in several texts [63–67]. The oscillations arise from the photoelectron wave being backscattered and interfering with the outgoing wave. If the two waves are in phase there will be constructive interference, a lower final state energy and a higher probability for absorption. If the two waves are out of phase then there will be destructive interference, higher final state energy and a lower probability for absorption. Thus as the incident photon energy increases so does the energy of the emitted photoelectron with consequential changes on its wavelength. Since the distance between the target atom and its neighbours is fixed there will be shifts in and out of phase and hence the observation of the EXAFS oscillations.

From the qualitative explanation in the preceding paragraph it should be clear that the frequency of the EXAFS oscillations contains information on the distance from the target atom from its neighbours. The intensity of the oscillations will depend on the type of atom, which is acting as the backscatterer, i. e. the higher the atomic weight the more intense the oscillations, and the number of backscattering atoms. To be more precise, after subtraction of the background absorption, the normalised absorption coefficient, $\chi(k)$, as a function of the photoelectron momentum, k, can be written in an equation of the form [65, 67]

$$\chi(k)=\sum_j \frac{N_j}{kR_j^2}|f_j(\pi)|\exp\left(-2\sigma_j^2k^2\right)\exp(-2R_j/\lambda)\sin(2kR_j+\psi_j+2\delta). \qquad (1)$$

Here N_j is the number of atoms (all of the same type) in shell j with backscattering factor $f_j(\pi)$ at a distance R_j from the central atom. The other terms in (1) are a Debye-Waller like factor σ_j expressing the mean square variation in R_j, the phase factors δ and ψ_j of the photoelectron wave that depend on the central and scattering atom, and λ the mean free path of the photoelectron. The Fourier transform of $k\chi(k)$ with respect to $\sin(2kR)$ or $\exp(-2ikR)$ yields a partial radial distribution function in real space with peak areas proportional to N_j. If the phase factors are known, either from theoretical calculations or model compounds (i. e. fitted from the EXAFS of chemically similar compounds to that under investigation but with known R_j and N_j), then the radial distances can be determined, typically out to 5 Å from the target atom. The uncertainty in R_j that can be achieved with EXAFS is about +/−0.01 Å. The determination of N_j is usually less accurate, about +/−20%, as it is strongly coupled to the Debye-Waller factor. $f_j(\pi)$ does not vary strongly with atomic number and the identification of the type of atoms in the shells is limited to differentiation between rows of the Periodic Table. The advantages of EXAFS over diffraction methods are that it does not depend on long range order, hence it can be used to study local environments in both crystalline and amorphous solids, and liquids, it is atom specific and can be sensitive to low concentrations of the target atom.

2.2 XANES

In the XANES region (sometimes referred to as the near-edge X-ray absorption fine structure, NEXAFS) the electron is excited from an inner orbital to an unoccupied orbital. In the case of a K-edge spectrum the transition is from the 1*s* to a *p* orbital. Since the electron in this region has a low kinetic energy it is subject to many multiple scattering events. Thus quantitative analysis of this region has been difficult. Nevertheless, useful qualitative information is available and is obtained by comparison with model compounds. First, the structure of the spectrum in this region can be used as a *fingerprint* of the local environment of the target atom. This is illustrated in Fig. 2 for the case of Cu(II) K-edge XANES. The spectra are clearly different for the octahedral and square planar geometries, and are the same for the two different octahedral complexes. The second application of XANES is to identify the oxidation state of an atom. As an element loses an electron and goes to a higher oxidation state, the shielding of the nucleus is reduced. Thus the remaining electrons are more strongly bound resulting in higher ionisation energies. Hence the absorption edge moves to a higher energy, typically about 5 eV per oxidation number. An example of this is shown for Co K-edge XANES in Fig. 3. The edge position is seen to shift to higher energies as the oxidation state increases from Co(0) to Co(II) and to Co(II/III). For a sample in which an element is in mixed oxidation states, it is

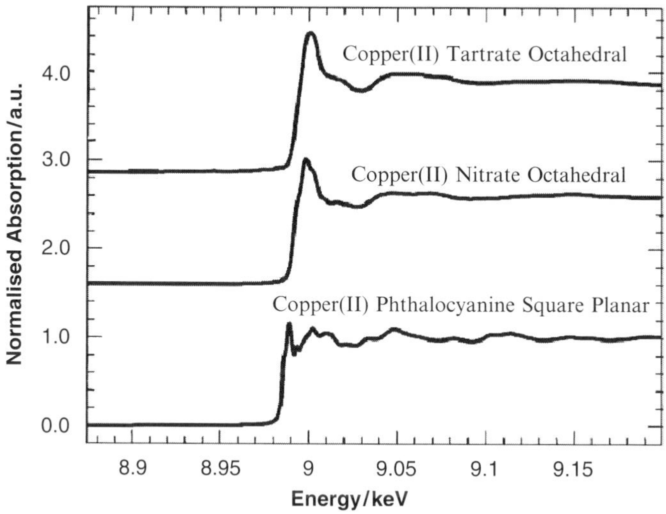

Fig. 2 The Cu K-edge XANES spectra for different Cu geometries (after [11])

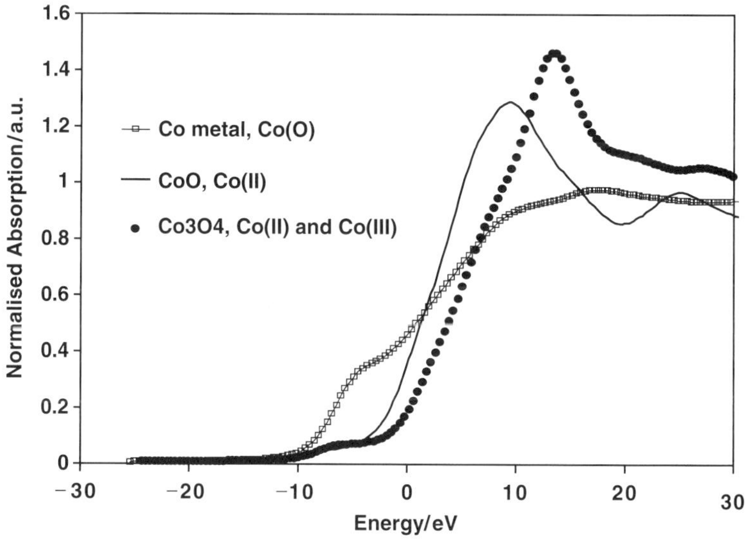

Fig. 3 The Co K-edge XANES spectra for Co in different oxidation states (after [68])

possible, by comparison with standard compounds of known oxidation state, to estimate the proportion of each state.

Recently there has been progress in the development of the theory of the XANES region, which along with the availability of ab initio computer codes for the calculation of electronic structure, allowing a treatment of the multiple scattering effects [69]. In the future these promise a unified quantitative approach to the analysis of EXAFS and XANES.

2.3. *Experiments and Data Analysis*

The set up of a typical XAFS experiment is shown schematically in Fig. 4. The requirement of a high intensity, tunable X-ray source means that the majority of experiments are now undertaken on synchrotron radiation sources. The experiment is basically simple. The white beam from the synchrotron is passed through a two crystal (usually silicon) monochromator, the wavelength being selected by the Bragg condition and step-wise rotation of the crystals allows a sweep of the X-ray energy. The intensity of the X-ray beam incident on the sample, I_o, is measured with an ion chamber filled with a gas mixture set to be 80% transmitting. For samples in which the target atom is concentrated (>1%), a transmission mode is employed and the transmitted intensity, I_t, is measured with a second ion chamber, in this case set to be 80% absorbing. The absorption coefficient is simply evaluated from log (I_o/I_t). The sample thickness for a transmission experiment is adjusted to give a step jump at the absorption edge of μ about 1, by using thin films or making pellets of powders with a non-adsorbing diluent (i. e. containing only light atoms, for example boron nitride, silica or polythene). For dilute samples a fluorescence mode is used, where the fluorescence X-rays that are emitted upon absorption of the incident X-rays are used to monitor the absorption coefficient. A scan using stepper motors to drive the monochromator will typically take the order of 30 min. However, scan times can be short as a few minutes using quick scanning EXAFS mode (QuEXAFS) [70–73], where the monochromator is moving continuously. For XANES spectra, normal scanning is usually employed to provide high energy resolution in the edge region. XAS spectra can be collected for most elements in the Periodic Table; however, for the lighter

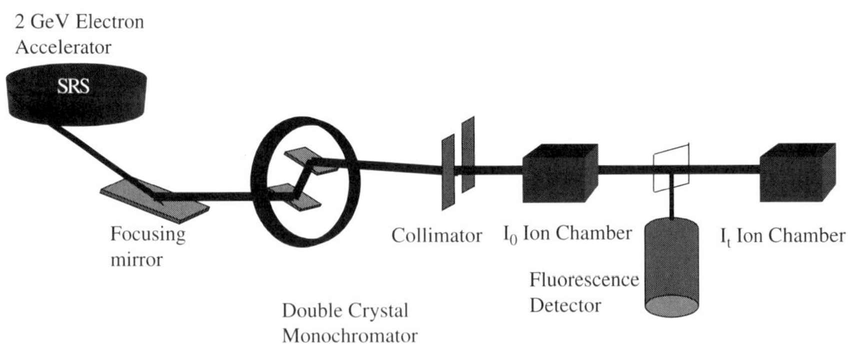

Fig. 4 Typical EXAFS station

elements (roughly those lighter than calcium) air absorption will be significant and the sample needs to be in a vacuum along with the X-ray beam.

For XAS experiments the beam must be highly monochromatic and free from harmonic components. der-sorting monochromators are usually employed, which allow higher harmonics to be rejected from the incident beam. A feature of many synchrotron XAS stations is the possibility to perform other types of experiment simultaneously with the XAS spectrum. Common combined experiments include XAS with X-ray powder diffraction (XRPD) [70, 71, 73–77], differential scanning calorimetry [78], IR/UV [72, 79, 80] and mass spectrometry [79, 81–83]. In addition it is possible to carry out a range of in situ experiments using furnaces [70, 71, 74], electrochemical cells [84–86] and catalytic reactions [73, 79, 81].

We will give here only brief summary of the data analysis and more details, such as the calculation of phase shifts, the treatment of multiple scattering on the EXAFS, errors, etc., can be found elsewhere [65, 87–90]. EXAFS data analysis is performed with interactive computer programmes such as the EXCALIB, EXBACK and EXCURVE codes developed at the Daresbury SRS [88] or the University of Washington codes, UWAXFS and FEFF [89, 90]. EXCALIB allows for correction of the monitors for background counts and converts monochromator angle to X-ray energy. EXBACK removes background X-ray absorption of the atom to produce the normalised absorption, $\chi(k)$. EXCURVE, the final step in the procedure is least-squares fitting the data to a model of the local structure with parameters such as N_j, R_j and A_j [$=2\sigma_j^2$] as variables. Typical graphical outputs from EXCURVE are shown in Fig. 5, which is for the Zr K-edge EXAFS of bulk ZrO_2. Figure 5 (top) is a plot of the experimental normalised absorption coefficient, $\chi(k)$, (depicted by a solid line) as a function of the X-ray wave vector, k, in Å^{-1}, and the value predicted by the theoretical model (depicted by the dashed line). Since the magnitude of the EXAFS oscillations decreases with increasing k it is usual to weight $\chi(k)$ by multiplying by k, k^2 or k^3 to emphasise the oscillations at high k. The iterative least-squares fitting in EXCURVE provides the 'best-fit' to the normalised absorption plot, the quality of the fit being measured by the sum of the deviations between the experiment and the model, usually expressed as a percentage, referred to as an *R-value*. Figure 5 (bottom) is a plot of the magnitude of the Fourier transform as a function of radial distance, R. It is important to note that the Fourier transformation has to be performed with a phase-shift, usually that for the atom in the first co-ordination shell. Thus for compounds it can be misleading to read off peak positions from these plots as exact radial distances. Tabulated output from EXCURVE contains the best-fit parameters, the errors on the parameters, the correlation matrix for the parameters and the quality of the fit. EXCURVE has the facility to isolate a peak in the Fourier transform and back transform to yield the normalised absorption, referred to as 'Fourier filtering'. This can be useful in the early stages of an analysis to determine the types of atoms in a co-ordination shell.

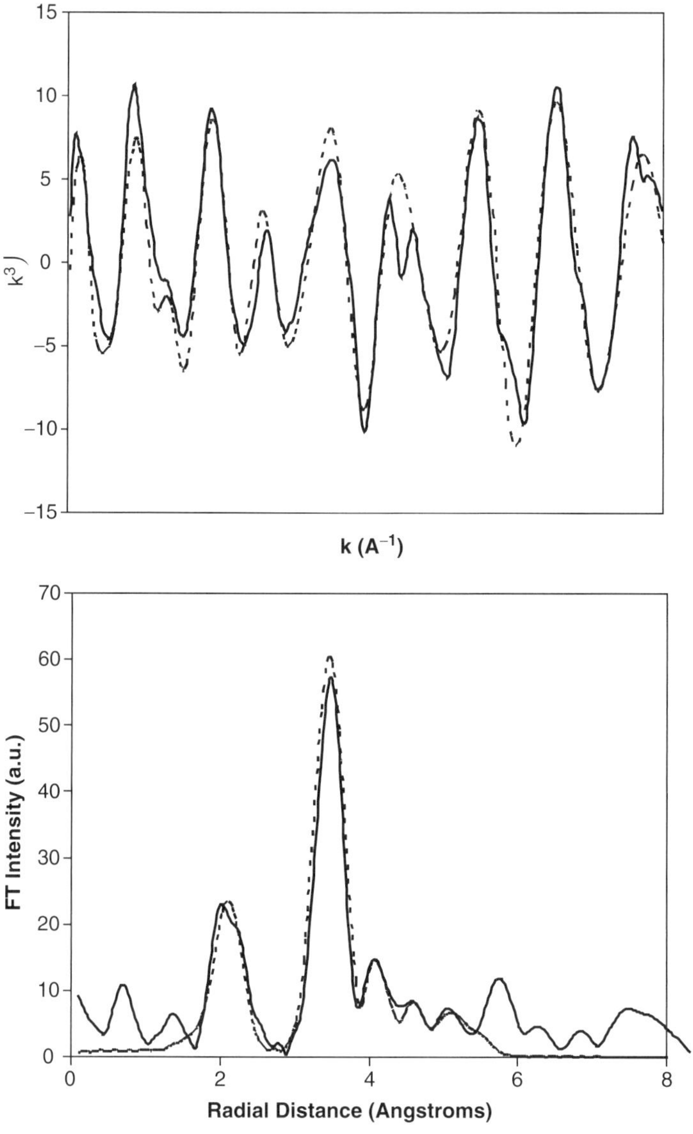

Fig. 5 The Zr K-edge EXAFS of bulk monoclinic ZrO_2: (top) the normalised EXAFS and (bottom) the Fourier transform that has been corrected with the phase shift of the first shell. The *solid line* is the experimental data; the *dashed line* is the theoretical fit

3 Applications to Nanocomposite Systems

We will now consider the application of XAS measurements to the various types of nanocomposites using illustrative examples from the literature.

3.1 Nanoparticles in/on an Inorganic Matrix

Areas attracting particular attention are systems where clusters of nanoparticles are coated/deposited on an inorganic solid. A good example is the catalytic system of nanoparticles gold (Au) clusters on titania (TiO_2), which has been used for CO oxidation. In a particularly thorough study [91], XAS was used to elucidate the effect of synthesis and reaction conditions on the structure and activity of the Au clusters (10 nm) supported on nanocrystalline (anatase, rutile and brookite) and mesoporous TiO_2. XAS experiments for the Au L_{III} edge were performed both ex situ (catalyst pre-treated in quartz reactor tube at given temperature and gas flow, and transferred to a cryogenic chamber for XAS studies) and in situ. XANES spectra were collected at higher temperatures, whilst EXAFS spectra were collected under He gas at liquid N_2 temperatures to minimise thermal disorders. The XANES studies during pre-treatment and reaction conditions showed that varying the TiO_2 crystalline structure does not affect the redox behaviour of the Au clusters. For all samples, the Au precursors were completely reduced to Au^0 after treatment in 4% H_2/He at 150°C. The distinction between Au^{3+} (in Au_2O_3) and Au^0 (in Au metal) is very clear in the XANES spectra, the former having a clearly defined and large white line (refer to Fig. 1 for illustration of the white line). The results showed that all samples were most active after reduction at 150°C, implying that the fully reduced state (i. e. Au^0) is the most active. Once reduced, no re-oxidation occurred, even after treatment in air at 150°C or 300°C. EXAFS spectra obtained for samples prepared using precursor solutions of aqueous $HAuCl_4.3H_2O$ with different pH showed that particles smaller than 4 nm were obtained when the pH was >6. The effect of Au loading on size and reducibility of Au clusters was studied using the mesoporous TiO_2. For lower Au loadings, a higher co-ordination number (CN) for the Au–Au correlation was obtained after treatment with 4% H_2/He at 200°C. Higher CN indicates larger metallic particles; however, the reasons were not clear as to why the catalyst containing the least amount of Au would have the bigger particles, but results for the particle sizes were in agreement with those obtained from TEM measurements.

Cation-doped TiO_2 is being explored as a photocatalyst and an interesting novel application is as anti-microbial agents [92]. The most widely investigated system is Ag-doped TiO_2. However, there are questions as to the exact nature of these systems. First it is not clear whether the metal ions have entered the TiO_2 lattice or reside on the surface. In addition, the oxidation state of the cation appears to be unknown. These problems have recently been investigated in XANES studies of Ag-doped TiO_2 films and powders [93, 94]. The samples were prepared by sol–gel methods. The results for the films are shown in Fig. 6a. It is clear that the Ag is present as Ag^+, from a simple comparison of the spectra with the standards, Ag metal, Ag_2O and AgO. The results for the powder samples are shown in Fig. 6b. In contrast to the films these samples were annealed at a range of temperatures up to 900°C. For samples at 300 and 600°C the Ag is again present as Ag^+. However, annealing at 900°C clearly causes a reduction to Ag^0, as can be seen from the

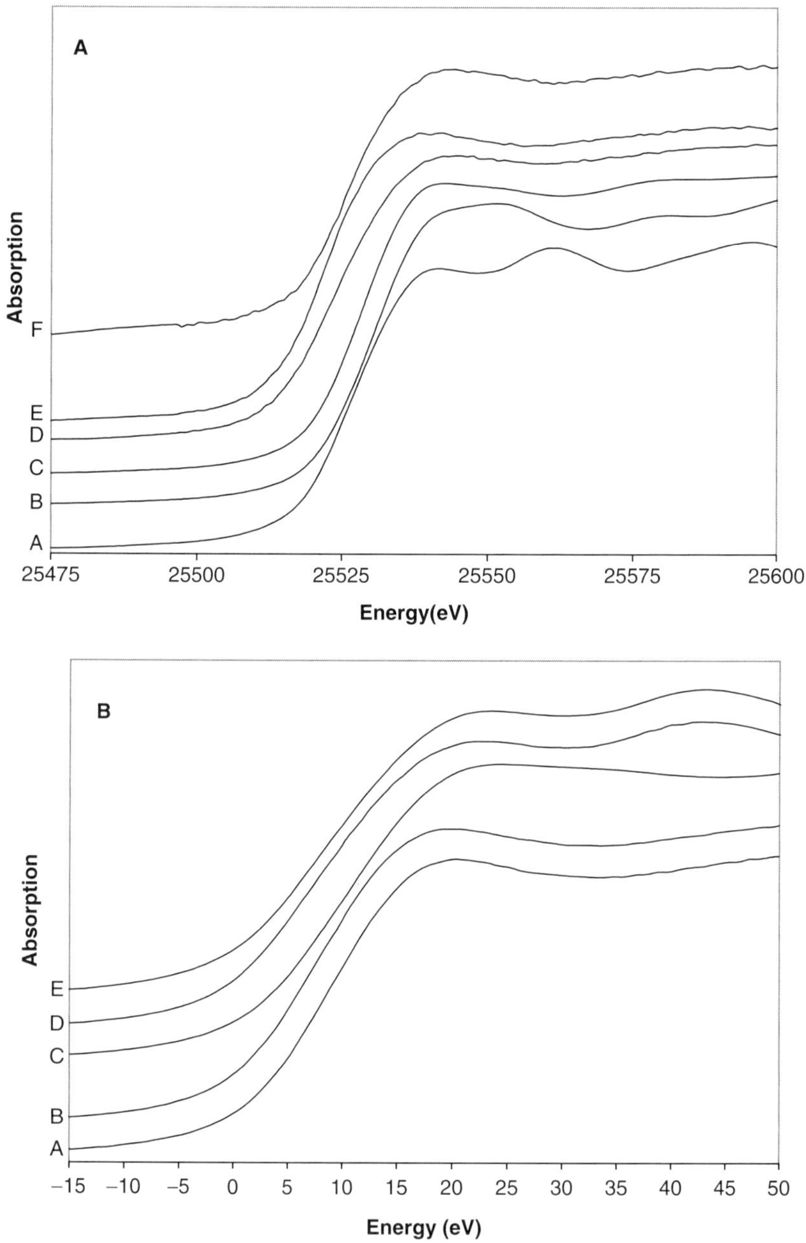

Fig. 6 (**a**) A comparison of the Ag K-edge XANES spectra of Ag metal powder (A), AgO (B), Ag_2O (C), 20% (D), 10% (E) and 5% (F) Ag-doped TiO_2 thin films. (**b**). A comparison of the Ag-K-edge XANES spectra of 1% Ag-doped TiO_2 powders annealed at 300°C (A), 600°C (B) and 900°C (D) with Ag_2O (C) and Ag metal powder (E)

similarity of the XANES spectra of the sample and Ag metal. In addition, the similarity of all the XANES spectra for the Ag-doped TiO_2 samples with the reference compounds also indicates that the silver species are on the surface of the TiO_2. Similar XANES studies have been reported for Pt-, Pd- and Au-doped TiO_2 [94].

An unusual example of nanocomposites is the presence of metal particles in ceramics. These can give rise to lustre, one of the most interesting medieval ceramic decorations and has been an intriguing problem for archaeologists. A rather common lustre is formed by the application of a mixture of copper and silver salts with iron-rich clay on a ceramic glaze and fired about 500°C to develop a gold-like decoration exhibiting metallic reflection. The earliest lustre decorated ceramics were probably made in Iraq in the early ninth century AD and followed the expansion of the Arabian culture through Spain and the rest of the western Mediterranean [95]. It has been demonstrated that lustre is a nanostructured thin layer, typically about 500 nm thick, formed by metallic copper and silver nanocrystals embedded in a glass matrix [96]. The traditional lustre technique is thus one of the oldest technologies known to achieve metal glass nanocomposites. A recent XAS study [97] was made of copper nanoparticles within a glass matrix, the structure on which lustre glazes are based on. Standard EXAFS at Cu K-edge showed that both metallic copper and copper oxide local sites are present in the matrix and detailed atomic arrangements of each were derived. The method provides no information as to whether either structure is involved in the luminescence processes. To further explore the luminescence process optically detected XAS (OD-XAS) at O K-edge and Cu L_3 edges were collected at 10–300 K. The results provided strong evidence that copper metal nanoparticles are responsible for the red emission and that the X-ray absorption processes of the metal particles (red emission) and silicate lattice (blue emission) are different. In a continuation of this work an EXAFS study was made of the colour variations in thirteenth century Hispanic lustre, based on copper–silver nanoparticles, on a potsherd [98]. Spectra were collected for Cu K- and the Ag L_3-edges using a microfocus beamline with a spot size of $20 \times 5\,\mu m$. The typical red and green colour variations in the sample were correlated with different copper/silver ratios. EXAFS data were used to determine the local atomic environment of the copper and revealed a corresponding variation in the metal to oxide ratio of the copper content, which was related to the visual effects.

Zeolites and related structures are important catalysts consisting of a three-dimensional network of metal–oxygen tetrahedra (or rarely octahedra) that provide the periodically sized microporous structure (with interlinked pores from 3 Å in size), in which the active sites are part of the structure. There are many variations based on Si–O, Al–O and P–O tetrahedra and the activity is due to a charge imbalance. For example, replacing one Si^{4+} atom by Al^{3+} causes a formal negative charge on the tetrahedron. This negative charge is then balanced by a proton or metal cation forming an acid site, which acts as the site for catalytic activity. The activity of these catalysts can be tuned by the addition of transition metal atoms and these can be either be part of the framework or deposited onto the framework as a supported species. XAS studies have played a very significant role in determining the local environment of these metal atoms and their role in catalytic reactions, particularly by using in situ measurements [99]. In general these systems would not strictly be classed as nanocomposites as the metal atoms are isolated as virtually single atoms in the small pores. However, there are a range of mesoporous analogues of the zeolites with pore diameters in the range 1.5–30 nm, which can incorporate nanoparticles, to form true nanocomposites. A good example is the incorporation of highly dispersed gold and platinum in SBA-15 (a mesoporous silica with pore size in the range 4.6–10 nm and wall thickness 3.1–6.4 nm) [100]. The metal was incorporated into the pores by impregnation of an ethanolic solution of $HAuCl_4$ or H_2PtCl_6. Pt L_3-edge XAS measurements were used, following the formation of the Pt/SBA-15 nanocomposite on reduction in a hydrogen flow at 573 K. In the XANES spectrum, the white line intensity decreased significantly, accompanied by a change of the shape of the spectrum after reduction, indicating the formation of Pt nanoparticles. This was confirmed by the EXAFS data, which indicated Pt nanoparticles 2.9 nm in size in the reduced sample, which is in good agreement with the values of 2.3 and 2.0 nm from XRD and TEM, respectively. After reduction the nanocomposites were exposed to air and the XAS spectra showed some evidence of partial oxidation.

3.2 Nanoparticles in/on a Polymer Matrix

Polymers, due to their ease of processing, are widely used as matrices for active species with particular chemical and/or physical properties, for example in biosensing and drug delivery [101], dental materials [102], electrochromic devices [103], ferroelectric composites [104] and non-linear optics [105]. A general review of use of polymer matrices in nanocomposites has recently been published [106]. Generally XAS studies of the polymer-based nanocomposites have been restricted to investigations of the inorganic additive.

A typical example is the Cr K-edge EXAFS study of non-linear optical polymeric film nanocomposites based on bis(arene)chromium complexes incorporated into a CN-containing matrix, polyacrylonitrile (PAN) [107]. Extraordinary high third-order non-linear optical susceptibilities of the ultra-fast electronic type have been demonstrated at various wavelengths for polymer films prepared from these composites. Electron diffraction patterns of the films and the presence of a non-zero electro-optical coefficient in the absence of an external poling field indicate the formation of self-organised structures. In CrPAN and $[CrPAN]^+$ $TCNE^-$(where TCNE is tetracyanoethylene) a single Cr–C distance was found in the EXAFS, showing that the arene rings are parallel and symmetric as in bis(benzene)chromium. The Cr–C distance in these materials is very similar to the distance of 2.1 Å in bis(benzene)chromium. In $[CrPAN]^+$ OH^-, more than one Cr–C distance was resolved, showing either that the two arene rings are at different distances or that they are tilted rather than parallel.

Another example is the XAS investigation of the interfacial structure of poly(phenylenevinylene)/TiO_2 (PPV/TiO_2) nanocomposite where data were collected for the Ti K-edge [108]. The pre-edge region of the XANES spectra displayed a distortion of octahedral geometry (TiO_6) in the surface of TiO_2 nanoparticles, while post-edge region of the XANES showed that PPV could modify the surface of the nano-TiO_2. The EXAFS showed a reduction in intensity compared to bulk and that the bond length (Ti–O) and co-ordination number decreased in nano-TiO_2. This was attributed to co-ordinatively unsaturated TiO_x surface species (where $x < 6$). It was proposed that in the PPV/TiO_2 nanocomposite, the PPV molecule can bond to the surface of the TiO_2 nanoparticles, leading to the interfacial structure being reconstructed. However, the detailed analysis of the EXAFS spectra appears to over-interpret the raw data with too much reliance on the accuracy of the derived co-ordination numbers.

A final example of this class is the investigation of cobalt aggregates electrochemically incorporated on composite polypyrrole films, materials with interesting magnetic behaviour [109]. In-situ Co K-edge fluorescence mode XAS measurements were collected at different stages of the electrochemical process. In addition, ex-situ XAS data were collected at the N and O K-edges, in total electron yield mode. The Co K-edge results, both XANES and EXAFS, showed that the reaction starts with the Co^{2+} ions entrapped in the polymeric matrix as a complex and the reduction of this complex leads to the formation of Co metal aggregates on the films. The N and O K-edge XANES results showed that the main interaction between Co metal aggregates and the polymer is via Co–N bonds, the N originating from the polypyrrole amine group.

3.3 *Nanoparticle–Nanoparticle Composites*

This class covers a very wide range of materials, as the criterion is simply that the two phases are on the nanoscale. A common variety is the case of composites of metals (or alloys) with an inorganic oxide. There is a range of potential applications for such systems, including catalysts, sensors, magnetic devices and coatings.

Nanocrystalline metal oxides such as MgO, SnO_2 and ZrO_2 are currently being investigated for a range of potential applications, such as gas-sensing, fuel cells and catalysis [110–112]. There are many routes employed to prepare nanocrystalline metal oxides including low temperature routes such as sol–gel [113–115] and ball milling [116–119]. However, many applications require the material to be stable at moderate to high temperatures (i. e. temperatures > 300°C), temperatures at which grain growth is pronounced. It has been shown that it is possible to restrict the high temperature grain growth of a range of nanocrystalline metal oxides by use of a 'sol–gel pinning' method [120–123] based on the principle of *Zener* pinning [124]. The materials are prepared with the inclusion of a small amount, generally 10–15%, of a second oxide phase, usually silica or alumina, which allows us to obtain particle of ~ 10 nm after calcining at 1,000°C [120–123].

However, while techniques such as XRD can be used to characterise the majority nanocrystalline metal oxide phase present, it cannot provide useful information regarding the pinning phase. The main questions that need to be answered are (a) has the pinning phase doped into the lattice of the nanocrystalline oxide or is it present as a separate phase and (b) has the pinning phase reacted with the nanocrystalline oxide? Since XAS provides information about the local environment of a specific element, it is an ideal technique to supply answers to these questions.

We have previously used XAS at the Si, Mg, Zr and Sn K-edges to study the local environment of nanocrystalline MgO, ZrO_2 and SnO_2 pinned with 15% SiO_2 (referred to as MgO–SiO_2, ZrO_2–SiO_2 and SnO_2–SiO_2) [120–123]. The K-edge XANES for Si are shown in Fig. 7 for crystalline SiO_2 (quartz), amorphous SiO_2 and MgO–SiO_2, ZrO_2–SiO_2 and SnO_2–SiO_2 samples calcined at 1,000°C. The average particle sizes for these samples are 11 nm for MgO–SiO_2, 10 nm for ZrO_2–SiO_2 and 8 nm for SnO_2–SiO_2. This figure shows that the spectra for quartz and amorphous silica are significantly different, with three peaks present at ~5 eV, 8.5 eV and 11.5 eV in the quartz spectrum, reflecting the more ordered environment around the Si atoms in this material, which are not observed in the amorphous spectrum. The spectra for the SnO_2–SiO_2 and ZrO_2–SiO_2 samples are very similar to those of the amorphous SiO_2, indicating that (a) the SiO_2 is present as amorphous silica and (b) there has been no reaction between the silica phase and the nanocrystalline metal oxide. Thus in these samples the silica phase is present as a discrete, amorphous phase and this was confirmed by the results of a ^{29}Si NMR study on the SnO_2–SiO_2 system [125].

However, it is clear that the spectrum of the MgO–SiO_2 sample is different from all of the other samples, with a peak present at ~10 eV that is not observed in either of the two silica standards, indicating that it is likely the silica has reacted with the MgO to form a new phase. In the Mg K-edge EXAFS of the sample there is a clear

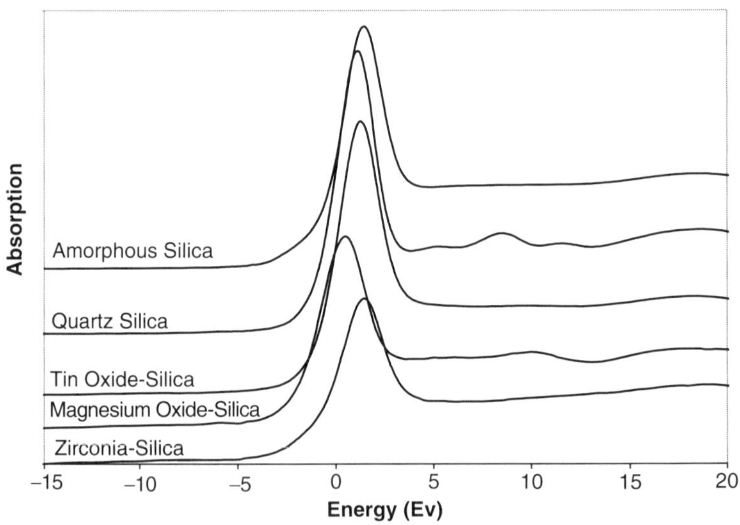

Fig. 7 Si K-edge XANES spectra for various silica standards and silica-pinned metal oxides. The spectra are offset on the vertical scale for clarity

reduction in the intensity of the EXAFS and Fourier transform spectra, compared to the bulk sample, as shown in Fig. 8. This corresponds to an ~15% decrease in the intensity of Fourier transform peak at 3 Å, indicating that ~15% of the magnesium atoms are not present as MgO, but as a different phase. Thus, both the Si K-edge XANES and the Mg K-edge EXAFS indicate that a discrete, second phase, which is not SiO_2, has formed and is acting to restrict the high temperature growth. ^{29}Si and ^{25}Mg NMR performed on the same sample [126] clearly showed that this discrete second phase was magnesium silicate. The three-way catalyst is used in

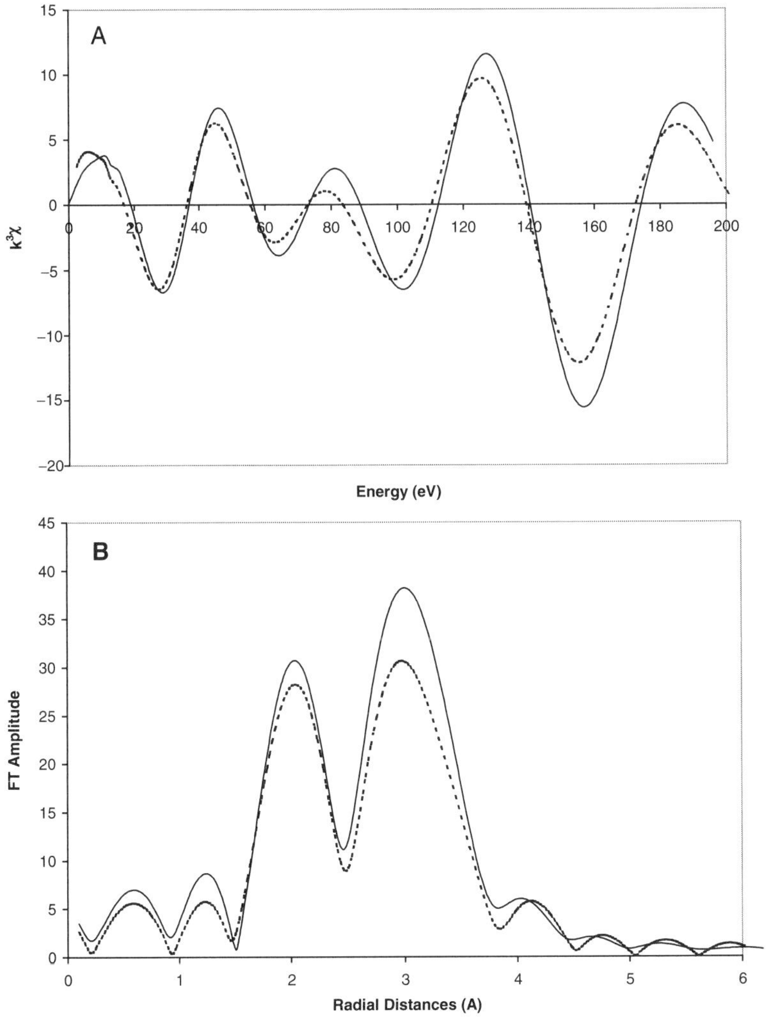

Fig. 8 The Mg K-edge EXAFS (**a**) of bulk magnesium oxide (*solid line*) and 11 nm MgO–SiO_2 (*dashed line*) and corresponding Fourier transform (**b**), which has been corrected with the phase shift of the first shell

automobiles to remove hazardous gases from the emissions by simultaneous conversion of oxides of nitrogen (NO_x), carbon monoxide (CO), hydrocarbons (HC) and formaldehyde (CH_2O) to N_2, CO_2 and water. Current commercial systems consist of Pt/Rh and CeO_2 coatings on an alumina substrate. There is considerable research effort aimed at replacing the expensive precious metal component with a cheaper alternative. Some success has been achieved by the use of nanocomposites, in particular those based on Cu and CeO_2 [127,128]. The activity of this system in CO oxidation is higher than any other combination of base metal catalyst or oxide support, and is even comparable to the activity of more expensive precious metal catalysts. The same catalytic system also has applications as the anode in solid oxide fuel cells as it permits the use of hydrocarbon fuels rather than just hydrogen [129, 130].

It is important to understand the role of metal and support in this particular combination but many experimental techniques are inapplicable due to the highly dispersed nature of the metal component; for example, the Cu or Cu oxides cannot be observed in the XRD pattern. Progress has been possible by combining the results of N_2O programmed oxidation with in situ EXAFS studies of Cu/CeO_2 [127] and Cu/SiO_2 [131] samples with a range of loadings. It is important to stress the need for in situ studies for this particularly reactive system, since small traces of O_2, CO_2 or H_2O adsorbed on the surface of Cu will significantly change the Cu EXAFS. The Cu K-edge EXAFS for a 9.1% loaded sample are shown in Fig. 9. The size of the CeO_2 nanocrystals was 4.5 nm, which meant the loading represents 0.57 Cu atoms per CeO_2 surface unit. Treatment with 4% N_2O/Ar at 100°C and 8% O_2/Ar at 300°C leads to oxidation as shown in Fig. 9. There is one major peak in the Fourier transforms at ~1.9 Å consistent with the Cu–O bond distance in CuO, showing the presence of Cu ions on the CeO_2 surface. Figure 9 shows that reduction in 5% H_2/Ar at 400°C leads to a first major peak at ~2.5 Å, which is consistent with the Cu–Cu bond distance of 2.55 Å in Cu metal, and the presence of small metallic clusters. Subsequent treatment with 4% N_2O/Ar only leads to oxidation above 300°C as shown in Fig. 9, where the Fourier transform changes from metallic to oxide-like between 260 and 400°C, the final state being an atomic dispersion of Cu ions. The differences in the state of Cu atom dispersion during reduction/oxidation was explained by a wetting/de-wetting transition on the CeO_2 surface

4 Conclusions

The examples discussed in this contribution demonstrate the unique role that XAS measurements can play in the study of nanocomposites. The technique can provide information on the local environment and oxidation state of an atom, with the particular advantages of being element specific, usable at low concentrations of target atom and not restricted to crystalline systems. The examples that have been used focus on the three general classes; however, they are equally applicable to biological systems.

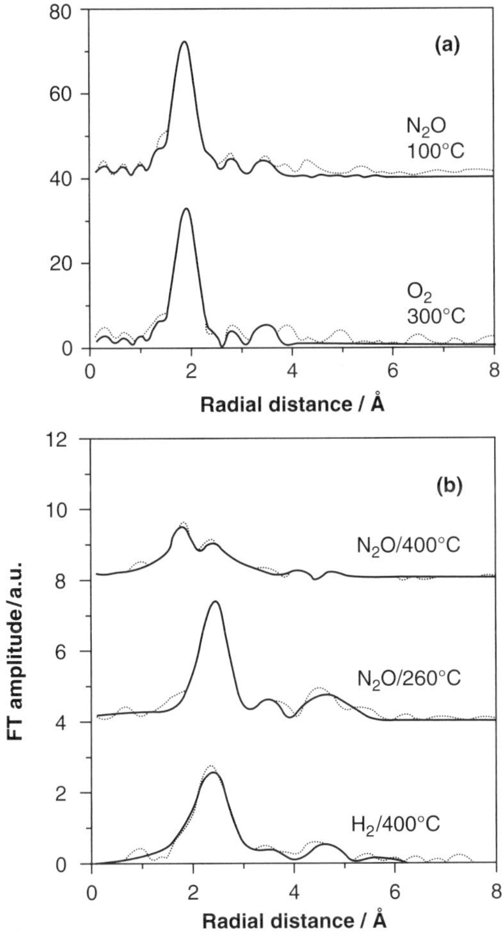

Fig. 9 The Fourier transforms of the Cu K-edge EXAFS for 9.1% Cu/CeO_2; (**a**) under oxidising conditions and (**b**) under H_2 and N_2O. *Solid lines* are the experimental data and *dashed lines* are the simulated fits

Although combined (e. g. XAS with XRD, etc.) and in situ measurements are now relatively routine, XAS techniques and their variations are still under continuous development. Microfocus XAS [132, 133] allows the mapping of species with a resolution of ~1 μm, which is particularly useful for composite samples. Time-resolved experiments, including Energy Dispersive EXAFS used to probe the reactivity of solids [83] are also possible allowing data to be collected on the millisecond timescale and the study of rapid changes or reactions can be explored. In addition, access to synchrotron facilities is widening with the commissioning of new facilities (e. g. Diamond in the UK [134], Soleil in France [135] and the Australian synchrotron [136]). Thus the use of XAS measurements in the characterisation of nanocomposites will only continue to expand.

Acknowledgements We wish to acknowledge the various students and co-workers at Kent who have worked on nanocrystalline materials. We would also like to express our thanks to various collaborators in our work on nanomaterials over recent years. First, we wish to thank Prof. Mark Smith and Mr. Luke O'Dell in the Physics Department at the University of Warwick for an extremely fruitful collaboration on nanocrystalline oxides. Second, we wish to thank the members of ALISTORE, the EU Network of Excellence working on nanomaterials for lithium battery applications. We are very grateful for help over several years from staff at the Daresbury SRS that has facilitated the work described in this Chapter, especially staff associated with the XAS beam lines: Ian Harvey, Bob Bilsborrow, Fred Mosselmans, Steve Fiddy, Shusaku Hayama and Alistair Lennie. Finally, we wish to acknowledge the financial support for our work on nanomaterials, especially the EPSRC (grants/S61881 and GR/S61898) and the EU (contract SES6-CT2003–503532, ALISTORE).

References

1. R. Roy, S. Komarneni and D.M. Roy, *Mater. Res. Soc. Symp. Proc.,* (1984), **32**, 347.
2. R. Roy, *Mater. Res. SOC. Annu. Mtg., Abstr.,* Boston, MA, 1982, p. 370.
3. R.A. Roy and R. Roy, *Mater. Res. Bull.,* (1984), **19**, 169.
4. D.W. Hoffman, R. Roy and S. Komarneni, *J. Am. Ceram. Soc.,* (1984), **67**, 468.
5. S. Komarneni, *J. Mater. Chem.,* (1992), **2**, 1219.
6. J.B. Park and R.S. Lakes, *Biomaterials: An Introduction*, (Plenum, New York, 1992).
7. B. Ji and H. Gao, *J. Mech. Phys. Solids*, (2004), **52**, 1963.
8. X. Li and P. Nardi, *Nanotechnology*, (2004), **15**, 211.
9. C.-S. Choi and Y.-W. Kim, *Biomaterials*, (2000), **21**, 213.
10. X.W. Sui and F.Z. Cui, *Mater. Sci. Eng. C*, (1999), **7**, 19.
11. J.L. Gardea-Torresdey, J.R. Peralta-Videa, G. de la Rosa and J.G. Parsons, *Co-ord. Chem. Rev.*, (2005), **249**, 1797.
12. S. Morris, S. Hanna and M.J. Miles, *Nanotechnology*, (2004), **15**, 1296.
13. R. Murugan and S. Ramakrishna, *Compos. Sci. Tech.,* (2005), **65**, 2385.
14. D.X. Cui and H.J. Gao, *Biotech. Prog.*, (2003), **19**, 683.
15. V. Thomas, D.R. Dean and Y.K. Vohra, *Curr. Nanosci.*, (2006), **2**, 155.
16. S.B. Mitra, D. Wu and B.N. Holmes, *J. Am. Dental Assoc.*, (2003), **134**, 1382.
17. P. Oelhafen, and A. Schuler, *Sol. Energ.* (2005), **79**, 110.
18. B. Li, L.D. Wang, B.N. Kang, P. Wang and Y. Qiu, *Sol. Energ. Mater. Sol. Cell.*, (2006), **90**, 549.
19. S. Yonemochi, A. Sugiyama, K. Kawamura, T. Nagoya and R. Aogaki, *J. Appl. Electrochem.*, (2004), **34**, 1279
20. Y.Z. Li and S.J. Kim, *J. Phys. Chem. B.*, (2005), **109**, 12309.
21. Y. Hu, Y. Tang and L. Song, *Polym. Adv. Tech.*, (2006), **17**, 235.
22. S.M. Lomakin and G.E. Zaikov, *Polym. Sci. B*, (2005) **47**, 9.
23. F. Bauer, R. Flyunt, K. Czihal, M.R. Buchmeiser, H. Langguth and R. Mehnert, *Macromol. Mater. Eng.*, (2006), **291**, 493.
24. X.G. Li, M.R. Huang, J.F. Zeng and M.F. Zhu, *Colloid. Surface. A*, (2004), **248**, 111.
25. R. Hauert and J. Patscheider, *Adv. Eng. Mater.*, (2000), **2**, 247.
26. S. Zhang, D. Sun, Y.Q. Fu and H.J. Du, *Surf. Coating. Tech.*, (2003), **167**, 113.
27. T. Mizutani, K. Arai, M. Miyamoto and Y. Kimura, *Progr. Org. Coating.*, (2006), **55**, 276.
28. R. Gangopadhyay and A. De, *Chem. Mater.*, (2000), **12**, 608.
29. G. M. do Nascimento, V.R.L. Constantino, R. Landers and M.L.A. Temperini, *Macromolecules*, (2004), **37**, 9373.
30. W.L. Qiu, X.H. Ma, Q.H. Yang, Y.B. Fu and X.F. Zong, *J. Power Sourc.*, (2004), **138**, 245.
31. H. Kawaoka, M. Hibino, H.S. Zhou and I. Honma, *J. Power Sourc.*, (2004), **125**, 85.
32. F. Croce, L. Persi, F. Ronci and B. Scrosati, *Solid State Ionics*, (2000), **135**, 47.

33. H.K. Liu, G.X. Wang, Z.P. Guo, J.Z. Wang and K. Konstantinov, *J. Nanosci. Nanotech.*, (2006), **6**, 1.
34. M.M.E. Jacob, E. Hackett and E.P. Giannelis, *J. Mater. Chem.*, (2003), **13**, 1.
35. C.S. Karthikeyan, S.P. Nunes, L.A.S.A. Prado, M.L. Ponce, H. Silva, B. Ruffmann and K. Schulte, *J. Membr. Sci.*, (2005), **254**, 139.
36. S.J. Han, Y.K. Yun, K.W. Park, Y.E. Sung and T. Hyeon, *Adv. Mater.*, (2003), **15**, 1922.
37. A.K. Cuentas-Gallegos, M. Lira-Cantu, N. Casan-Pastor and R. Gomez-Romero, *Adv. Funct. Mater.*, (2005), **15**, 1125.
38. Z.X. You, I. Balint and K. Aika, *Appl. Catal. B*, (2004), **53**, 233.
39. W. Oelerich, T. Klassen and R. Bormann, *J. Alloys Comp.*, (2001), **315**, 237.
40. B. Skarman, D. Grandjean, R.E. Benfield, A. Hinz, A. Andersson and L.R. Wallenberg, *J. Catal.*, (2002), **211**, 119.
41. A.A. Athawale and S.V. Bhagwat, *J. Appl. Polym. Sci.*, (2003), **89**, 2412.
42. G. Zhao, J.J. Feng, Q.L. Zhang, S.P Li and H.Y. Chen, *Chem. Mater.*, (2005), **17**, 3154.
43. X.R. Zhai, W.Z. Wei, J.X. Zeng, X.Y. Liu and S.G. Gong, *Anal. Lett.*, (2006), **39**, 913.
44. K.H. An, S.Y. Jeong, H.R. Hwang and Y.H. Lee, *Adv. Mater.*, (2004), **16**, 1005.
45. Y. Ma, N. Li, C. Yang and X.R. Yang, *Colloid. Surface. A*, (2005), **269**, 1.
46. M.Z. Atashbar, D. Banerji, S. Singamaneni and V. Bliznyuk, *Mol. Cryst. Liq. Cryst.*, (2005), **427**, 529
47. X. Lu and M.A. Winnik, *Chem. Mater.*, (2001), **13**, 3449.
48. P. Innocenzi and B. Lebeau, *J. Mater. Chem.*, (2005), **15**, 3821.
49. L.L. Beecroft and C.K. Ober, *Chem. Mater.*, (1997), **9**, 1302.
50. T. Ichikawa, N. Hanada, S. Isobe, H.Y. Leng and T. Ichikawa, *Mater. Trans.*, (2005), **46**, 1.
51. R. Gensler, R. Groppel, V. Muhrer and N. Muller, *Part. Part. Syst. Char.*, (2002), **19**, 293.
52. M. Takafuji, S. Ide, H. Ihara and Z.H. Xu, *Chem. Mater.*, (2004), **16**, 1977.
53. G.C. Hadjipanayis, *J. Magn. Magn. Mater.*, (1999), **200**, 373.
54. S. Calvin, M.M. Miller, R. Goswami, S.F. Cheng, S.P. Mulvaney, L.J. Whitman and V.G. Harris, *J. Appl. Phys.*, (2003), **94**, 778.
55. I. Gourevich, H. Pham, J.E.N. Jonkman and E. Kumacheva, *Chem. Mater.*, (2004), **16**, 1472.
56. S.G. Lu, B.R. Li, C.L. Mak and K.H. Wong, *J. Inorg. Mater.*, (2004), **19**, 1231.
57. T.C. Merkel, B.D. Freeman, R.J. Spontak, Z. He, I. Pinnau, P. Meakin and A.J. Hill, *Science*, (2002), **296**, 519.
58. K.E. Gonsalves, L. Merhari, H.P. Wu and Y.Q. Hu, *Adv. Mater.*, (2001), **13**, 703.
59. R. Prost and B. Yaron, *Soil Sci.*, (2001), **166**, 880.
60. J.M Thomas and G. Sankar, *Acc. Chem. Res.*, (2001), **34,** 71.
61. F. Farges, G.E. Brown and J.J. Rehr, *Phys. Rev. B*, (1997), **56**, 1809.
62. C. Prestipino, P.L. Solari and C. Lamberti, *J. Phys. Chem. B*, (2005), **109**, 13132
63. B.K. Teo and D.C. Joy, eds. "*EXAFS Spectroscopy: Techniques and Applications*," (Plenum Press, New York, 1980).
64. T.M. Hayes and J.B. Boyce, *Solid State Phys.*, (1982), **37**, 173.
65. D.C. Koningsberger and R. Prins, eds. "*X-Ray Absorption*," (Wiley, New York, 1988).
66. C.R.A. Catlow and G.N. Greaves, eds. "*Applications of Synchrotron Radiation*," (Blackie, Glasgow, 1990).
67. P.A. Lee and J.B. Pendry, *Phys. Rev. B*, (1975), **11**, 2795.
68. A.V. Chadwick, S.L.P. Savin, R. Alcántara, D. Fernández Lisbona, P. Lavela, G.F. Ortiz and J.L. Tirado, *Chem. Phys. Chem.*, (2006), **7**, 1086.
69. J.J. Rehr and A.L. Ankudinov, *Co-ord. Chem. Rev.*, (2005), **249**, 131
70. A.J. Dent, M. Oversluizen, G.N. Greaves, M.A. Roberts, G. Sankar, C.R.A. Catlow and J.M. Thomas, *Physica B*, (1995), **208** & **209**, 253.
71. G. Sankar, P.A. Wright, S. Natarajan, J.M. Thomas, G.N. Greaves, A.J. Dent, B.R. Dobson, C.A. Ramsdale and R.H. Jones, *J. Phys. Chem.*, (1993), **97**, 9550.
72. V. Briois, D. Lutzenkirchen-Hecht, F. Villain, E. Fonda, S. Belin, B. Griesebock and R. Frahm, *J. Phys. Chem. A*, (2005), **109**, 320.
73. A.J. Dent, *Top. Catal.*, (2002), **18**, 27.

74. S.R. Davis, A.V. Chadwick and J.D. Wright, *J. Phys. Chem. B*, (1997), **101**, 9901
75. B.S. Clausen, H. Topsoe and R. Frahm, *Adv. Catalysis*, (1998), **42**, 315.
76. 0B.S. Clausen, L. Grabaek, G. Steffensen, P.L. Hansen, and H. Topsoe, *Catal. Lett.*, (1993), **20**, 23.
77. J.M. Thomas and G.N. Greaves, *Catal. Lett.*, (1993), **20**, 337.
78. S. Ansell and G.W. Neilson, *Biophys. Chem.*, (2004), **107**, 229.
79. M.A. Newton, B. Jyoti, A.J. Dent, S.G. Fiddy and J. Evans, *Chem. Commun.*, (2004), 2382.
80. G.C. Shen and M. Ichikawa, *J. Chem. Soc. Faraday Trans.*, (1997), **6**, 1185.
81. J.D. Grunwaldt, S. Hannemann, C.G. Schroer and A. Baiker, *J. Phys. Chem. B*, (2006), **110**, 8674.
82. J.D. Grunwaldt, L. Basini and B.S. Clausen, *J. Catal.*, (2001), **200**, 321.
83. J. Evans and M.A. Newton, *J. Mol. Catal. A*, (2002), **182**, 351.
84. A. Deb and E.J. Cairns, *Fluid Phase Equil.*, (2006), **241**, 4.
85. O. Haas, A. Deb, E.J. Cairns and A. Wokaun, *J. Electrochem. Soc.*, (2005), **152**, A191.
86. W.S. Yoon, C.P. Grey, M. Balasubramanian, X.Q. Yang and J. McBreen, *Chem. Mater.*, (2003), **15**, 3161.
87. J.E. Penner-Hahn, *Co-ord. Chem. Rev.*, (1999), **190–192**, 1101.
88. N. Binsted, J.W. Campbell, S.J. Gurman and P.C. Stephenson, *SERC Daresbury Program Library*, Daresbury Laboratory, Warrington, Cheshire WA4 4AD, UK, 1992; A.J. Dent and J.F.W. Mosselmans, *Guides to EXCALIB, EXBACK and EXCURV92*, Darebury Laboratory, Warrington, Cheshire WA4 4AD, UK, 1995.
89. E.A. Stern, M. Newville, B. Ravel, Y. Yacoby and D. Haskel, *Physica B*, (1995), **209**, 117.
90. M. Newville, *J. Synch. Rad.*, (2001), **8**, 322.
91. V. Schwartz, D.R. Mullins, W.F. Yan, B. Chen, S. Dai and S.H. Overbury, *J. Phys. Chem. B*, (2004), **108**, 15782.
92. A. Mills and S. LeHunte, *J. Photochem. Photobiol. A: Chem.*, (1997), **108**, 1.
93. K. Page, R. Palgrave, I.P. Parkin, M. Wilson, S.L.P. Savin and A.V. Chadwick, *J. Mater. Chem.*, (2007), **17**, 95.
94. K.E. Rammutla, S.L.P Savin, M.G. Matshaba, A.V. Chadwick and P.E. Ngoepe, *Phys. stat. solidi (c)*, (2007), **4**, 765.
95. A. Caigher-Smith, *"Lustre Pottery: Technique, Tradition and Innovation in Islam and the Western World,"* (Faber & Faber, London, UK, 1985).
96. J. Roqué, T. Pradell, J. Molera and M. Vendrell-Saz, *J. Non-Cryst. Solids*, (2005), **351**, 568.
97. J. Roque, N.R.J. Poolton, J. Molera, A.D. Smith, E. Pantos and M. Vendrell-Saz, *Phys. Stat. Solidi. B*, (2006), **243**, 1337.
98. A.D. Smith, T. Pradell, J. Roqué, J. Molera, M. Vendrell-Saz, A.J. Dent and E. Pantos, *J. Non-Cryst. Solids*, (2006), **352**, 5353.
99. J.M. Thomas and G. Sankar, *J. Synchrotron Rad.*, (2001), **8**, 55.
100. C.-M. Yang, P.-H. Liu, Y.-F. Ho, C.-Y. Chiu and K.-J. Chao, *Chem. Mater.* (2003), **15**, 275.
101. S. Geetha, C.R.K. Rao, M. Vijayan and D.C. Trivedi, *Anal. Chim. Acta*, (2006), **568**, 119.
102. M.S. Soh, A. Sellinger and A.U.J. Yap, *Curr. Nanosci.*, (2006), **2**, 373.
103. H.J. Byker, *Electrochimica Acta*, (2001), **46**, 2015.
104. C.J. Dias and D.K. Das-Gupta, *IEEE Transactions on Dielectrics and Electrical Insulation*, (1996), **3**, 706.
105. A.W. Harper, S. Sun, L.R. Dalton, S.M. Garner, A. Chen, S. Kalluri, W.H. Steier and B.H. Robinson, *J. Opt. Soc. Am. B: Opt. Phys.*, (1998), **15**, 329.
106. F. Hussain, M. Hojjati, M. Okamoto and R.E. Gorga, *J. Composite Mater.* (2006), **40**, 1511.
107. L.G. Klapshina, I.S. Grigoryev, T.I. Lopatina, V.V. Semenov, G.A. Domrachev, W.E. Douglas, B.A. Bushuk, S.B. Bushuk, A.Y. Lukianov, A.V. Afanas'ev, R.E. Benfield and A.I. Korytin, *New J. Chem.*, (2006), **30**, 615.
108. J. Zhang, Z.Y. Wu, X. Ju, B.J. Wang, Q.S. Li, T.D. Hu, K. Ibrahim and Y.N. Xie, *Opt. Mater.*, (2003), **21**, 573.

109. N. Watanabe, J. Morais, S.B.B. Accione, A. Morrone, J.E. Schmidt and M.C.M. Alves, *J. Phys. Chem. B.*, (2004), **108**, 4013.
110. Th. Becker, S. Ahlers, Chr. Bosch-v.Braunmühl, G. Müller and O. Kiesewetter, *Sensor. Actuator. B*, (2001), **77**, 55.
111. S. Park, R.J. Gorte and J.M. Vohs, *Appl. Catal. A*, (2002), **200**, 55.
112. K.J. Klabunde, J. Stark, O. Koper, C. Mohs, D.G. Park, S. Decker, Y. Jiang, I. Lagadic and D. Zhang, *J. Phys. Chem.*, (1996), **100**, 12142.
113. M. Schneider and A. Baiker, *Catal. Rev. Sci. Eng.*, (1995), **37**, 515.
114. B.L. Cushing, V.L. Kolesnichenko and C.J. O'Connor, *Chem. Rev.*, (2004), **104**, 3893.
115. X. Bokhimi, A. Morales, T. Lopez and J. Gomez, *Solid State Chem.*, (1995), **115**, 411.
116. D. Michel, E. Gaffet and P. Berthet, *Nanostruct. Mater.*, (1995), **6**, 667.
117. T.D. Shen, C.C. Koch, T.L. McCormick, R.J. Nemanich, J.Y. Huang and J.G. Huang, *J. Mater. Res.*, (1995), **10**, 139.
118. S. Indris, D. Bork and P, Heitjans, *J. Mater. Synth. Proc.*, (2000), **8**, 245.
119. D. Michel, I. Mazerolles, P. Berthet and E. Gaffet, *Eur. J. Solid State Inorg. Chem.*, (1995), **32**, 673.
120. S.L.P. Savin, A.V. Chadwick, L.A. O'Dell and M.E. Smith, *J. Phys. Condens. Matter*, (2006), **18**, L163.
121. S.L.P. Savin and A.V. Chadwick, *Rad. Eff. Latt. Def. Solids*, (2003), **158**, 73.
122. S.L.P. Savin, A.V. Chadwick, L.A. O'Dell and M.E. Smith, *Solid State Ionics*, (2006), **177**, 2519.
123. S.L.P. Savin, A.V. Chadwick, L.A. O'Dell and M.E. Smith, *Phys. Stat. Solidi (c)*, (2005), **2**, 661.
124. C. Zener, quoted in C.S. Smith, *Trans. Metall. Soc.*, (1948), **175**, 15.
125. L.A. O'Dell, S.L.P. Savin, A.V. Chadwick and M.E. Smith, *Nanotechnology*, (2005), **16**, 1836.
126. L.A. O'Dell, S.L.P. Savin, A.V. Chadwick and M.E. Smith, *Faraday Dissuss.*, (2007), **134**, 83.
127. A. Tschöpe, J. Markmann, P. Zimmer, R. Birringer and A.V. Chadwick, *Chem. Mater.*,(2005), **17**, 3935.
128. B. Skårman, D. Grandjean, R.E. Benfield, A. Hinz, A. Andersson and L. Reine Wallenberg, *J. Catal.*, (2002), **211**, 119.
129. S.D. Park, J.M. Vohs and R.J. Gorte, *Nature*, (2000), **404**, 265.
130. H. Kim, S. Park, J.M. Vohs and R.J. Gorte, *J. Electrochem. Soc.* (2001), **148**, A693.
131. G. Sankar, J.M. Thomas, D. Waller, J.W. Couves, C.R.A. Catlow and G.N. Greaves, *J. Phys. Chem.* (1992), **96**, 7485.
132. A.M. Flank, G. Cauchon, P. Lagarde, S. Bac, M. Janousch, R. Wetter, J.M. Dubuisson, M. Idir, F. Langlois, T. Moreno and D. Vantelon, *Nucl. Instr. Meth. Phys. Res. B*, (2006), **246**, 269
133. H. Oyanagi, *Jpn. J. Appl. Phys.*, (1993), **32**, 861.
134. http://www.diamond.ac.uk
135. http://www.synchrotron-soleil.fr
136. http://www.synchrotron.vic.gov.au

Dynamical Aspects of Nanocrystalline Ion Conductors Studied by NMR

P. Heitjans[1], Sylvio Indris[1,2], and M. Wilkening[1]

1 Introduction

In recent years, nanocrystalline materials have considerably attracted the interest of the materials research community [1–4]. Single-phase materials with an average particle diameter of less than 50 nm exhibit new or at least enhanced chemical and physical properties when compared to their coarse-grained counterparts. For example, they show new mechanical [5–8], electrical [9–13], magnetic [14–19], optical [20–24], catalytic [25, 26], and/or thermodynamic [27–29] features. In ion conducting nanocrystalline materials, an enhancement of the diffusivity of small cations and anions like Li^+ and F^-, or even larger anions like O^{2-}, is often observed [2–4, 30–41]. The diffusivity is additionally influenced by admixing an ionic insulator to the conducting phase [4, 42–47].

Nuclear magnetic resonance (NMR) spectroscopy (e. g. [48, 49]) is very useful for the study of atomic diffusion in ion conductors [4]. Up to now nearly exclusively Li^+ and F^- nanocrystalline conductors were investigated by NMR. We review NMR studies both on single- and two-phase nanomaterials. In our laboratory, mainly the measurement of spin-lattice relaxation rates and the analysis of NMR line shapes were applied so far to extract information about ionic self-diffusion. Quite recently, field-gradient NMR has also been used [46], which, contrary to the techniques mentioned, traces long-range diffusion. Further NMR techniques with considerable potential particularly in the slow-motion regime are spin-alignment echo (e. g. [50–52]) as well as two-dimensional exchange NMR (e. g. [53, 54]). These methods have so far only been applied to microcrystalline ion conductors but are likely to be extended to nanocrystalline materials in the near future. For example, we have started to probe the Li diffusivity in nanocrystalline $LiTaO_3$ by ^7Li spin-alignment echo NMR giving diffusion parameters comparable to the

[1] Institute of Physical Chemistry and Electrochemistry, and Center for Solid State Chemistry and New Materials, Leibniz University Hannover, Callinstr. 3a, 30167 Hannover, Germany

[2] Present address: Forschungzentrum Karlsruhe, Institute of Nanotechnology, 76201 Karlsruhe, Germany

P. Knauth and J. Schoonman (eds.), *Nanocomposites: Ionic Conducting Materials and Structural Spectroscopies*.

results obtained from dc-conductivity (for the latter comparison see e. g., recent NMR studies on microcrystalline $Li_4Ti_5O_{12}$ [55] and Li_7BiO_6 [56]).

In a first approach, single-phase nanocrystalline materials can be simply described as heterogeneous materials consisting of structurally ordered grains and interfacial regions being expected to show structural disorder [4, 57, 58]. Because of the small particle size, the volume fraction of the interfacial regions in these materials is drastically increased [1, 4, 27, 59] as compared to the corresponding coarse-grained (microcrystalline) materials with an average particle diameter in the μm range. The interfacial regions seem to be responsible for fast ionic diffusion. This heterogeneous structure gives rise to heterogeneous diffusion in nanomaterials [44, 45]. We will show that the highly mobile ions in the interfacial regions can be differentiated from the slowly diffusing ones in the crystalline grains via their different NMR relaxation behavior. Thus, it is possible to probe diffusion parameters of the ions in the interfacial regions separately from those in the grains. Up to now we have investigated the F^- and Li^+ transport properties in a series of single-phase nanocrystalline ion conductors like CaF_2 [33, 34, 48, 60], BaF_2 [61], $LiNbO_3$ [30, 31, 57, 58], Li_xTiS_2 [62], and Li_2O [44] as well as in nanocrystalline composites like $Li_2O : B_2O_3$ [44, 63] and $Li_2O : Al_2O_3$ [45]. The structure of the latter ones, which are so-called dispersed ionic conductors, is sketched in Fig. 1. Instead of one kind of interface like in single-phase nanomaterials three types of interfaces can be identified in two-phase composites. These include contacts between (a) the ionic

Fig. 1 Sketch of a composite material built of ionic conductor (*light grey areas*) and insulator grains (*dark grey areas*). The network of interfaces consists of three different types, which are represented by *dotted*, *broken*, and *full lines*. The latter ones highlight the insulator-conductor heterocontacts (from [4])

conductor grains (dotted lines), (b) the insulator grains (broken lines), and (c) the ion conductor and the insulator grains (full lines). In the composites studied here, the enhancement of Li dc-conductivity [32] is due to the presence of insulator-conductor contacts in the material. This dispersed ion conductor effect was first observed for microcrystalline LiI : Al_2O_3 by Liang [47]. The increase of the Li diffusivity may have different origins, e. g., the formation of space charge layers, an enhanced concentration of structural defects or the formation of new phases. In general, the heterocontacts build a network of fast diffusion pathways, which lead to a pronounced maximum in the total ionic conductivity as a function of the insulator content and which can be described by percolation theory. This was shown for example via dc-conductivity measurements on nanocrystalline Li_2O : B_2O_3 composites [32]. Similar results were also found for CuBr : TiO_2 [64–66]. Recently, also the composite LiI : Al_2O_3 in its nanostructured form was studied by various methods including ^{7}Li NMR [46].

2 NMR Methods: Line Shape and Spin-Lattice Relaxation Spectroscopies

2.1 *Influence of Diffusion on NMR Resonance Lines*

The NMR line width $\Delta\nu$ is related to the spin-spin relaxation rate $1/T_2$ quantifying the decay of a transverse magnetization $M_{y'}(t)$, i. e., perpendicular to the external magnetic field $\boldsymbol{B}_0$. A transverse magnetization can be generated simply with a 90° radio-frequency pulse. y' denotes that $M_{y'}(\mathrm{t})$ is sampled as free induction decay (FID) in the rotating reference frame. In the simplest case, the transients $M_{y'}(t)$ follow either exponential or Gaussian-shaped functions depending on the sample temperature and diffusivity of the nuclei under investigation. In the latter case, also a Gaussian-shaped NMR resonance line is obtained after Fourier transformation of $M_{y'}(t)$. At low temperatures, the FIDs of solids proceed on a very short time scale compared to those measured in liquids. Accordingly, broad NMR lines with a width of some tens kHz are obtained. This is due to the fact that the nuclei in a solid are much less mobile than in the liquid state. They are exposed to slightly different dipolar magnetic fields leading to a distribution of resonance frequencies around the mean value $\omega_0/2\pi$ being the resonance frequency that is related to B_0 via $B_0 = \omega_0/\gamma$ with γ being the gyromagnetic ratio of the probe nuclei. In this temperature range, which is called the *rigid-lattice* regime, the jump rate $1/\tau$ of the ion is much smaller than the line width $\Delta\nu$. In this T range, $\Delta\nu$ and thus $1/T_2$ are independent of temperature. Usually, the rigid-lattice value of $1/T_2$ is of the order of 10^4 s^{-1}.

At higher temperatures, however, the nuclei start moving and local magnetic fields are averaged. With increasing diffusivity, the transverse relaxation process is slowed down and, thus, the FIDs proceed on a longer time scale similar to the case that is present in liquids. Consequently, the corresponding NMR line starts to narrow. The beginning of this *motional narrowing* indicates the onset of ion motions with residence times τ of the order of the inverse line width $1/\Delta\nu$. At even higher temperatures, i. e., in the regime of *extreme narrowing*, the line width reaches once again a temperature-independent value due to inhomogeneities of the external magnetic field B_0.

In a heterogeneous material with two distinct reservoirs of fast and slow diffusing spins – as it may be the case in a nanocrystalline ion conductor – it should be possible to distinguish between these two species via their different spin-spin relaxation behavior (see below).

2.2 Influence of Diffusion on NMR Spin-Lattice Relaxation

The primary observable in an NMR experiment is the magnetization $\boldsymbol{M}$ being proportional to the sum of the nuclear magnetic moments of the sample. In thermal equilibrium $\boldsymbol{M}$ is parallel to the external magnetic field $\boldsymbol{B}_0$, i. e., $\boldsymbol{M} = \boldsymbol{M}_0$. Using an external radio frequency field, the direction of the magnetization with respect to B_0 can be changed, i. e., nuclear spins can be "reversed". The transition probability will be maximum when this external radio frequency ω_1 is in resonance with the Larmor precession frequency of the nucleus given by $\omega_0 = \gamma B_0$. Immediately after excitation with a radio frequency pulse, the spin-system will relax to its state of thermal equilibrium. This spin-lattice relaxation process can be recorded when the recovery of the magnetization. vector $\boldsymbol{M}$ along the axis defined by the magnetic field $\boldsymbol{B}_0$ is monitored as a function of waiting (or delay) time t. Usually, an inversion or a saturation recovery pulse sequence is used for this purpose [48]. In the simplest case $M(t)$ follows an exponential

$$M(t)/M_0 = 1 - \exp(-t/T_1), \tag{1}$$

with $1/T_1$ being the spin-lattice relaxation rate, which is a function of temperature. Besides other effects, the relaxation process is induced by internal fluctuating fields due to the temperature-dependent motion of the nuclei. These fields may be dipolar magnetic from neighboring nuclei as well as quadrupolar electric from local electric field gradients. The fluctuations of an internal field are described by a correlation function $G(t)$. According to the model introduced by Bloembergen, Purcell and Pound (BPP) [67], $G(t)$ is assumed to be a simple exponential

$$G(t) = G(0)\exp(-|t|/\tau_c). \tag{2}$$

τ_c is the correlation time that is within a factor of the order of unity equal to the mean residence time τ between two successive jumps of the nucleus. The Fourier transform of $G(t)$, to which $1/T_1$ is related, is the spectral density function $J(\omega)$. Spin-lattice relaxation becomes effective when $J(\omega)$ has intensities at the resonance frequency, see e. g. [48].

$$1/T_1 \propto J(\omega_0) = G(0)\frac{2\tau_c}{1+(\omega_0\tau_c)^2} \tag{3}$$

The temperature dependence of τ_c in (3) is typically given by an Arrhenius relation

$$\tau_c = \tau_0 \exp(E_A / (k_B T)), \tag{4}$$

where τ_0 is the preexponential factor and E_A the activation energy of the diffusion process. Thus, for a given Larmor frequency ω_0, $1/T_1$ vs. reciprocal temperature T passes through a maximum at a specific temperature that is determined by the condition $\omega_0 \tau_c \approx 1$.

Contrary to the prediction of the BPP model (3), in real materials usually asymmetric diffusion induced rate peaks ln $1/T_1(1/T)$ are obtained, i. e., the slope at temperatures above the maximum is larger than that below. The slope of the high-T flank of the peak is related to long-range diffusion parameters. In this region $\omega_L \tau_c \ll 1$ holds, i. e., several jumps are sampled during one period of spin-precession. The low-T flank on the other hand reflects short-range diffusion because below the maximum $\omega_0 \tau_c \gg 1$ holds. It may be influenced by correlation effects due to structural disorder and/or Coulomb interactions of the moving particles, cf. [68], for example. In such cases the slope of the low-T flank is reduced.

In analogy to spin-spin relaxation measurements in samples with two distinct regions of different diffusivity of the nuclei, one should be able to identify these different species via their individual spin-lattice relaxation rates, too. In such a case the overall magnetization transient $M(t)$ (eq. 1) might be composed of two exponentials with distinct relaxation rates.

3 NMR Line Shapes of Nanocrystalline Ion Conductors – Case Studies

3.1 *Single-Phase Materials*

The transport characteristics of chemically homogeneous materials are exemplarily studied by NMR on nanocrystalline lithium oxide, Li_2O. Nanosized Li_2O with an average particle diameter of about 20 nm was prepared by high-energy ball-milling of the coarse-grained microcrystalline starting material. For details of sample preparation see [44]. It crystallizes in the antifluorite structure (space group *Fm*3*m*), cf. Fig. 2. Li resides in sites with perfect tetrahedral symmetry. Thus, although ^{7}Li is a quadrupole nucleus with a spin-quantum number $I = 3/2$, no significant contributions arising from quadrupolar interactions to the NMR spectra are expected. In the whole temperature range covered here (140 K – 500 K) no distinct ^{7}Li satellite transitions in the NMR spectra were observed for microcrystalline Li_2O. The ^{7}Li NMR spectra of Li_2O are mainly governed by homonuclear dipole–dipole interactions and consist of a single resonance line, only. Quadrupolar contributions may arise from impurity ions or due to local defects as well as to (larger) regions with disordered structure. At temperatures above 700 K Xie at al. observed a small but well-defined quadrupolar powder pattern for pure Li_2O ascribed to the formation of thermal vacancies [69]. However, also for the nanocrystalline sample, the ^{7}Li NMR spectra, being recorded at temperatures up to 473 K, do not show any indication of quadrupole

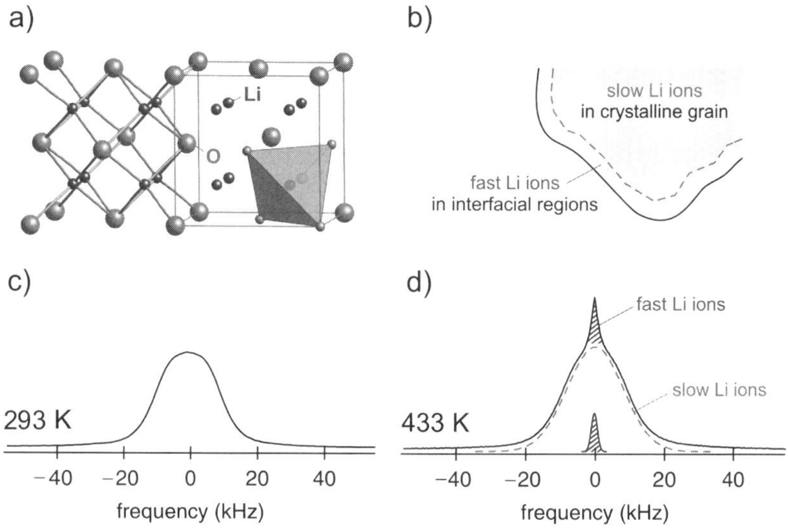

Fig. 2 (**a**) Crystal structure of Li_2O. Li ions reside in tetrahedral sites with perfect symmetry. Thus, there is no electric field gradient with which the quadrupole moment of the ^{7}Li nucleus may interact. (**b**) Simplified view of a nanocrystal consisting of a crystallographically ordered grain and an interfacial region with fast diffusion pathways. (**c**) and (**d**) ^{7}Li NMR spectra of nanocrystalline Li_2O with an average particle diameter of about 20 nm at two temperatures. The spectra were recorded at 155.5 MHz. The two-component line shape that appears at higher T reflects the heterogeneous structure of the nanocrystalline ion conductor

intensities next to the central line. In Fig. 2c, the ^{7}Li NMR spectrum of nanocrystalline lithium oxide is presented. The spectrum was recorded at 293 K. At this temperature, it is identical to the one which is obtained for the coarse-grained microcrystalline counterpart. However, in contrast to the microcrystalline material, at higher temperatures, i. e., starting at about 433 K, a two-component line shape shows up (Fig. 2d). The broad NMR line, which represents the less mobile ions in the crystalline grains, is now superimposed by a sharp NMR line due to the much more mobile charge carriers in the interfacial regions. Thus, the heterogeneous structure of a nanocrystalline material – consisting of crystalline grains and a large volume fraction of interfacial regions – is reflected by the shape of the NMR line. The narrower the NMR line, the more mobile are the ^{7}Li nuclei in Li_2O.

Another example where Li ions in the interfacial regions can be detected by NMR is nanocrystalline lithium niobate, $LiNbO_3$, which was prepared by high-energy ball-milling of the coarse-grained source material [58, 70]. The single- and microcrystalline forms are rather poor ionic conductors. However, reducing the particle size by mechanical treatment to about 20 nm [30, 31, 58] and, thus, introducing a large volume fraction of grain boundaries and interfacial regions has a dramatic effect on Li diffusivity in the case of $LiNbO_3$. Its Li conductivity increases by about six orders of magnitude [70]. In contrast to lithium oxide, in $LiNbO_3$ the Li nuclei are exposed to an electric field gradient at the nuclei sites. Thus, according to the corresponding alteration of the Zeeman levels, the ^{7}Li NMR signal consists of a central line and a rather broad contribution of satellite intensities as expected for a powder sample [58]. Both contributions can be distinguished (Fig. 3, see the magnification of the spectrum on the right-hand side of this figure). Increasing the temperature to 463 K the quadrupole contribution of the ^{7}Li NMR line remains nearly unchanged whereas the central line is partly narrowed. As in the case of Li_2O it is superimposed at higher temperatures by a narrow NMR line which takes about 45 % of the total area under the overall signal. Altogether, a three-component line shape evolves at temperatures above 293 K. The interfacial effect in nanocrystalline $LiNbO_3$ is dominant in the overall ionic transport – at least at temperatures below about 600 K [58, 70]. At higher temperatures also the Li ions residing in the crystalline grains become mobile [30, 31, 57, 58]. Below 500 K Li diffusion in a purely amorphous $LiNbO_3$ sample is nearly identical to that present in the ball-milled material. Comparative ^{7}Li NMR spin-lattice relaxation as well as conductivity measurements on different forms of $LiNbO_3$ have shown that the interfacial regions of nanocrystalline $LiNbO_3$, which was prepared by high-energy ball-milling, are of amorphous structure [45, 58]. This was corroborated by recent EXAFS and XRD measurements [58]. EXAFS data revealed that about 50% of the Li ions in nanocrystalline $LiNbO_3$ reside in the interfacial regions [58, 71]. This is in good agreement with the results obtained by NMR line shape analysis. Using HR-TEM micrographs, the amorphous structure of nanocrystalline $LiNbO_3$

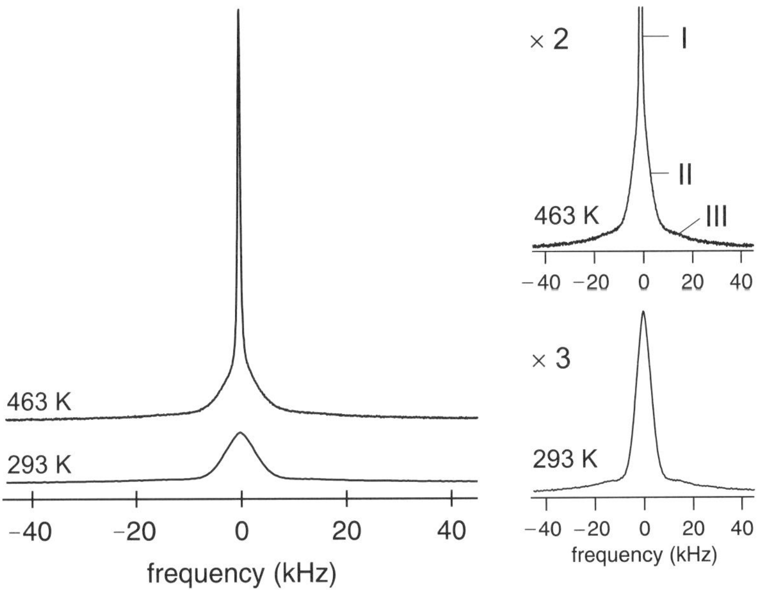

Fig. 3 ^{7}Li NMR spectra of nanocrystalline $LiNbO_3$ at 293 K and 463 K. The sample was prepared by high-energy ball milling of the coarse grained source material for 16 h. The spectra on the right show the magnification of those on the left. At $T = 293$ K the ^{7}Li NMR spectrum is composed of a central line (II) and a quadrupolar powder pattern (III). Both contributions are dipolarly broadened. Increasing T to 463 K the central line is partly narrowed so that a three-component line shape arises whose narrow line (I) reflects highly mobile Li ions in the interfacial regions that have amorphous structure in the case of $LiNbO_3$

prepared by ball-milling and probed by NMR spectroscopy was made visible [58]. In Fig. 4, an HR-TEM micrograph of $LiNbO_3$ ball-milled for 32 h is shown. The heterogeneous nature of the ball-milled material consisting of a crystalline and an amorphous phase is clearly visible. The nanocrystallites are surrounded by an amorphous grain boundary region of about 2 nm thickness (indicated in Fig. 4 by two solid lines). Contrary to that, in nanocrystalline $LiNbO_3$, which was prepared chemically using a sol–gel route, amorphous interfacial regions are virtually absent [58]. Although the Li diffusivity is somewhat enhanced compared to that in the microcrystalline sample, the conductivity does not reach the values found for the ball-milled nanomaterial. Thus, the small increase of Li diffusivity in the sol–gel prepared sample may be ascribed to space-charge effects [43, 72] rather than to the structural disorder being the cause in ball-milled $LiNbO_3$.

Apart from Li^+ in Li conductors, also the dynamics of small anions were studied in nanocrystalline CaF_2 and BaF_2 by ^{19}F NMR so far, see [48, 61]. Nanocrystalline CaF_2, which was prepared by inert-gas condensation, was historically the first example of a nanocrystalline ceramic on which hopping dynamics was investigated by solid state NMR [60]. Similar to nanocrystalline Li_2O, a two-component line shape shows up. Quadrupolar interactions are absent for nuclei with a spin-quantum number of $I = 1/2$ like ^{19}F. The motionally narrowed NMR line of the spectra shown in Fig. 5, whose centre of gravity coincides with that of the broad component, again

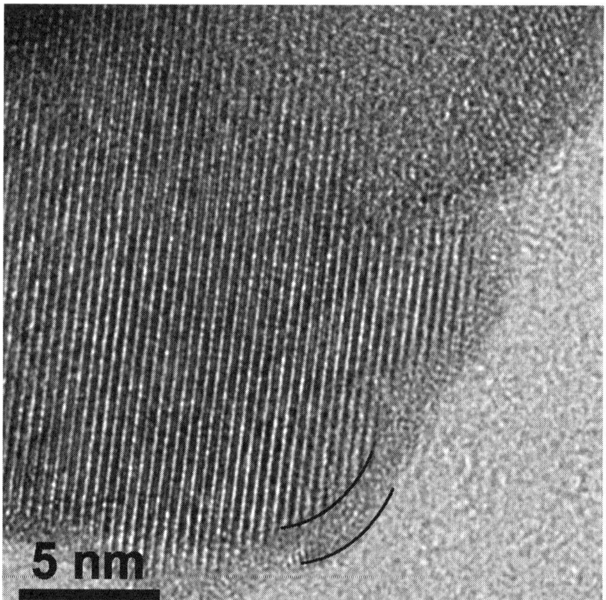

Fig. 4 HR-TEM micrograph of nanocrystalline $LiNbO_3$ prepared by ball-milling. The average particle size was about 20 nm. An amorphous region of about 2 nm thickness covers the crystalline grains (from [58])

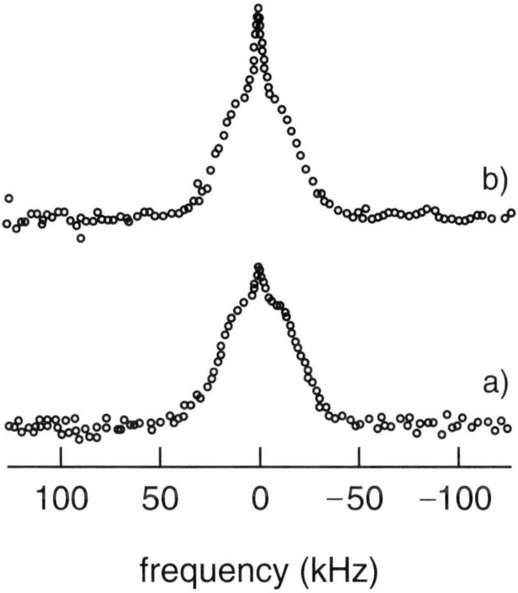

Fig. 5 NMR line shapes of ^{19}F (spin-1/2 nucleus) in nanocrystalline CaF_2 recorded at $T = 400\,K$ (**a**) and 440 K (**b**) both at a resonance frequency of 24 MHz. As in the case of lithium oxide (Fig. 2d) a two-component line shape is observed. At 440 K, the narrowed line has attained 10% of the signal intensity (after [48])

represents highly mobile ions (here F^- anions) in the interfacial regions. The same characteristics were also obtained for nanocrystalline BaF_2, which was prepared by ball-milling [61]. Thus, the heterogeneous structure observed by NMR seems to be independent of the method used to prepare nanocrystalline CaF_2 and BaF_2. In contrast to nanocrystalline $LiNbO_3$, where the enhanced Li diffusivity is due to an amorphous layer covering the nanocrystalline grain, in CaF_2 a space charge layer [72] adjacent to the grain boundaries results in enhanced vacancy concentration leading to the observed enhancement of the overall ionic transport [34]. The same explanation may hold also for the enhanced diffusivity in other ionic conductors, see e. g. [43].

3.2 Two-Phase Materials

In chemically heterogeneous nanocrystalline materials like $Li_2O : Al_2O_3$ [45] and $Li_2O : B_2O_3$ [44] composites, the same features of NMR line shapes can be observed as in the single-phase materials, however, much more pronounced. The composites were prepared by mixing nanocrystalline powders of α-Al_2O_3 as well as B_2O_3 with nanocrystalline Li_2O for a few minutes in a high-energy ball mill using a vial set made of α-Al_2O_3. The whole preparation was done in inert gas atmosphere in order to eliminate any influence of water vapor. Before mixing, the materials were separately milled for about 16h. The average particle size determined via XRD line broadening and using the Scherrer equation was about 20 nm [73]. Mixing the Li conductor Li_2O with the insulator material for a short time, e. g. for several minutes only, does not lead to a mechanochemical reaction of the two components as is verified by XRD analysis (Fig. 6) [73]. Therefore, this step of the preparation procedure can be regarded just as a compaction process. However, after mechanical treatment for about 4 h, new XRD lines show up in the XRD pattern that belong to $Li_2B_4O_7$ (marked with crosses). Abrasion of the milling vial can also be detected via the appearance of XRD peaks characteristic of Al_2O_3. However, this small amount of alumina has negligible impact on the NMR data presented here, as can be clearly seen when studying the changes of the NMR line shapes for different insulator contents with $x = 0.2$, 0.5, and 0.8, see Fig. 8.

Static ^{7}Li NMR spectra of the nanocrystalline $x\,Li_2O : (1-x)\,Al_2O_3$ composites with $x = 0.5$ are shown in Fig. 7. The NMR line width (full width at half maximum) of the microcrystalline composite is homogenously narrowed with increasing temperature, whereas the corresponding line shapes of the nanocrystalline counterpart exhibit a heterogenous narrowing as in the case of single phase Li_2O. Incipient line narrowing indicates the onset of slow Li motions. Usually it is observed when the ionic jump rate reaches a value in the kHz range. As mentioned in Sect. 2.1, this means that local magnetic fields, to which the ^{7}Li nuclei are exposed, are averaged due to the onset of ion hopping. Thus, the relatively broad distribution of local resonance frequencies at the lowest temperature (rigid-lattice line width) is increasingly reduced to a single (mean) value for all nuclei when ionic diffusion becomes fast enough. Interestingly, the area fraction of the narrow NMR line, and thus the number of highly mobile charge carriers, depends on temperature (Fig. 7, right-hand side).

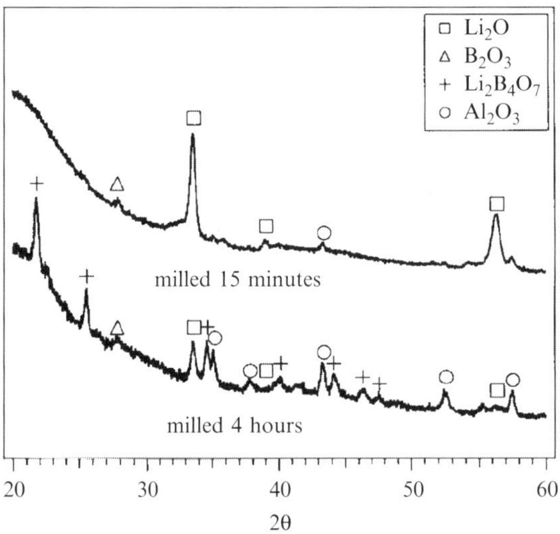

Fig. 6 XRD pattern of the mixture of Li_2O and B_2O_3 after different milling times. Formation of $Li_2B_4O_7$ (*crosses*) is observed for milling times longer than 4 h (from [73])

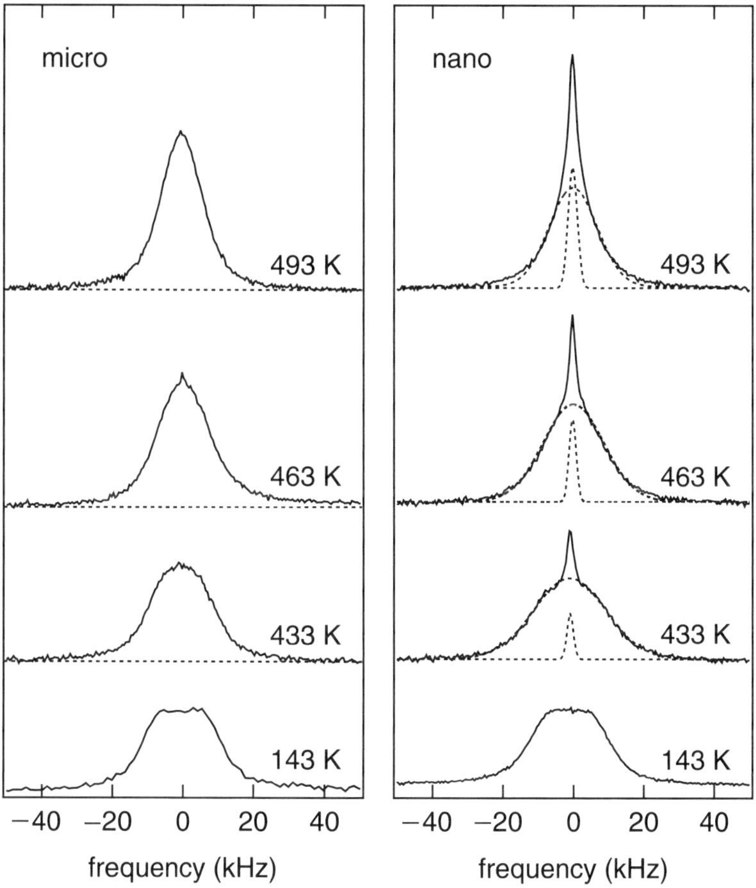

Fig. 7 Comparison of ^{7}Li NMR spectra of micro- and nanocrystalline Li_2O recorded for various temperatures. NMR spectra were measured at a radio frequency of 58.8 MHz. The particle size of high-energy ball-milled lithium oxide was about 20 nm (from [45])

At 433 K, the narrow contribution of the overall NMR line is about 4%. Furthermore the area fraction of the narrow contribution to the NMR signal depends on the amount of insulator added to the ion conducting phase. Similar results were obtained for composites with B_2O_3 as insulator phase. In Fig. 8, line shapes of $(1-x)$ $Li_2O : x$ Al_2O_3 composites with different insulator contents are shown as an example, see [49]. Each spectrum was recorded at 433 K. Increasing the alumina content leads to an increase of mobile charge carriers in the composites. In contrast to the single-phase materials, in nanocrystalline composite systems, contacts between adjacent Li_2O and insulator particles are present. In a simple explanation, the insulator being an oxide with a certain basicity is adsorbing Li ions. Consequently, cation vacancies are generated in the conductor phase, so that Li diffusion is enhanced in a region along the grain boundaries. Thus, the additional number of

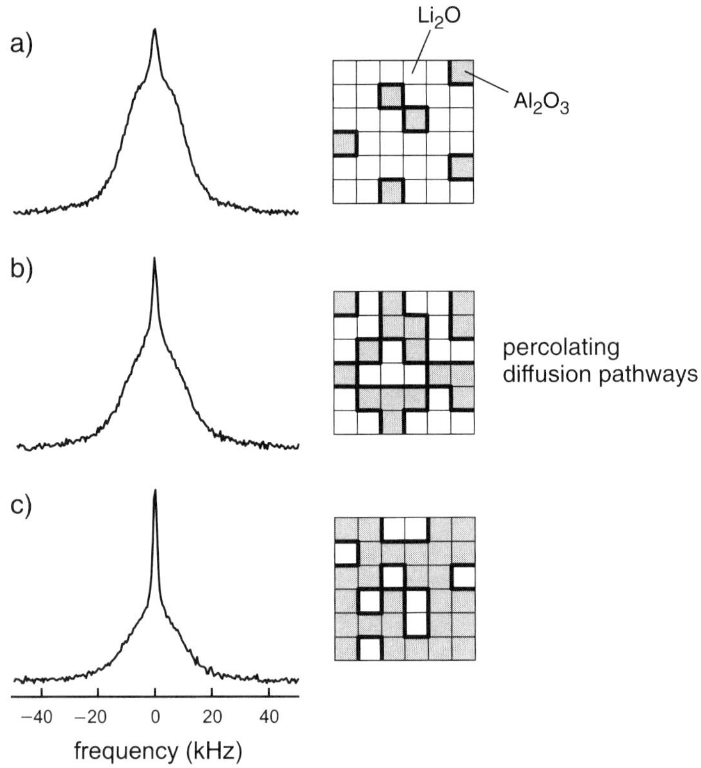

Fig. 8 ^{7}Li NMR spectra of nanocrystalline $Li_2O : Al_2O_3$ composites with different amounts of alumina: (**a**) $x = 0.2$, (**b**) $x = 0.5$, and (**c**) $x = 0.8$. Spectra were recorded at 433 K and 58.8 MHz. The higher the number of conductor–insulator contacts with respect to the total Li_2O content, the larger the area fraction of the mobile ions. Heterocontacts between insulator and conductor particles are represented in the schemes on the right by *bold lines* (from [49])

mobile charge carriers with respect to those in the single-phase material are located in the interfacial regions adjacent to the insulator grains.

In Fig. 8, the ^{7}Li NMR spectra are compared with schemes of composites with different insulator contents x, where the interfacial regions between unlike particles are represented by bold lines. The fraction of bold lines relative to the Li_2O content, and thus the fraction of mobile ions, increases with insulator content. In other words, the homointerfaces between adjacent Li_2O particles are continuously replaced by heterointerfaces between neighboring Li_2O and Al_2O_3 particles being faster diffusion pathways for the Li ions. Remarkably, impedance spectroscopy measurements show a maximum of the dc-conductivity at compositions with $x \approx 0.5$ as explicitly shown for the $(1-x)Li_2O : xB_2O_3$ composites [32]. This observation can be ascribed to the formation of highly diffusive percolation pathways by the heterointerfaces at intermediate insulator contents (cf. Fig. 8). It should be noted that the two methods probe diffusion on different length scales. In the temperature range examined here, NMR is sensitive to short-range Li motions in the $Li_2O : Al_2O_3$ composites, whereas dc-conductivity measurements probe Li diffusivity on a long-range length scale and, thus, are sensitive to percolation pathways being *not* detectable by the low-T ^{7}Li NMR measurements reported here.

4 NMR Relaxation in Nanocrystalline Conductor–Insulator Composites

4.1 Heterogeneous Spin-Spin Relaxation

The ^{7}Li NMR spectra shown in Figs. 7 and 8 were obtained after Fourier transformation of the corresponding FIDs that are displayed in Fig. 9. All FIDs are composed of two contributions, namely a fast decaying part being detectable below $t = 50\,\mu$s and a slowly decaying one for acquisition times larger than $50\,\mu$s. The corresponding spin-spin relaxation rates differ by about one order of magnitude at, e. g., 433 K and $x = 0.5$. Fourier transformation of the first part of the FID expectedly results in a broad NMR line with a width of about 20 kHz, whereas the second one yields a narrow NMR line whose width is less than 3 kHz. Thus, the two parts can be attributed to Li ions in the grains and interfacial regions, respectively. The temperature dependence of the spin-spin relaxation rates of

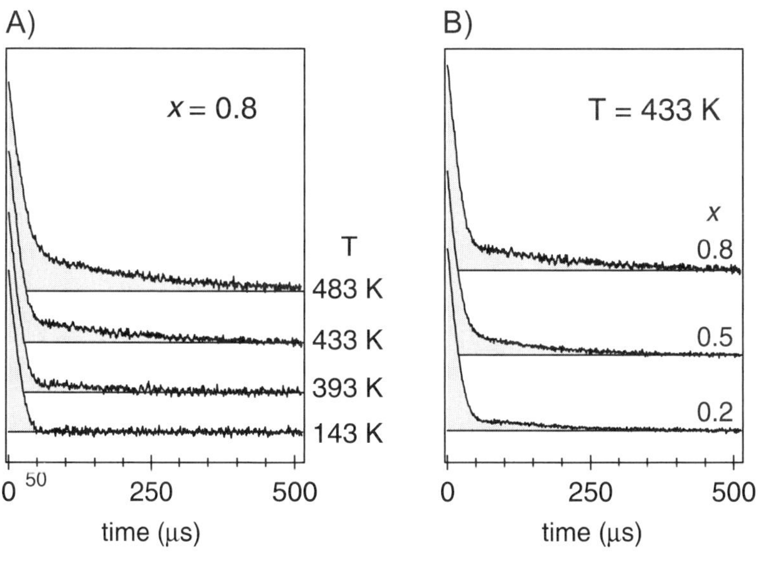

Fig. 9 ^{7}Li NMR-free inductions decays of nanocrystalline $Li_2O : Al_2O_3$ composites recorded at 58.8 MHz for different temperatures T (**a**, $x = 0.8$) and insulators contents x (**b**, $T = 433$ K) (from [45])

both Li species and, thus, their dynamics, can be studied independently. Plotting the area fraction of the slowly and fast decaying part of the FID as a function of waiting time in the frame of a saturation recovery experiment used to measure longitudinal relaxation, two individual rates $T_{1\,\mathrm{slow}}^{-1}$ and $T_{1\,\mathrm{fast}}^{-1}$ are obtained [44, 45]. The corresponding transients $M_{\mathrm{slow}}(t)$ and $M_{\mathrm{fast}}(t)$ can be well described with a single-exponential function each. In Fig. 10 the rates $T_{1\,\mathrm{slow}}^{-1}$ and $T_{1\,\mathrm{fast}}^{-1}$ of ^{7}Li in $(1-x)\mathrm{Li_2O} : x\mathrm{Al_2O_3}$ with $x = 0.5$ (full symbols) are shown in an Arrhenius plot [45]. They differ by one order of magnitude in the temperature range measured here. The results represent the low-temperature flanks of the diffusion-induced $T_{1\,\mathrm{slow}}^{-1}(T)$ and $T_{1\,\mathrm{fast}}^{-1}(T)$ peaks, respectively, for which in each case $T_1^{-1} \propto \tau^{-1} \propto \exp(-E_A/k_B T)$ is valid [48]. Surprisingly, the corresponding activation energies of both spin ensembles are similar ($E_{A,\mathrm{slow}} \approx E_{A,\mathrm{fast}}$) and amount to about 0.32 eV. For samples with $x = 0.2$ and $x = 0.8$ nearly the same results were obtained ($E_A \neq f(x)$). Furthermore, E_A seems to be not only independent of x but also of the kind of insulator added because very similar results were obtained for composites with boron oxide, $\mathrm{B_2O_3}$, as insulator component [44]. For comparison, the results of that nanocomposite are shown in Fig. 10, too.

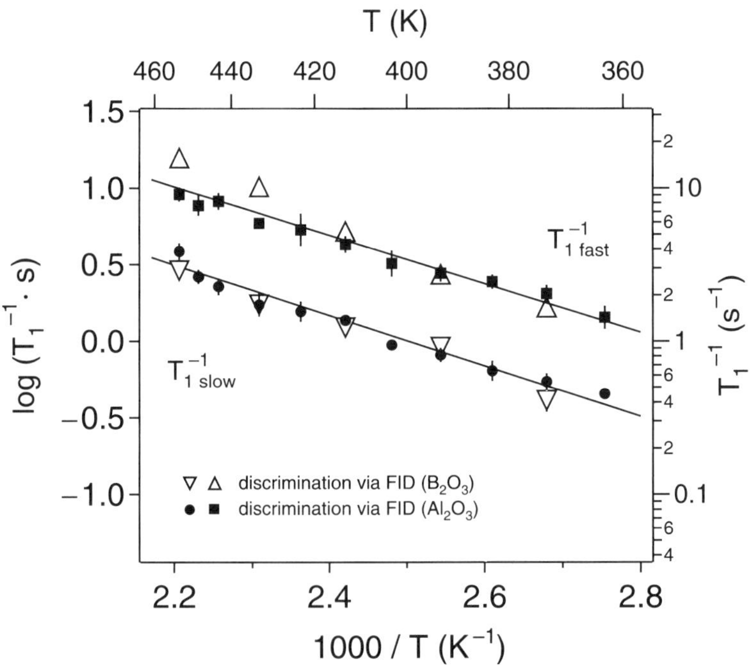

Fig. 10 ^{7}Li NMR spin–lattice relaxation rates of the nanocrystalline composites $(1-x)\mathrm{Li_2O} : x\mathrm{Al_2O_3}$ (*full symbols*) and $(1-x)\mathrm{Li_2O} : x\mathrm{B_2O_3}$ (*open symbols*), respectively ($x = 0.5$). The two different Li species were separated from each other via their individual T_2^{-1} rates (from [45])

4.2 Heterogeneous Spin-Lattice Relaxation

Another way of discriminating the two Li species in the nanocrystalline $(1-x)Li_2O : xAl_2O_3$ and $(1-x)Li_2O : xB_2O_3$ composites is the measurement of the *overall* magnetization transients $M(t)$ that are obtained after integrating the total area under the FID as a function of delay time t (see Sect. 2.2). Magnetization transients $M(t)$ of the nanocrystalline composite $(1-x)Li_2O : xAl_2O_3$ with $x = 0.8$ are shown in Fig. 11 for several temperatures, see [45]. The transients $M(t)$ of the analogous microcrystalline composites follow a single-exponential function with the rate $T_{1\,\mathrm{micro}}^{-1}$, as expected when quadrupole interactions are absent, whereas those of the nanocrystalline composites can be described only with a combination of two exponentials (cf. solid lines in Fig. 11).

$$\begin{aligned} M(t) &= M_{\mathrm{slow}}(t) + M_{\mathrm{fast}}(t) \\ &= M_{0\,\mathrm{slow}}(t)\left[1-\exp\left(-\frac{t}{T_{1\,\mathrm{slow}}}\right)\right] + M_{0\,\mathrm{fast}}(t)\left[1-\exp\left(-\frac{t}{T_{1\,\mathrm{fast}}}\right)\right] \end{aligned} \tag{5}$$

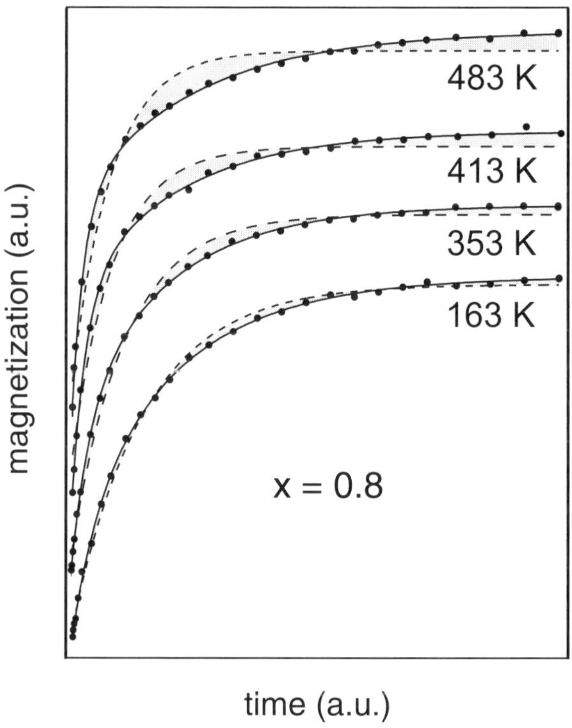

Fig. 11 ^{7}Li spin–lattice relaxation NMR transients $M(t)$ of the nanocrystalline composite $(1-x)Li_2O : xAl_2O_3$ with $x = 0.8$ recorded at 58.8 MHz. With increasing temperature T, the deviation from single-exponential behavior (*dashed lines*) increases. *Solid lines* represent fits according to a sum of two exponential functions with the decay constants $T_{1\,\mathrm{slow}}^{-1}$ and $T_{1\,\mathrm{fast}}^{-1}$, respectively (from [45])

The deviation from single-exponential time behavior increases with increasing temperature as shown exemplarily for the sample with $x = 0.8$ in Fig. 11. This behavior correlates with the result obtained from the analysis of the static ^{7}Li NMR spectra (see Fig. 7). The number fraction of fast Li ions represented by the narrow NMR component also increases with T. Furthermore, this fraction depends on x. Consistently, with increasing x the deviations from single-exponential time behavior of $M(t)$ increase with the amount of insulator added. The latter observation is highlighted in Fig. 12 where magnetization transients recorded at the same temperature but for different compositions ($x = 0.2$, 0.5, and 0.8) are shown. On the right-hand side of Fig. 12, the logarithm of $(M_0 - M(t))/M_0$ is plotted vs. t, so that single-exponential transients will give a straight line in this representation. This is clearly fulfilled for the microcrystalline composites where the negligible volume fraction of interfacial regions does not influence the overall Li dynamics. In this representation, the deviations from single-exponential time behavior of the nanocrystalline composites can be easily visualized. At short delay times, the fast relaxation process increasingly becomes important at higher insulator contents. Expectedly, the slopes of the long-time tail of the transients, i. e., above $t = 1$ s at 433 K, are similar for *all* composites. Thus, this part of the nanocrystalline composites can be ascribed to the slow Li ions in the crystalline Li_2O grains ($T_{1\,\text{micro}}^{-1} = T_{1\,\text{slow}}^{-1}$). Between 360 K and 460 K the rates $T_{1\,\text{slow}}^{-1}$ and $T_{1\,\text{fast}}^{-1}$ obtained from the overall

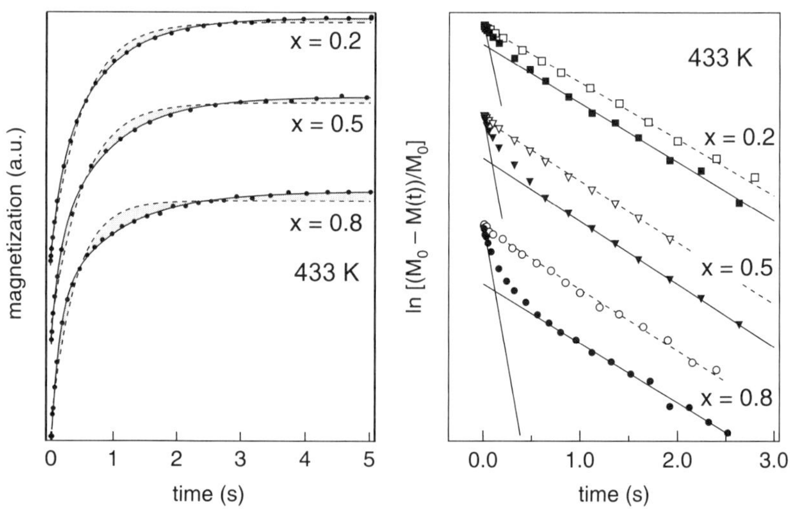

Fig. 12 *Left*: ^{7}Li NMR spin–lattice relaxation transients $M(t)$ recorded at 58.8 MHz of a nanocrystalline composite $(1 - x)$ Li_2O : $x Al_2O_3$ with $x = 0.2$, 0.5, and 0.8. The larger the insulator content x the more obvious becomes the deviation from single-exponential time behavior (*dashed lines*). *Solid lines* represent fits according to (5). *Right*: Linearization of the transients shown on the left, i. e., logarithm of $(M_0 - M(t))/M_0$ vs. delay time t (*filled symbols*). *Open symbols* show transients of the corresponding microcrystalline composites giving a straight line through the origin and representing single-exponential time behavior (from [45])

transients $M(t)$ differ by about one order of magnitude. In Fig. 13, the rates $T_{1\,\mathrm{slow}}^{-1}$ and $T_{1\,\mathrm{fast}}^{-1}$ for the nanocrystalline composite with $x = 0.5$ obtained via the two different methods for discrimination (see also Sect. 4.1) are compared with those from the microcrystalline ones of the same composition. Both methods give nearly the same rates $T_{1\,\mathrm{fast}}^{-1}$ of the ions in the interfacial regions. Comparing the $T_{1\,\mathrm{slow}}^{-1}$ values, the differences are somewhat larger. As mentioned above, the rates $T_{1\,\mathrm{slow}}^{-1}$ and $T_{1\,\mathrm{micro}}^{-1}$ obtained by analyzing the $M(t)$ transients are very similar.

Similar to the situation when discriminating the two Li species via their different spin-spin relaxation rates, the temperature dependence of the spin-lattice relaxation rates shown in Fig. 13 reveal the same activation energy for all composites. E_A amounts to 0.32(2) eV. Almost the same results were obtained for the analogous composite system with boron oxide used as insulator [44]. Consequently, the enhanced diffusivity is due to an increase of the preexponential factor in the corresponding Arrhenius relation. In view of the unaltered activation energy, one may suppose that the structure in the vicinity of the hopping ions in the grains and in the interfacial regions of high-energy ball-milled Li_2O is very similar so that the

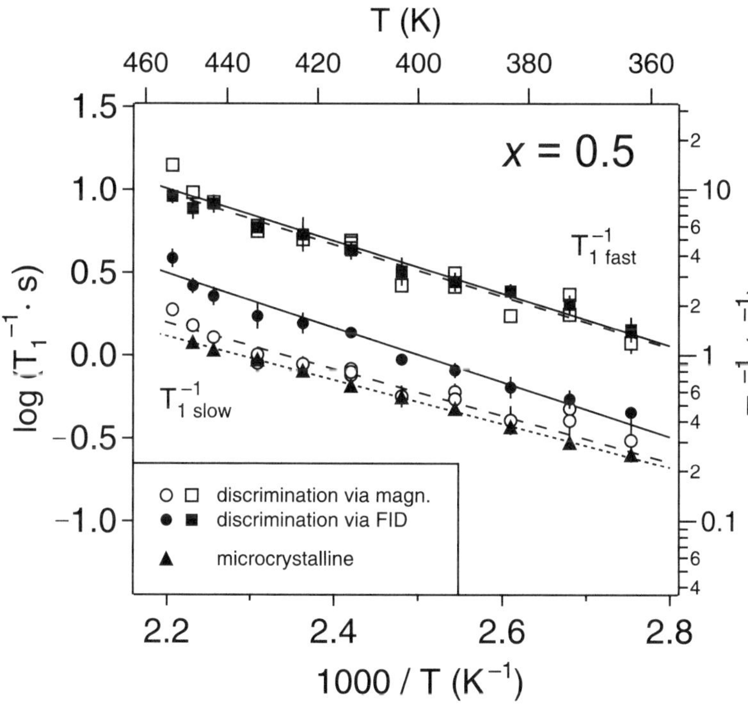

Fig. 13 Separated ^{7}Li NMR spin–lattice relaxation rates of the nanocrystalline $(1-x)\,Li_2O : xAl_2O_3$ composite with $x = 0.5$. The rates were obtained either by the discrimination method via the double-exponential magnetization transients (see (5)) or via the two-component shape of the FIDs (see Sect. 4.1). For comparison, the relaxation rates of the microcrystalline counterpart of the same composition are also shown (from [45])

hopping ions have to overcome similar barriers. By contrast, the interfacial regions in nanocrystalline $LiNbO_3$ prepared by ball-milling seem to have amorphous structure and show a much smaller activation barrier (about 0.3 eV) than in the corresponding coarse-grained material (0.9 eV). Nevertheless, the diffusion pathways in Li_2O near the interfacial regions are much faster than those in the interior of the grains. This is consistent with the fact that both rates $T_{1\,\mathrm{slow}}^{-1}$ and $T_{1\,\mathrm{fast}}^{-1}$ are independent of x, whereas the number fraction A_f of highly mobile ions clearly increases with the amount of insulator x added. Increasing the number fraction of insulator–conductor contacts leads to a higher concentration of defects, which might be simply explained by adsorption of Li ions on the surface of the insulator grains or by migration processes of Li ions into the insulator phase. Consequently, similar structures of the diffusion pathways near the interfacial regions as in the grains in nanocrystalline Li_2O but higher defect concentrations near interfaces between insulating and conducting phases would result in similar activation energies but an increased preexponential factor and, therefore, in enhanced diffusivity.

5 Summary

Materials with grains in the nanometer regime can exhibit considerably different properties from those with particle diameters in the μm range. The large volume fraction of interfacial regions in nanocrystalline materials may influence the transport properties in such a way that the overall diffusivity is significantly enhanced. Examples for single-phase Li and F conductors are $LiNbO_3$ and CaF_2, respectively. An additional degree of freedom in the design of fast ion conductors is introduced when a nanocrystalline ion conductor is mixed with a nanocrystalline insulator forming a composite material like $(1 - x)Li_2O : xX_2O_3$ (X = B, Al). In these composite materials, a further increase of the diffusivity is found.

^{7}Li NMR line shape studies on both single- and two-phase materials revealed a heterogeneous structure consisting of crystalline grains and interfacial regions. This is in contrast to the microcrystalline counterparts appearing to be homogeneous when studied with NMR techniques due to the very small volume fraction of interfaces. In the case of nanocrystalline composites temperature dependent ^{7}Li spin-spin as well as spin-lattice relaxation NMR investigations revealed clearly the heterogeneous dynamics of slow and fast Li ions located in the nanosized grains and interfacial regions, respectively. In contrast to that, the diffusion behavior of the coarse-grained materials can be described with a single species of (slow) Li ions. Interestingly, the number fraction of fast Li ions of the nanocrystalline material is significantly enhanced by increasing the content x of nanosized insulator grains. That means that interfaces between different phases (Li_2O and X_2O_3) generate more fast Li ions than those between adjacent Li_2O grains. As the activation energy is not changed, the increased diffusivity can be ascribed to a change in the Arrhenius preexponential factor. In summary, the results for the composites studied up to now with different ion insulators are very similar and seem to reflect a generic behavior, at least for oxides.

References

1. H. Gleiter: Acta Mater. **48**, 1 (2000).
2. P. Knauth, H. L. Tuller: Solid State Ionics **136–137**, 1215 (2000).
3. A. V. Chadwick: Diffusion Fundamentals **2**, 44 (2005).
4. P. Heitjans, S. Indris: J. Phys. Condens. Matter **15**, R1257 (2003).
5. J. Karch, R. Birringer, H. Gleiter: Nature **330** 556 (1987).
6. S. Yip: Nature **391** 532 (1998).
7. J. Baller, J. K. Krüger, R. Birringer, C. Proust C: J. Phys. Condens. Matter **12** 5403 (2000).
8. U. Betz, H. Hahn: Nanostruct. Mater. **12** 911 (1999).
9. H. L. Tuller: J. Electroceram. **1**, 211 (1997).
10. A. M. George, J. Íñiguez, L. Bellaiche: Nature. **413**, 54 (2001).
11. C. Demetry, X. Shi: Solid State Ionics **118**, 271 (1998).
12. C.-W. Nan, S. Holten, R. Birringer, H. Gao, H. Kliem, H. Gleiter: Phys. Status Solidi **a 164**, R1 (1997).
13. H. Wittmer, S. Holten, H. Kliem, H. D. Breuer: Phys. Status Solidi **a 181**, 461 (2000).
14. H. Mamiya, I. Nakatani, T. Furubayashi: Phys. Rev. Lett. **80**, 177 (1998).
15. T. Jonsson, J. Mattsson, C. Djurberg, F. A. Khan, P. Nordblad, P. Svedlindh: Phys. Rev. Lett. **75**, 4138 (1995).
16. S. J. Stewart, R. A. Borzi, G. Punte, R. C. Mercader, F. Garcia: J. Phys. Condens. Matter **13**, 1743 (2001).
17. J. Garcia-Otero, M. Porto, J. Rivas, A. Bunde: Phys. Rev. Lett. **84**, 167 (2000).
18. J. P. Chen, C. M. Sorensen, K. J. Klabunde, G. C. Hadjipanayis: Phys. Rev. **B 51**, 11527 (1995).
19. D. Zhang, K. J. Klabunde, K. J. Sorensen, G. C. Hadjipanayis: Phys. Rev. **B 58**, 14167 (1998).
20. H. Weller, H. M. Schmidt, U. Koch, A. Fojtik, S. Baral, A. Henglein, A. Kunath, K. Weiss, E. Dieman: Chem. Phys. Lett. **124**, 557 (1986).
21. X. Peng, L. Manna, W. Yang, J. Wickham, E. Scher, A. Kadavanich, A. P. Alivisatos: Nature **404**, 59 (2000).
22. M. Shim, P. Guyot-Sionnest: Nature **407**, 981 (2000).
23. C.-W. Nan, R. Birringer, W. Krauss, H. Gao, H. Gleiter: Phys. Status Solidi A **162**, R3 (1997).
24. R. Schmechel, M. Kennedy, H. von Seggern, H. Winkler, M. Kolbe, R. A. Fischer, L. Xaomao, A. Benker, M. Winterer, H. Hahn: J. Appl. Phys. **89**, 1679 (2001).
25. A. L. Linsebigler, G. Lu, J. T. Yates Jr: Chem. Rev. **95**, 735 (1995).
26. J. Y. Ying, T. Sun: J. Electroceram. **1**, 219 (1997).
27. H. Gleiter: Prog. Mater. Sci. **33**, 223 (1989).
28. J. Weissmüller, H. Ehrhardt: Phys. Rev. Lett. **81**, 1114 (1998).
29. K. T. Jacob, K. P. Jayadevan, R. M. Mallya, Y. Waseda: Adv. Mater. **12**, 440 (2000).
30. D. Bork, P. Heitjans: J. Phys. Chem. B **102**, 7303 (1998).
31. D. Bork, P. Heitjans: J. Phys. Chem. B **105**, 9162 (2001).
32. S. Indris, P. Heitjans, H. E. Roman, A. Bunde, Phys. Rev. Lett. **84**, 2889 (2000).
33. W. Puin, P. Heitjans: Nanostruct. Mater. **6**, 885 (1995).
34. W. Puin, S. Rodewald, R. Ramlau, P. Heitjans, J. Maier: Solid State Ionics **131**, 159 (2000).
35. U. Brossmann, R. Würschum, U. Södervall, H.-E. Schaefer: Nanostruct. Mater. **12**, 871 (1999).
36. P. Knauth: J. Solid State Electrochem. **147**, 115 (2002).
37. Y.-M. Chiang, E. B. Lavik, I. Kosacki, H. L. Tuller, J. Y. Ying: J. Electroceram. **1**, 7 (1997).
38. A. Tschöpe, E. Sommer, R. Birringer: Solid State Ionics **139**, 255 (2001).
39. J. Lee, J. H. Hwang, J. J. Mashek, F. O. Mason, A. E. Miller, R. W. Siegel: J. Mater Res. **10**, 2295 (1995).
40. C.-W. Nan, A. Tschöpe, S. Holten, H. Kliem, R. Birringer: J. Appl. Phys. **85**, 7735 (1995).
41. G. Li, L. Li, S. Feng, M. Wang, L. Zhang, X. Yao: Adv. Mater. **11**, 146 (1999).
42. J. Maier: Prog. Solid State Chem. **23**, 171 (1995).
43. J. Maier: Nat. Mater. **4**, 805 (2005).
44. S. Indris, P. Heitjans: J. Non-Cryst. Solids **307**, 555 (2002).

45. M. Wilkening, S. Indris, P. Heitjans: Phys. Chem. Chem. Phys. **5**, 2225 (2003).
46. H. Maekawa, Y. Fujimaki, H. Shen, J. Kawamura, T. Yamamura: Solid State Ion. **177**, 2711 (2006).
47. C. C. Liang: J. Electrochem. Soc. **120**, 1289 (1973).
48. P. Heitjans, A. Schirmer, S. Indris: NMR and β-NMR Studies of Diffusion in Interface-dominated and Disordered Solids. In: *Diffusion in Condensed Matter – Methods, Materials, Models*, ed. by P. Heitjans, J. Kärger (Springer, Berlin Heidelberg, 2005) pp. 367–415.
49. P. Heitjans, S. Indris, M. Wilkening: Diffusion Fundamentals **2**, 45 (2005).
50. F. Qi, T. Jörg, R. Böhmer: Solid State Nucl. Magn. Reson. **22**, 484 (2002).
51. F. Qi, C. Rier, R. Böhmer, W. Franke, P. Heitjans: Phys. Rev. B **72**, 104301 (2005).
52. M. Wilkening, W. Küchler, P. Heitjans: Phys. Rev. Lett. **97**, 065901 (2006).
53. Z. Xu, J. F. Stebbins: Science **270**, 1332 (1995).
54. M. Wagemaker, A. P. M. Kentgens, F. M. Mulder: Nature **418**, 397 (2002).
55. M. Wilkening, R. Amade, W. Iwaniak, P. Heitjans: Phys. Chem. Chem. Phys. **9**, 1239 (2007).
56. M. Wilkening, C. Mühle, M. Jansen, P. Heitjans: J. Phys. Chem. B **111**, 8691 (2007).
57. M. Wilkening, D. Bork, S. Indris, P. Heitjans: Phys. Chem. Chem. Phys. **4**, 3246 (2002).
58. P. Heitjans, M. Masoud, A. Feldhoff, M. Wilkening: Faraday Discuss. **134**, 67 (2007).
59. R. W. Siegel: Nanophase Materials. In: *Encyclopedia of Applied Physics*, vol. 11, ed. by G. L. Trigg, E. H. Immergut, E. S. Vera, W. Greulich (VCH, New York, 1994) pp. 173–200.
60. W. Puin, P. Heitjans, W. Dickenscheid, H. Gleiter: NMR Study of Dynamic Processes in Nanocrystalline CaF_2. In: *Defects in Insulating Materials*, ed. by O. Kanert, J. Spaeth (World Scientific, Singapore, 1993) pp. 137–139.
61. P. Heitjans, S. Indris: J. Mater. Sci. **39**, 5091 (2004).
62. R. Winter, P. Heitjans: J. Phys. Chem. B **105**, 6108 (2001).
63. S. Indris, P. Heitjans, H. E. Roman, A. Bunde: Defect Diffus. Forum **194–199**, 935 (2001).
64. J.-M. Debierre, P. Knauth, G. Albinet: Appl. Phys. Lett. **71**, 1335 (1997).
65. P. Knauth, G. Albinet, J.-M. Debierre: Ber. Bunsenges. Phys. Chem. **102**, 945 (1998).
66. P. Knauth: J. Electroceram. **5**, 111 (2000).
67. N. Bloembergen, E. M. Purcell, R. V. Pound: Phys. Rev. **73**, 679 (1948).
68. A. Bunde, W. Dieterich, P. Maass, M. Meyer: Ionic Transport in Disordered Materials. In: *Diffusion in Condensed Matter – Methods, Materials, Models*, ed. by P. Heitjans, J. Kärger (Springer, Berlin Heidelberg, 2005) pp. 367–415.
69. Z. H. Xie, M. E. Smith, J. H. Strange, C. Jaeger: J. Phys. Condens. Matter **7**, 2479 (1995).
70. M. Masoud, P. Heitjans: Defect Diffus. Forum **237–240**, 1016 (2005).
71. A. V. Chadwick, M. J. Pooley, S. L. P. Savin: Physica status solidi C **2**, 302 (2005).
72. J. Maier: Solid State Ionics **23**, 59 (1987).
73. S. Indris, D. Bork, P. Heitjans: J. Mater. Synth. Process. **8**, 245 (2000).

Mössbauer Spectroscopy and New Composite Electrodes for Li-ion batteries

Pierre-Emmanuel Lippens and Jean-Claude Jumas

1 Introduction

Lithium-ion batteries have become one of the most promising power sources for portable equipment because of their high specific energy and working voltage. Many studies have been devoted to negative electrodes, which are actually carbon-based materials, in order to improve their electrochemical performances, especially the energetic capacities and the safety. For example, the theoretical specific and volumetric capacities of graphite are 372 Ah kg^{-1} and about 800 Ah L^{-1}, respectively [1]. Different families of new compounds have been proposed, especially tin- or silicon-based materials that form alloys with lithium and have specific and volumetric theoretical capacities higher than about 1,000 Ah kg^{-1} and 7,000 Ah L^{-1}, respectively. These materials differ by the chemical nature of the elements and Li insertion mechanisms [2–10]. The study of such complex mechanisms requires different experimental tools, especially at the atomic scale, since commonly used techniques such as X-ray diffraction (XRD) fail to characterize small particles or amorphous phases.

Mössbauer spectroscopy allows characterizing materials at the atomic scale and is a very efficient tool to study charge/discharge process of electrode materials, in situ by using specific test cells or ex situ by extracting the material from the cell at a given point of the charge/discharge curve. It allows to study the bulk in transmission geometry (TMS) as well as the surface or interface in scattering geometry by Conversion Electron Mössbauer Spectroscopy (CEMS). Each absorber atom contributes to the absorption spectrum independently from all other atoms. There is therefore no restriction concerning a domain size like in XRD. The technique allows therefore the study of nanostructured materials or nanocomposites. Mössbauer spectroscopy is further very useful because it can give information not only on static properties (such as crystal structure, magnetic properties, valence state, bonding) but also on dynamic properties.

AIME, Institut Charles Gerhardt, UMR 5253 CNRS, Université Montpellier II, CC15, Place Eugène Bataillon, F-34095 Montpellier Cedex 5, France

P. Knauth and J. Schoonman (eds.), *Nanocomposites: Ionic Conducting Materials and Structural Spectroscopies.*

In this chapter, we first give a brief description of the Mössbauer spectroscopy, since this technique is well documented [11–15]. Then, we show different applications to new negative electrode materials for Li-ion battery. These materials are powders and they are mixed with binder and carbon black to form composite electrodes. In most cases, they are transformed into nanocomposites during the first discharge (lithiation). Most of the results concerns tin- and antimony-based materials, since these two elements can be studied with Mössbauer spectroscopy. We also show the use of ^{57}Fe local probe in titanates that do not contain a Mössbauer isotope.

2 The Mössbauer Spectroscopy

2.1 The Mössbauer Effect

The technique is also known as "recoil-free resonant absorption of gamma radiation." The source is a radioactive material containing the desired mother isotope. Its decay populates the excited nuclear state of the Mössbauer isotope, which, in turn, emits the gamma radiation used for spectroscopy by transition to its nuclear ground state. This radiation can then be absorbed by a nucleus of the same isotope in the absorber, which is usually the sample that must be characterized (Fig. 1).

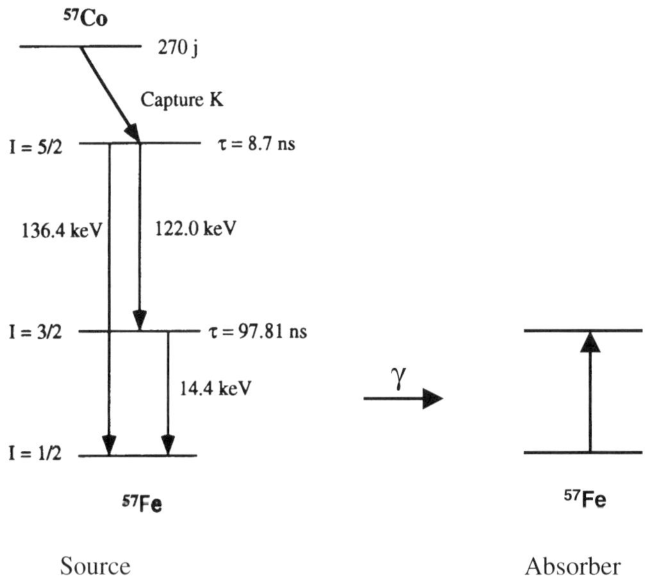

Fig. 1 Resonant absorption of gamma radiation by ^{57}Fe. The mother isotope is ^{57}Co, which transforms to ^{57}Fe by K capture. Transition to the nuclear ground state of ^{57}Fe produces the 14.4 keV gamma radiation used for Mössbauer spectroscopy

Typical gamma energies are in the range 5–200 keV. Emission and absorption lines are extremely narrow (10^{-5} to 10^{-17} eV).

The recoil energy transferred to the nucleus by the absorption or emission process (conservation of energy and momentum) lies in the range 10^{-1} to 10^{-4} eV. Emission lines are shifted to lower energies, absorption lines to higher energies by an amount equal to the recoil energy. As this energy is several orders of magnitude greater than the line width, absorption and emission lines are generally out of resonance. From this follows that only recoil-free absorption/emission processes (i. e., without excitation of phonons in the source or the absorber) contribute to the Mössbauer effect. In this case, the recoil energy is absorbed by the crystal lattice as a whole and not by a single atom. This implies that Mössbauer spectroscopy is limited to solid samples. Some isotopes with particularly high gamma energies (^{121}Sb: E = 37.15 keV, ^{125}Te: E = 35.46 keV) require that source and absorber be cooled to liquid helium temperature to increase the fraction of recoil-free absorption/emission processes.

The extremely narrow spectral lines cannot be resolved by gamma detectors. Instead, the energy of the gamma-rays from the source is slightly varied by the Doppler effect. The energy of gamma-rays emitted from a nucleus moving with a velocity v along the gamma-ray propagation direction is shifted by a first-order linear Doppler effect

$$E_D = v/c\ E_y. \tag{1}$$

2.2 The Hyperfine Interactions

Hyperfine interactions result from electric and magnetic interactions between the nucleus and electronic charges. The effect on the nuclear energy is small but can be easily detected by Mössbauer spectroscopy. One can distinguish three main interactions that influence the shape of the Mössbauer spectra. The first one is the Coulomb interaction between the nuclear and the electronic charge distributions over the finite size nucleus. This leads to a shift of the nuclear energy levels that depends on the nuclear spin. As a result the value of the nuclear transition energy, E_n (where n denotes the nuclear transition which is for example 1/2–3/2 for ^{119}Sn), depends on the electron distribution, $\rho(r)$, within the nucleus. The Mössbauer isomer shift, δ, is defined as the difference between the values of E_n for the absorbing material and for a material of reference. It can be written as

$$\delta = \alpha\left(\rho(0) - \rho_{\text{ref}}(0)\right) \tag{2}$$

where $\rho(0)$ and $\rho_{\text{ref}}(0)$ are the electron density at the Sn nucleus of absorbing and reference materials, respectively, and α is only dependent on the nucleus:

$$\alpha = \frac{Ze^2}{6\varepsilon_0}\Delta\left\langle r^2\right\rangle, \tag{3}$$

where $\Delta\langle r^2\rangle$ is the difference between the average squared nuclear radius of the excited and ground states, e is the electron charge, and ε_0 is the permittivity of vacuum. Thus, α is a constant for a given isotope and the value of δ is a measure of the electronic density at the nucleus of the Mössbauer isotope relative to a reference material. The value of $\rho(0)$ can be evaluated from first principles calculations [16–18] or can be related to the valence electron populations from some simple models [19–20]. In the latter case, variations of $\rho(0)$ are found to be very sensitive to changes in s-valence electrons but the p- and d-valence electrons have a significant indirect effect. For example in the case of ^{119}Sn, a tight-binding model [19] gives

$$\rho(0) = 6.6 + 39.6\, N_{5s} - 3.7\, N_{5p}, \tag{4}$$

where N_{5s} and N_{5p} are the Sn 5s and 5p electron numbers as defined within the tight-binding approach. In the case of ^{119}Sn, $\alpha > 0$, which means that δ varies as $\rho(0)$. The expressions (2) and (4) shows that δ strongly depends on N_{5s} and have very different values for Sn(IV) ($N_{5s} = 0$), Sn(0) ($N_{5s} = 1$) and Sn(II) ($N_{5s} = 2$). In most cases, the Sn 5p electrons only provide small changes in the values of δ compared to those due to Sn 5s electrons.

The second interaction results on the nonspherical nuclear charge when the nuclear angular moment quantum number is greater than 1/2. This is the quadrupole interaction between the nuclear quadrupole moment and the electric fields gradients due to asymmetric electron distribution around the nucleus. In the simple case of 1/2–3/2 nuclear spin transition, this leads to the splitting of the 3/2 nuclear levels and a simple expression of the energy difference between these two levels, called quadrupole splitting, is given by

$$\Delta = \frac{1}{2} eQV_{zz} \left(1 + \frac{\eta^2}{3}\right)^{1/2}, \tag{5}$$

where Q is the nuclear quadrupole moment of the excited state, V_{zz} is the main component of the diagonalized tensor of the electric field gradients, and η is the asymmetry parameter defined by

$$\eta = \left|\frac{V_{xx} - V_{yy}}{V_{zz}}\right| \text{ with } |V_{zz}| > |V_{yy}| > |V_{xx}|$$

Thus, the quadrupole splitting is directly related to the electric field gradients at the nucleus and provide information on the asymmetry of the charge distribution. The electric field gradients can be evaluated from first principles calculations [16–19] or from point charge models [18]. The effects of the two electric interactions on the nuclear levels are summarized in Fig. 2.

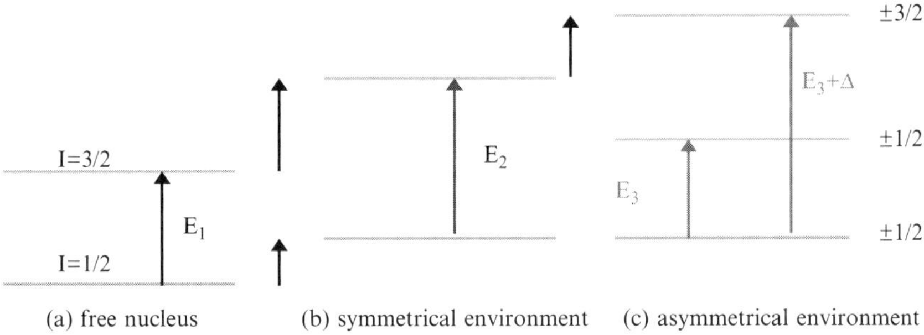

Fig. 2 Nuclear transitions in ^{119}Sn for absorption. (**a**) free nucleus, (**b**) nucleus surrounded by a symmetrical electron distribution (isomer shift: $\delta = E_2 - E_2$(reference material)), and (**c**) nucleus surrounded by an asymmetrical electron distribution (quadrupole splitting: Δ)

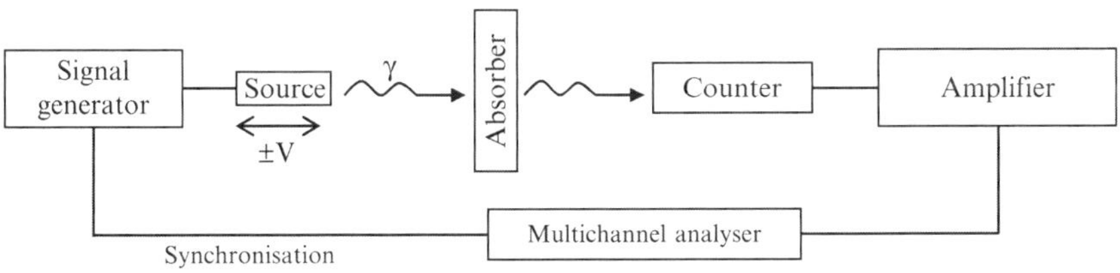

Fig. 3 Experimental setup for transmission Mössbauer spectroscopy

The effect of magnetic interactions is more complex and is not discussed here. It completely raises the degeneracy of nuclear states, which are split into $2I + 1$ substates. In the case of ^{57}Fe, this gives 2 states for $I = 1/2$ and 4 states for $I = 3/2$. Thus, there are 6 possible nuclear transitions by taking into account the selection rule.

2.3 The Experimental Setup

The typical experimental setup for Mössbauer spectroscopy in transmission geometry is shown in Fig. 3. The source is mounted on a loudspeaker-like electromechanical system that is excited by a signal generator to a periodic movement in a constant acceleration mode. The gamma-rays are counted by a gamma detector. The result is stored in a multichannel analyser synchronised by the signal generator.

The spectrum is typically presented as the relative transmission (in %) vs. the velocity (in mm s^{-1}), which can easily be converted to energy by using (1).

In contrast to the conventional transmission geometry described above, which probes the bulk of a material, Mössbauer spectroscopy in scattering geometry is surface sensitive. The deexcitation from the excited nuclear state in the absorber (after resonant absorption of a photon from the source) takes place via one of the following

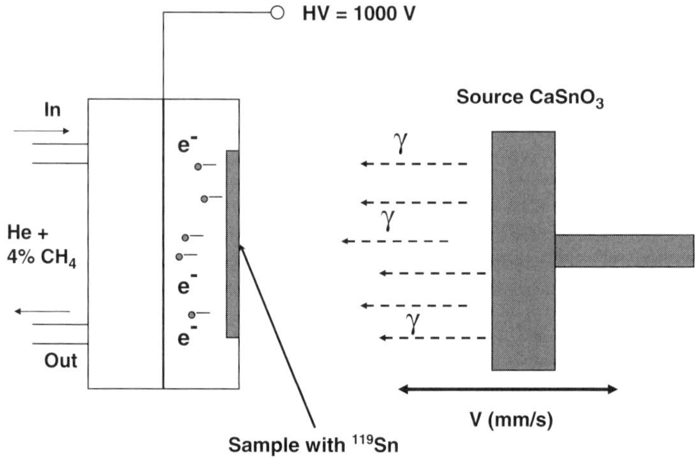

Fig. 4 Experimental setup for Conversion Electron Mössbauer Spectroscopy

three channels: (a) re-emission of a gamma photon, (b) absorption of this photon by an electron from an inner shell, which is then ejected, (c) filling of the core hole remaining after (b) by electron transitions from higher levels with emission of characteristic X-ray lines. With a specific electron counter (Fig. 4) for detection of the conversion electrons from process (b), the probing depth corresponds to the mean free path of the electrons in the material, i. e., typically several tens of nm.

2.4 Preparation of the Samples

^{57}Fe, ^{119}Sn, and ^{121}Sb Mössbauer spectra were recorded by transmission in the constant acceleration mode using an EG&G spectrometer, equipped with a cryostat to work at low temperature (down to 4 K). The absorbers were prepared by mixing powder samples and Apiezon grease inside the glove box, and sealed with parafilm to avoid contact with air. The sources were ^{57}Co in Rh matrix, ^{119m}Sn in $BaSnO_3$, and ^{121m}Sn in $BaSnO_3$, respectively. For ^{121}Sb, both source and absorber were simultaneously cooled down to 4 K in order to increase the fraction of recoil-free absorption and emission processes. The ^{57}Fe and ^{119}Sn spectra were fitted to Lorentzian profiles by least-squares method in order to determine the experimental values of the isomer shift δ and the quadrupole splitting Δ. The ^{121}Sb Mössbauer parameters δ, the quadrupole coupling eQV_{zz} (Q is the nuclear quadrupole moment and V_{zz} is the electric field gradient), and the asymmetry parameter η were obtained from the experimental data by using the fitting program of Rubenbauer and Birchall [21]. Isomer shift values are given relative to αFe for ^{57}Fe, $BaSnO_3$ for ^{119}Sn and InSb for ^{121}Sb. Measurements were performed in situ and ex situ at several depths of discharge and charge. In the latter case, the cells were opened inside the glove box and the electrode materials containing the active material were placed on specific sample holder transparent for the γ rays.

3 Characterization of Negative Electrodes

3.1 *βSn-Based Electrodes*

The binary phase diagram of Li-Sn shows the existence of seven Li_xSn compounds: Li_2Sn_5, LiSn, Li_7Sn_3 Li_5Sn_2, $Li_{13}Sn_5$, Li_7Sn_2, and $Li_{22}Sn_5$, although the exact composition of the later compound is still discussed [22–24]. Thus, the electrochemical reactions of tin-based materials with lithium, at very low lithium rate, are expected to give these Li_xSn compounds. The experimental potential curves obtained during the first discharge (lithiation) and charge (delithiation) of βSn at low lithium rate of C/70(1Li per βSn/70h) show 3 well-defined plateaus at different voltages (Fig. 5) [25].

The plateaus are due to biphasic reactions that could be related to the successive formation of the Li_xSn compounds. The expected reactions can be written

$$2/5\ \mathrm{Li} + \beta\mathrm{Sn} \rightarrow \mathrm{Li}_{2/5}\mathrm{Sn} \tag{6}$$

$$3/5\ \mathrm{Li} + \mathrm{Li}_{2/5}\mathrm{Sn} \rightarrow \mathrm{LiSn} \tag{7}$$

$$4/3\ \mathrm{Li} + \mathrm{LiSn} \rightarrow \mathrm{Li}_{7/3}\mathrm{Sn} \tag{8}$$

$$1/6\ \mathrm{Li} + \mathrm{Li}_{7/3}\mathrm{Sn} \rightarrow \mathrm{Li}_{5/2}\mathrm{Sn} \tag{9}$$

$$1/10\ \mathrm{Li} + \mathrm{Li}_{5/2}\mathrm{Sn} \rightarrow \mathrm{Li}_{13/5}\mathrm{Sn} \tag{10}$$

$$9/10\ \mathrm{Li} + \mathrm{Li}_{13/5}\mathrm{Sn} \rightarrow \mathrm{Li}_{7/2}\mathrm{Sn} \tag{11}$$

$$9/10\ \mathrm{Li} + \mathrm{Li}_{7/2}\mathrm{Sn} \rightarrow \mathrm{Li}_{22/5}\mathrm{Sn}. \tag{12}$$

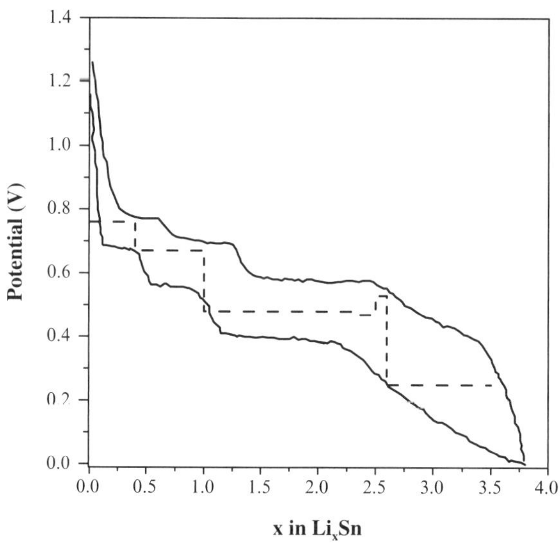

Fig. 5 Comparison between experimental [25] and theoretical potential for βSn

First principles calculations of the potential [25, 26] clearly show that the experimentally observed plateaus are due to the three first reactions (6)–(8). Quantitative differences are mainly due to the accuracy of the calculations and the experimental polarization effects. Insertion of more than about 2.3 Li leads to a continuous decrease of the potential up to about 3.8 Li. The calculated values of the potential for the reactions (9) and (10) are close to the value obtained for (8) but this concerns a small range for x (<0.3 Li) and this cannot be distinguished from experiments. Finally, the potential for (11) is close to the average potential obtained experimentally but the interpretation of this part is more complex as discussed previously [25]. The Li_xSn compounds formed during the electrochemical reactions are often difficult to characterize from conventional techniques such as XRD due to the small size of the particles and the poor crystallinity. In this case, the application of the Mössbauer spectroscopy is very helpful but requires accurate Mössbauer spectra as references.

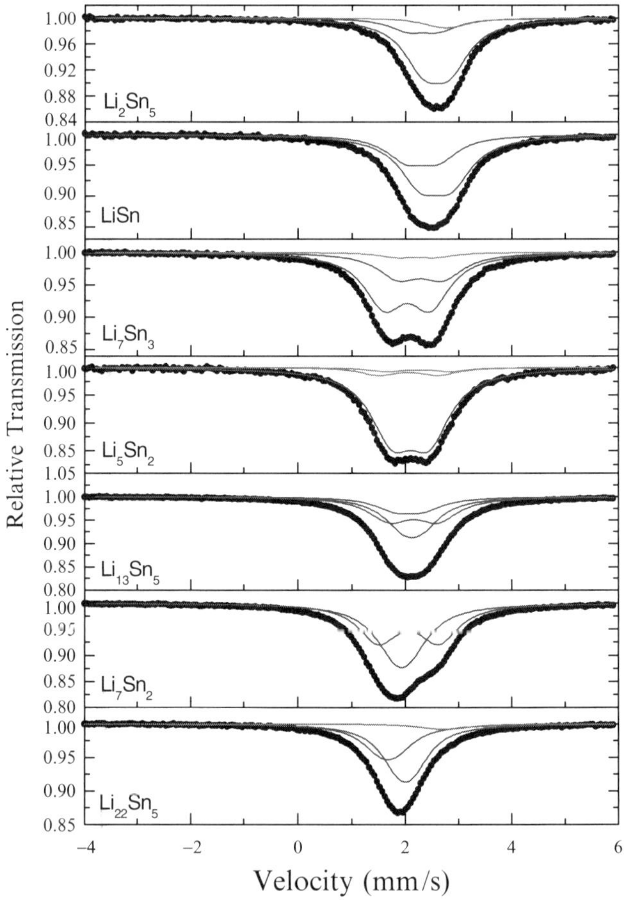

Fig. 6 Mössbauer spectra of the Li_xSn compounds

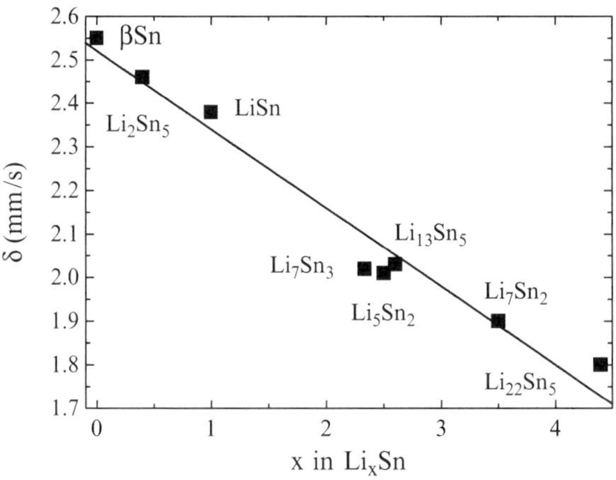

Fig. 7 Correlation between the averaged values of the isomer shift, δ, and *x* in Li_xSn

Li_xSn compounds were synthesized from both ceramic route [27] and mechanical milling [28]. In the two cases, the Mössbauer spectra showed the existence of impurities but it has been possible to obtain rather accurate values of δ and Δ for the different tin crystallographic sites in Li_xSn (Fig. 6).

The values of δ, averaged over the different tin sites, can be related to the composition of the Li_xSn compounds by a simple linear correlation curve as shown in Fig. 7. This is due to the linear dependency of δ with the Sn 5s and 5p electronic populations that change linearly with *x*. However, it is difficult to distinguish the three compounds Li_7Sn_3, Li_5Sn_2, and $Li_{13}Sn_5$ from the only average values of δ due to their close compositions and the values of Δ must be also taken into account [28]. This is confirmed in Fig. 6, where the shape of the spectra are very different for these three compounds.

Although βSn has a large theoretical capacity of about 1,000 Ah kg^{-1}, the particles cannot be used in electrode because of the poor cyclability. This is due to the large volume variations during the lithiation/delithiation processes that produce the pulverization of the electrodes and the loss of electrical contacts between the particles. Many experimental approaches have been proposed to overcome this problem and some of them are described below that include Mössbauer characterizations.

3.2 *Tin Oxides*

Tin oxide materials like SnO were proposed about a decade ago to solve the problems of volume variations [29, 30]. During the first discharge, the oxygen atoms react with lithium to form electrochemically inactive Li_2O particles that are expected to buffer the volume variations of Li_xSn. Thus, the beginning of the first discharge can be considered as an electrochemical synthesis of nanocomposites formed with βSn and Li_2O nanoparticles obtained from the reaction:

$$SnO + 2Li \rightarrow Sn + Li_2O. \tag{13}$$

The potential curve of SnO at C/20 rate shows a plateau at about 1 V, which is due to the transformation of the pristine material into Sn(0)-based material (Fig. 8). The Mössbauer spectra obtained for 1 and 2 Li at room temperature are rather complex but clearly show the increase of the βSn component and the decrease of the SnO component in agreement with the expected mechanism (Fig. 9). However, there is an additional component at about $\delta = 1$ mm s^{-1} that could be assigned to Sn atoms bonded to both Sn and O atoms in a region close to the interface between

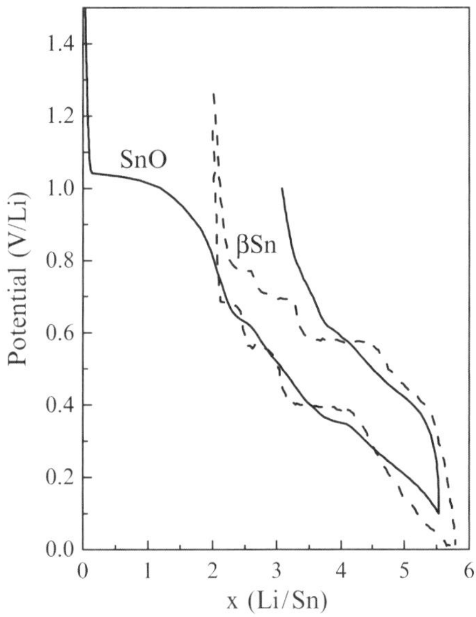

Fig. 8 Potential curves of SnO (*solid line*) and βSn (*dashed line*). In the latter case, the curve is shifted by 2 Li for comparison

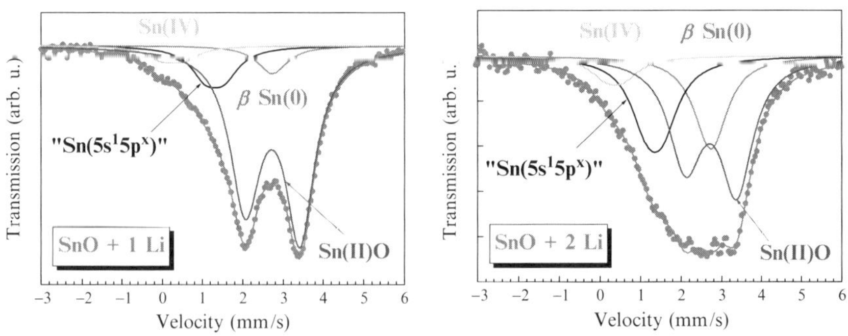

Fig. 9 Mössbauer spectra of SnO with 1 and 2 Li (*dots*) and the different Mössbauer components

βSn and Li_2O nanoparticles. The observed rather small contribution of βSn compared to those of the other two species, especially at 2 Li, is related to the small value of the Lamb-Mössbauer factor of βSn ($f \approx 0.04$) compared to those of the Sn(IV) and Sn(II) oxides, which should be in the range 0.1–0.4.

Comparison between the potential profiles of SnO and βSn, but shifted by 2 Li, shows similar variations although the plateaus are not well formed in the former case (Fig. 8). This could be due to kinetic effects. The room temperature ^{119}Sn Mössbauer spectra recorded at different steps of the discharge were qualitatively discussed in previously published papers but a quantitative analysis can be performed from comparison with theoretical Mösbbauer spectra obtained with first principles calculations of the hyperfine parameters. As an example, we consider the analysis of the spectrum obtained for the reaction of 3.5 Li into SnO. The electrochemical insertion of 3.5 Li is expected to form, at first βSn from the reaction (13), then LiSn from the reactions (6, 7) and finally, should transform LiSn into $Li_{7/3}Sn$ as suggested by the reaction:

$$LiSn + 1/2Li \rightarrow 3/8\ Li_{7/3}Sn + 5/8\ LiSn. \qquad (14)$$

The ^{119}Sn Mössbauer data form a complex spectrum that cannot be unambiguously interpreted by considering the usual fitting procedure of the experimental data (Fig. 10). The spectra of LiSn and $Li_{7/3}Sn$ were calculated from the LAPW method [17, 31, 32] by considering the experimental atomic positions of Sn and Li. The theoretical Mössbauer spectrum of SnO + 3.5 Li obtained from the summation of the calculated spectra according to the reaction (14) is favorably compared to the experimental data. This clearly indicates the existence of the two phases, LiSn and $Li_{7/3}Sn$, with the correct relative amounts and suggests that our analysis of the Mössbauer data is consistent with the existence of the stable phases.

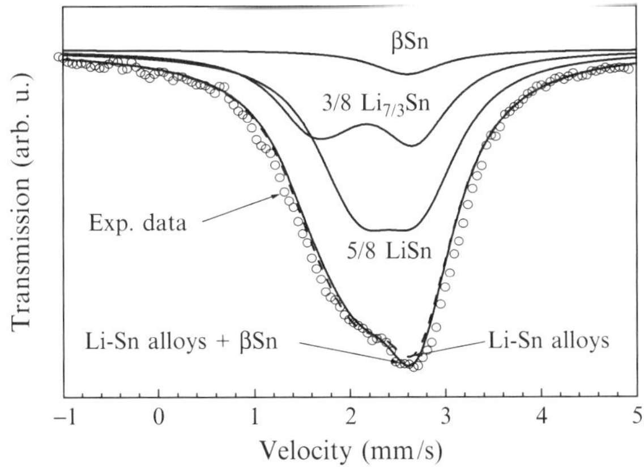

Fig. 10 Comparison between experimental (*circles*) and theoretical (*solid line*) ^{119}Sn Mössbauer spectra for SnO + 3.5 Li. The LiSn and Li_7Sn_3 theoretical subspectra are also shown

Although a better cyclability is obtained with SnO when compared with βSn, the electrochemical performances are not as good as carbon based materials. This is partially due to the existence of the irreversible insertion of Li at the beginning of the first discharge and to the coalescence of the βSn particles during the cycling.

SnO_2 was also considered as active material in negative electrodes. The potential curve is similar to that of SnO but with a larger irreversible part because of the transformation of Sn(IV)O_2 into βSn(0) that requires 4 Li. Similar mechanisms were also found for SnS and SnS_2 but with the formation of Li_2S instead of Li_2O [33].

3.3 *Tin Composite Oxides*

Tin composite oxides (TCO) were first proposed by Idota et al [34] and were widely studied by different groups [35–38]. In such composite material, the Sn atoms are dispersed into an oxide glass, typically a borophosphate amorphous material. The potential curve is similar to that of SnO except that the potential value of the irreversible part is higher. For the optimized composition $SnB_{0.6}P_{0.4}O_{2.9}$ the potential is about 1.6 V (Fig. 11).

Comparison between the Mössbauer spectra of TCO and SnO shows that both δ and Δ increase from SnO ($\delta = 2.7$ mm s^{-1}, $\Delta = 1.3$ mm s^{-1}) to TCO ($\delta = 3.1$ mm s^{-1}, $\Delta = 2.1$ mm s^{-1}) (Fig. 12). This is due to changes in the Sn local environment from asymmetric four coordinated Sn in SnO to distorted sixfold coordinated Sn in TCO. In the two cases, the Sn oxidation state is 2 and there is an irreversible transformation of Sn(II) into Sn(0) for the first two lithium and then reversible Sn(0)/Li_xSn reactions. Although the Sn atoms are dispersed within the pristine material, there is still coalescence of the Sn(0)-based particles and rather poor cycling performances.

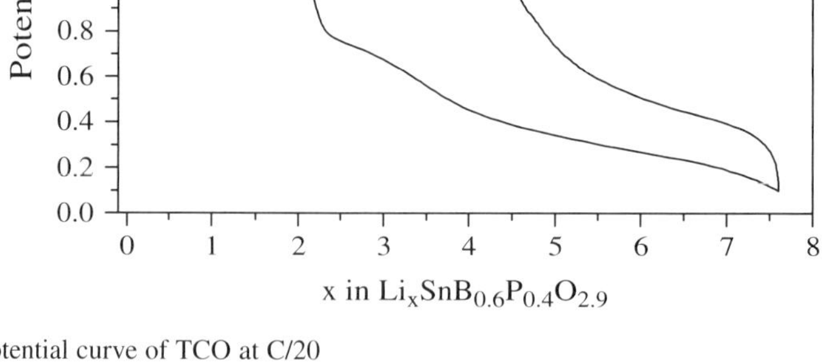

Fig. 11 Potential curve of TCO at C/20

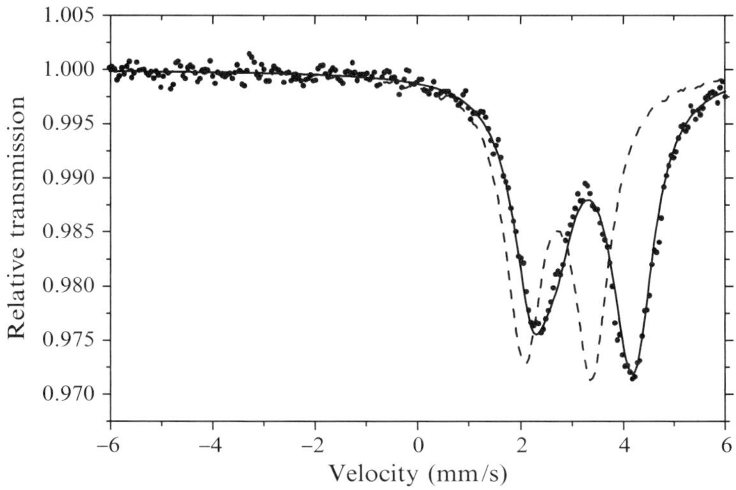

Fig. 12 Mössbauer spectra of SnO (*dashed line*) and TCO (*solid line*)

3.4 Tin Composites

A more recent approach to avoid the coalescence of tin particles concerns the synthesis of tin composite materials composed by βSn and crystalline oxides (BPO_4 or $CaSiO_3$) particles obtained from ceramic route (microsized) or sol–gel method (nanosized) [39–41]. During the synthesis there is a reaction between tin and oxide particles and the formation of an amorphous interface between them that improves the cohesion of the composite. The conversion electrons Mössbauer spectroscopy (CEMS), which is sensitive to tin atoms within a surface layer of several ten nanometers, has been used to characterize the interface between the particles. The Mössbauer spectrum obtained in transmission mode for Sn:BPO_4 shows similar contributions for both βSn and Sn(II), but due to the differences between the Lamb Mössbauer factors of these two species, more of 90% of the composite is formed by βSn (Fig. 13). The CEMS mainly shows an increase of the contribution of Sn(II), which corresponds to about 30% of the Sn sites at the particle surfaces (Fig. 13). This clearly indicates that Sn(II) are located at the interface between the βSn and BPO_4 particles. The Mössbauer parameters of the Sn(II) atoms at the interface are close to those found in TCO, which indicates the formation of an amorphous interface due to the reaction of βSn with BPO_4.

During the first discharge, the potential curve (Fig. 14) shows a small plateau at about 1.5 V and a strong decrease for the first 0.5 Li that corresponds to the reduction of Sn(II) into Sn(0) as confirmed by Mössbauer spectroscopy and then three plateaus that are similar to those observed with βSn. These plateaus are also observed during cycling, which clearly indicates the good reversibility of the βSn–Li_xSn (with $x \approx 3.5$) transformation and a small irreversible reaction compared to TCO. This confirms that such composite materials avoid the aggregation of tin particles.

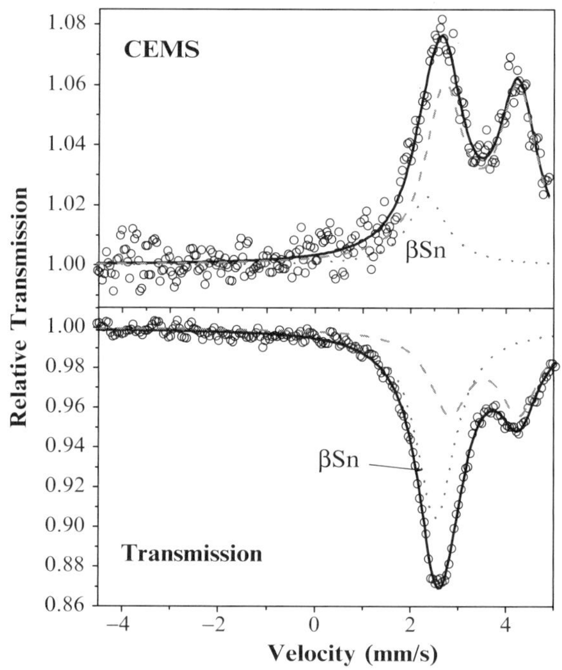

Fig. 13 Mössbauer spectra obtained in transmission and emission modes for $Sn:BPO_4$

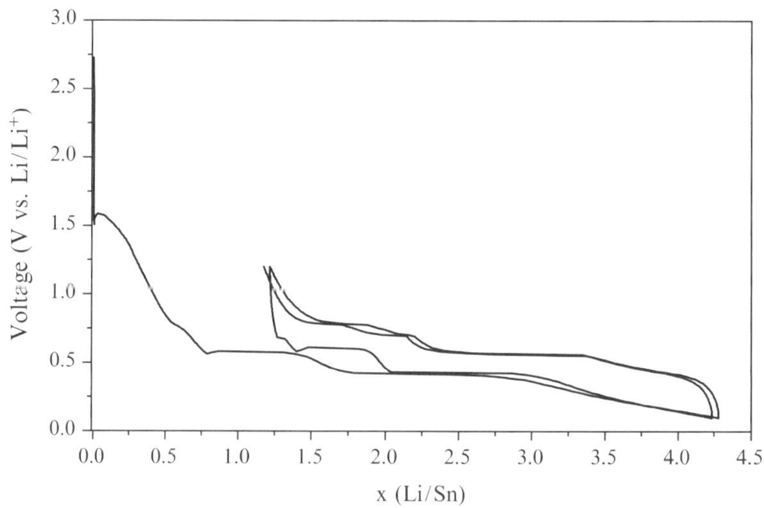

Fig. 14 Potential curve of $Sn:BPO_4$ at C/20

3.5 *Tin Based Intermetallic Compounds*

Tin intermetallic compounds [42–48] M_xSn, where M is a metal that does not alloy with Li, have been used to form nanocomposite electrodes during the first electrochemical discharge. Their functioning is based on displacement reactions in which M_xSn forms a lithium alloy Li_ySn and nanoparticles M following the reaction:

$$M_xSn + yLi \rightarrow Li_ySn + xM$$

The inactive matrix formed by these metallic nanoparticles may buffer the volume expansion of the active phase. There are many studies on the use of transition metals for M (Fe, Co, Ni, Cu) and the interest has recently risen due to the commercialization by Sony of the tin-based (Co-Sn-C) composite anode [49] with improved cyclability.

As a first example, we consider Ni_3Sn_4. During the first discharge, the potential curve shows a decrease of the potential in the range 0–4 Li and a plateau at very low potential (10 mV) in the range 4–12 Li (Fig. 15).

The Mössbauer spectrum at 4 Li is identical to that of Ni_3Sn_4, suggesting that Li does not react with the active material in the first range. The Mössbauer spectra obtained for different amounts of Li on the plateau show the existence of both Ni_3Sn_4 and Li_7Sn_2 (Fig.16) in agreement with the reaction:

$$Ni_3Sn_4 + 14Li \rightarrow 2Li_7Sn_2 + 3Ni. \quad (15)$$

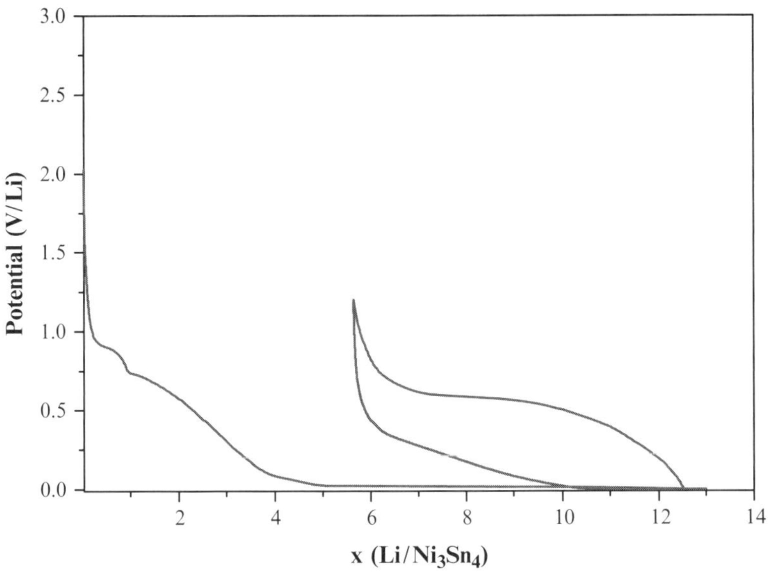

Fig. 15 Potential curve at C/20

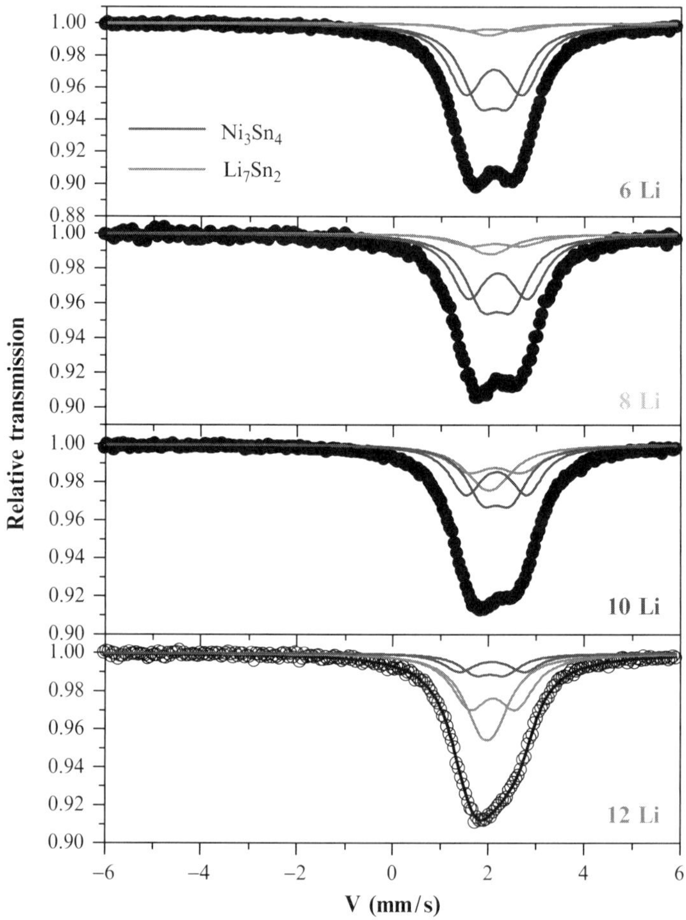

Fig. 16 Mössbauer spectra of Ni_3Sn_4 + xLi with x= 6, 8, 10, 12

Thus, at the end of the discharge the active material is a nanocoposite composed of Li_7Sn_2 and Ni nanoparticles. During the first charge, the delithiation leads to the formation of amorphous nanoparticles with composition close to Ni_3Sn_4 but with a small amount of lithium that could be within the nanoparticles.

A similar mechanism was also observed for $CoSn_2$ with the following reaction:

$$CoSn_2 + 7Li \rightarrow Li_7Sn_2 + Co. \tag{16}$$

In that case in situ Mössbauer spectroscopy was used during the first discharge (lithiation) and charge (delithiation) (Fig. 17). The relative amounts of $CoSn_2$ and Li_7Sn_2 have been estimated from the relative areas of the Mössbauer sub-spectra of these two compounds, including the effect of the Lamb-Mössbauer factor (Fig. 18) [50]. This clearly confirms the transformation of $CoSn_2$ into a nanocomposite formed by Co and Li_7Sn_2 nanoparticles.

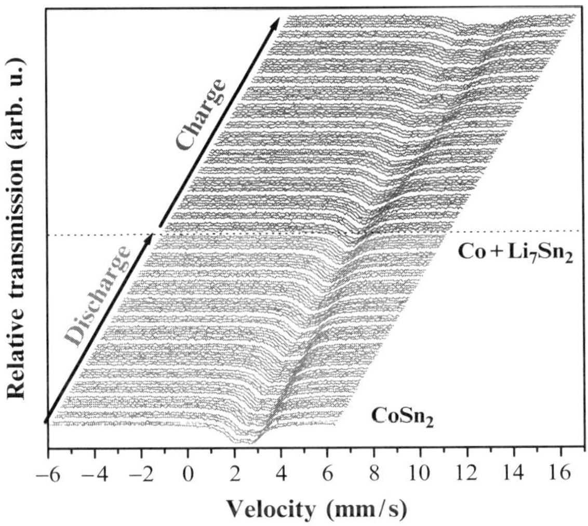

Fig. 17 In situ Mössbauer spectra of $CoSn_2$ + xLi

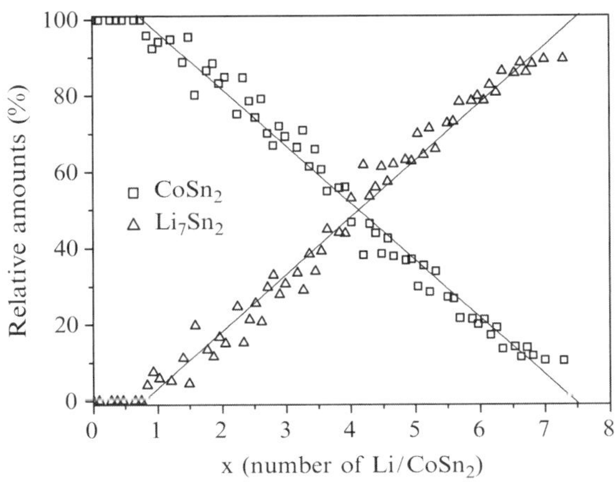

Fig. 18 Relative amounts of $CoSn_2$ and Li_7Sn_2 during the first discharge of $CoSn_2$ obtained from Mössbauer spectroscopy

3.6 *Antimony-Based Intermetallic Compounds*

The Li–Sb binary phase diagram shows the existence of Li_2Sb and Li_3Sb [51]. Experimental works based on XRD suggest that Li insertion in transition metal antimonides generally leads to the formation of small Li_3Sb particles at the end of the first discharge followed by reversible reactions involving alloying/dealloying mechanisms [52–58]. Because of the small size and partial amorphization of the particles, application of ^{121}Sb Mössbauer spectroscopy is of great interest. Unfortunately, the interpretation of the experimental data is rather difficult because of the manifold nuclear transitions (5/2 ↔ 7/2), the existence of different active Mössbauer sites, and the difficulty to obtain pure Li_2Sb and Li_3Sb crystalline phases as Mössbauer reference materials. Thus, we have combined experimental and theoretical evaluation of the isomer shift and quadruple coupling in order to improve the accuracy of the analysis. First, some reference materials were considered for the calibration of the isomer shift (Fig. 19) and quadruple coupling [18]. The observed linear correlation for the isomer shift as a function of the electronic density at the nucleus confirms the accuracy of the LAPW calculations and allows predicting the isomer shift from the calculated electron density at the nucleus when no experimental data are available. For example, we can predict the following theoretical values: $\delta(Li_2Sb) = 0.2\,mm\,s^{-1}$, δ(cubic Li_3Sb) = $1.1\,mm\,s^{-1}$ and δ(hexagonal Li_3Sb) = $1.5\,mm\,s^{-1}$ (relative to InSb). From the theoretical results, it can also be shown that variations of the isomer shift of Li_xSb are linearly correlated to x ($0 < x < 3$). Calculations of the atomic charges show that the observed decrease of the isomer shift from $1.1\,mm\,s^{-1}$ (Li_3Sb) to $-3\,mm\,s^{-1}$ (Sb) is mainly due to the increase in the Sb 5s electron population and to a less extent to the decrease in the Sb 5p electron population.

The electrochemical potential curve of $CoSb_3$ shows a first irreversible discharge at about 0.6 V followed by reversible reactions. At the end of the first discharge the Mössbauer spectra are found to be different depending on the lithium rate (Fig. 20). The spectrum obtained at C/70 (1 Li (per $CoSb_3$)/70 h) rate is formed by a sharp peak with the isomer shift $\delta = 1\,mm\,s^{-1}$ in good agreement with the calculated value for cubic Li_3Sb. This confirms the complete transformation of $CoSb_3$ into Li_3Sb and Co nanoparticles at low lithium rates. The Mössbauer spectrum at C/15 rate reflects the existence of at least two types of antimony sites, and among them one can be assigned to cubic Li_3Sb as expected from the basic mechanism. The other site does not seem to correspond to $CoSb_3$ as it should be for partial lithiation and is tentatively attributed to a more complex, maybe metastable, Li_xCoSb_y nanophase.

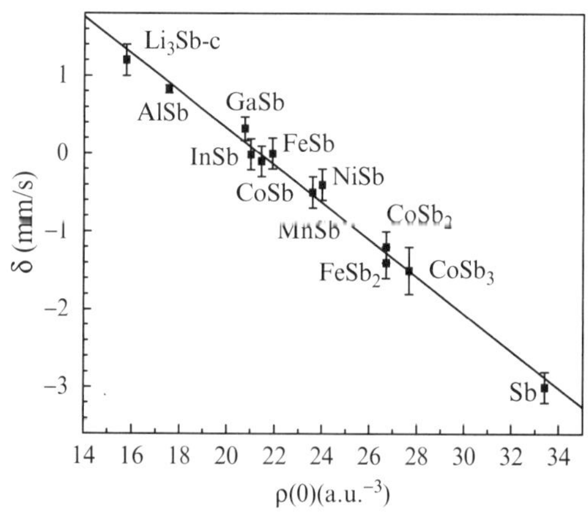

Fig. 19 Experimental values of the [121]Sb isomer shift (relative to InSb) as a function of the calculated density at the nucleus. The line is the linear interpolation to the experimental data

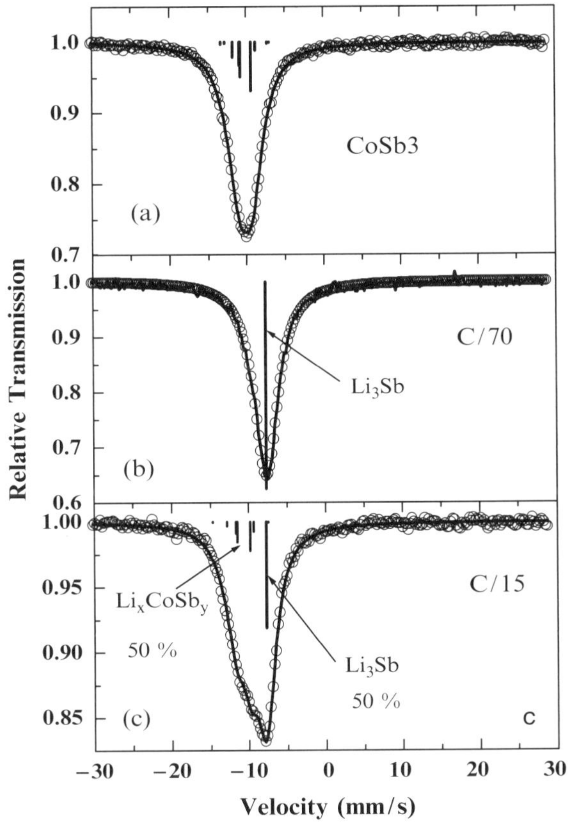

Fig. 20 [121]Sb Mössbauer spectra of $CoSb_3$ (**a**), at the end of the first discharge at C/70 rate (**b**), and at C/15 rate (**c**). The velocity scale is relative to $BaSnO_3$

The electrochemical curves for the metal antimonides MnSb, FeSb, CoSb, and NiSb are shown in Fig. 21. The value of the potential of the main plateau at the first discharge mainly depends on the nature of the M–Sb chemical bonds since all the compounds have the same NiAs structure. For MnSb and $MnSb_2$, there is a second plateau that reflects the existence of the intermediate phase LiMnSb [53, 57].

The Mössbauer spectra of NiSb (Fig. 22a) and CoSb (Fig. 23a) are formed by a broad and symmetrical peak. Thus, the determination of the sign of the quadrupole coupling from the only experimental data by considering the usual fitting procedure of Rubenbauer and Birchall [21] is not possible. However, a positive electric field gradient is obtained from LAPW calculations, which allows to unambiguously evaluate the Mössbauer parameters from the experimental data: $\delta = -0.1\,mm\,s^{-1}$ (CoSb), $-0.4\,mm\,s^{-1}$ (NiSb), $eQV_{zz} = -6.3\,mm\,s^{-1}$ (CoSb), $-5.8\,mm\,s^{-1}$ (NiSb).

At the end of the first discharge and with C/70 rate, the Mössbauer spectrum of NiSb is formed by a single peak at $1.1\,mm\,s^{-1}$ that can be assigned to cubic Li_3Sb and by a small asymmetry that is attributed to a small amount of unreacted NiSb (<10%). At very low Li rate of C/110, the Mössbauer spectrum of CoSb is

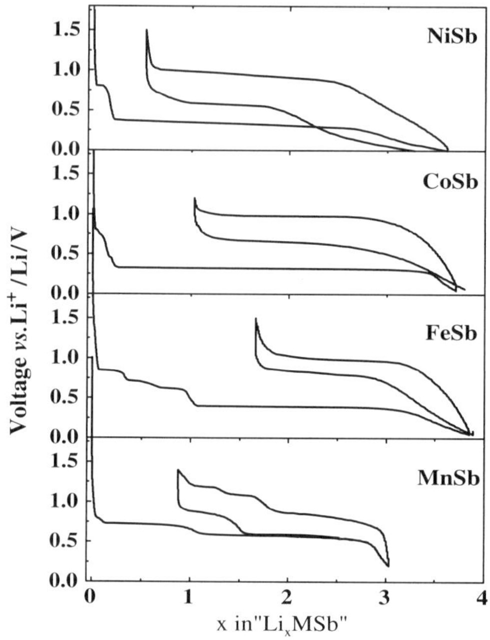

Fig. 21 Discharge/charge curves for MnSb, FeSb, CoSb, and NiSb

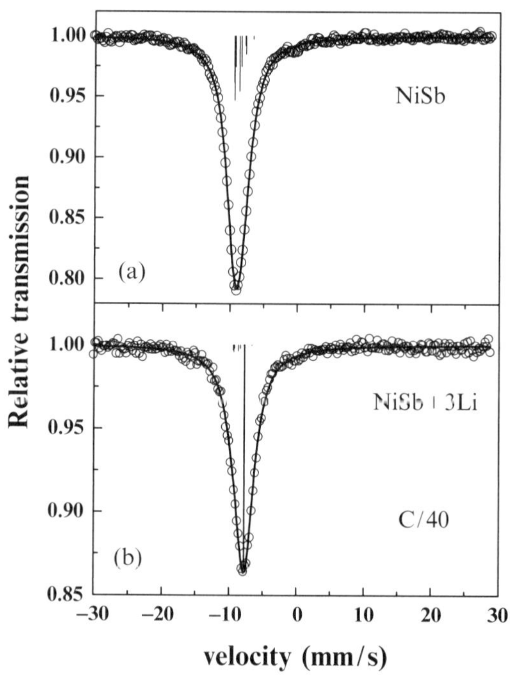

Fig. 22 ^{121}Sb Mössbauer spectra of NiSb (**a**) and at the end of the first discharge at C/40 (**b**). The velocity scale is relative to $BaSnO_3$

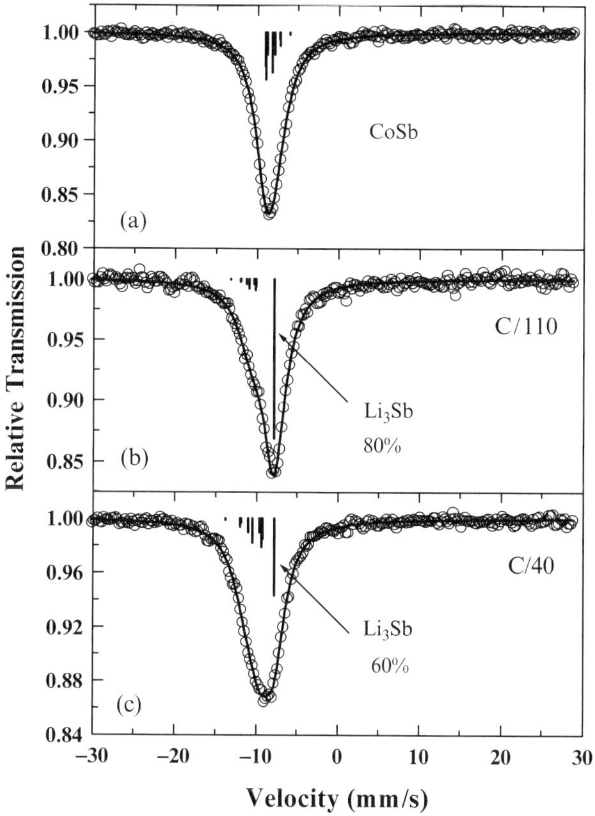

Fig 23 ^{121}Sb Mössbauer spectra of CoSb (**a**) and at the end of the first discharge at C/110 (**b**), and C/40 (**c**). The velocity scale is relative to $BaSnO_3$

asymmetrical, indicating the existence of about 20% CoSb at the end of the first discharge. The difference between the lithium reactions with NiSb and CoSb can be tentatively assigned to differences in Co–Sb and Ni–Sb bonds or to the different textures of the materials. It is worth noting that this effect is even more important at higher Li rate as shown for CoSb at C/40 (Fig. 23). In this case, the fitting of the experimental data by considering two Sb sites clearly shows the existence of Li_3Sb (60%) but the other site cannot be assigned to CoSb if we consider the experimental values of the Mössbauer parameters ($\delta = -1.5\,\text{mm s}^{-1}$, $eQV_{zz} = 9.5\,\text{mm s}^{-1}$). This could be due to the existence of metastable nanophases.

3.7 ^{57}Fe as Local Probe for Ti Oxides

Another example is a study of the Li intercalation in the spinel $Li_4Ti_5O_{12}$. Among the new anode materials, the spinel titanate $(Li)_3[LiTi_5]O_{12}$ allows to insert three lithium atoms per formula unit at a potential of 1.55 V leading to a theoretical

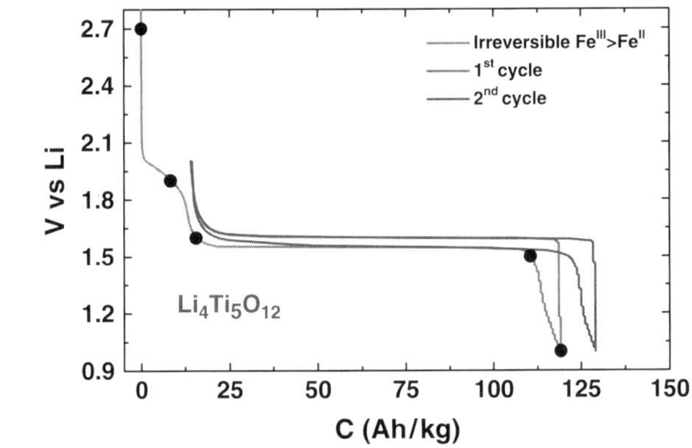

Fig. 24 Discharge/charge curve of pristine iron-doped $Li_4Ti_5O_{12}$ upon Li-insertion. Mössbauer experiments were performed at different discharge depths labeled a, b, c, d

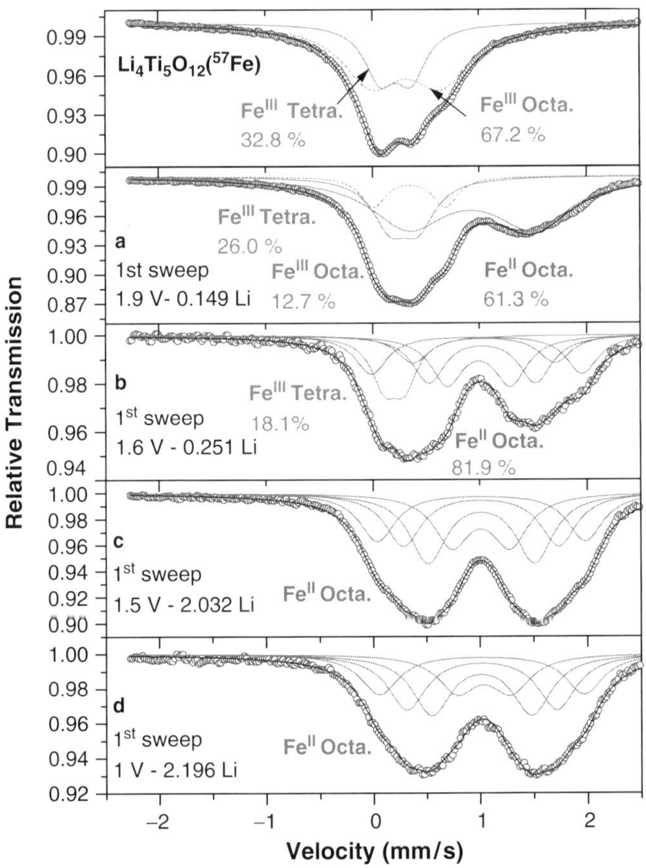

Fig. 25 ^{57}Fe Mössbauer spectra of pristine iron doped $Li_4Ti_5O_{12}$ before Li-insertion and after various discharge depths

capacity of 175 Ah kg^{-1} [59]. Recently, a new type of lithium-cell based on the combination of this high voltage spinel anode with a high voltage spinel cathode has been evidenced [60].

This compound contains no Mössbauer element. ^{57}Fe has therefore been introduced during the synthesis procedure via a precursor ($FeCl_3$•$6H_2O$) prepared from enriched ^{57}Fe iron powder. About 5 at.% of Ti has been substituted to obtain a pristine iron-doped $Li_{4.25}Ti_{4.75}Fe_{0.25}O_{12}$ whose typical discharge/charge curves are shown in Fig. 24.

^{57}Fe TMS shows that Fe dopants have been introduced in the spinel lattice as Fe^{III} on both octahedral ($\delta = 0.32$ mm s^{-1}, $\Delta = 0.65$ mm s^{-1}) and tetrahedral sites ($\delta = 0.22$ mm s^{-1}, $\Delta = 0.29$ mm s^{-1}) in an approximate 2:1 ratio (Fig. 24). During the first discharge the Fe^{III} ions are reduced into Fe^{II} (Fig. 25a–d), which can be explained by the migration of the iron atoms from tetrahedral to octahedral sites ($\delta = 0.9$–1.0 mm s^{-1} and $\Delta = 0.5$–1.94 mm s^{-1}). At the end of the first discharge (1 V), all the Fe atoms are located in octahedral sites and the different observed quadruple splitting are related to the probability to find Ti, Fe, and Li atoms in the neighbourhood of the probed iron atom. This mechanism corresponds to a complete spinel–NaCl phase transition according to the reaction:

$$(Li_3)_{8a}[LiTi_5]_{16_d}O_{12} + 3Li \rightarrow [Li_6]_{16_c}[LiTi_5]_{16d}O_{12}$$

As shown recently, ^{119}Sn can also be used to probe the local environment of Ti in TiO_2 nanoparticles [61] and to follow changes in the Sn local electronic structure during lithiation/delithiation [62].

4 Conclusion

Tin-based negative electrode materials for Li-ion batteries generally form nanocomposites at the end of the first electrochemical discharge that contain Li_xSn nanoparticles ($x \approx 3$–4) and other types of nanoparticles depending on the composition of the pristine material. Because of the small size and the poor crystallinity of the nanoparticles, the Mössbauer spectroscopy is an interesting alternative to the commonly used characterization tools such as XRD. We have shown in this paper that the mechanisms leading to the transformation of pristine materials into nanocomposites can be explained by using ^{119}Sn Mössbauer spectroscopy, since accurate values of the hyperfine parameters were evaluated for the Li_xSn reference compounds. In addition, quantitative information such as the composition of the electrodes can be obtained by combining Mössbauer experimental data and first-principles calculations or by using in situ experiments. We have also shown that interface between the particles of the composites can be characterized with the CEMS. In the case of antimony-based intermetallic materials the interpretation of the ^{121}Sb Mössbauer spectra is more complex. However, they clearly show that nanocomposites obtained at the end of the first discharge do not only contain Li_3Sb and metal nanoparticles but also some lithiated metastable phases that reduce the electrochemical performances. Finally, the use of Mössbauer probes, as illustrated for titanates, allows to characterize materials that do not contain a Mössbauer isotope.

References

1. J.R. Dahn, T. Zheng, Y. Liu, J.S. Xue, Science **270**, 590 (1995)
2. R.A. Huggins, "Handbook of Battery Materials", J.O. Besenhard, ed., Wiley VCH, Weinheim (1999).
3. R.A. Huggins, J. Power Sources **81**, 13 (1999).
4. J.O. Besenhard, J. Yang, M. Winter, J. Power Sources **68**, 87 (1997).
5. O. Mao, R.L. Turner, I.A. Courtney, B.D. Fredericksen, M.I. Buckett, L.J. Krause, J.R. Dahn, Electrochem. Solid State Lett. **2**, 3 (1999).
6. J. Yang, Y. Takeda, N. Imanishi, J.Y. Xie, O. Yamamoto, Solid State Ionics **96**, 77 (2000).
7. C.S. Johnson, J.T. Vaughey, M.M. Thackeray, T. Sarakonsri, S. Hackney, K. Edström, J.O. Thomas, Electrochem. Commun. **2**, 595 (2000).
8. J.M. Tarascon, M. Morcrette, L. Dupont, Y. Chabre, C. Payen, D. Larcher, V. Pralong, J. Electrochem. Soc. **150**, 732 (2003).
9. S. Grugeon, S. Laruelle, L. Dupont, J.M. Tarascon, Solid State Sciences **5**, 895 (2003).
10. J. Chouvin, J. Olivier-Fourcade, J.C. Jumas, B. Simon, O. Godiveau, Chem. Phys. Lett. **308**, 413 (1999).
11. T.C. Gibb, Principles of Mössbauer Spectroscopy, Chapman and Hall, London (1977).
12. U. Gonser, Topics in Applied Physics: Mössbauer Spectroscopy, Springer Verlag, Berlin (1981).
13. G. Long, Mössbauer Spectroscopy Applied to Inorganic Chemistry, vol. IX, Plenum Press, New York (1984).
14. P.E. Dickson, F.J. Berry, Mössbauer Spectroscopy, Cambrige University Press, Cambrige (1986).
15. P.E. Lippens, J.C. Jumas, J.M.R. Génin, Proceedings of ICAME 2005, Springer, The Netherlands (2007).
16. P. Blaha, K. Schwarz, P. Herzig, Phys. Rev. Lett. **54**, 1192 (1985).
17. P.E. Lippens, J. Olivier-Fourcade, J.C. Jumas, Hyp. Int. **126**,137 (2000).
18. P.E. Lippens, J. Olivier-Fourcade, J.C. Jumas, Hyp. Int. **141/142**, 303 (2002).
19. P.E. Lippens, Phys. Rev. B **60**, 4576 (1999).
20. P.E. Lippens, Solid State Commun. **113**, 399 (2000).
21. K. Rubenbauer, T. Birchall, Hyp. Int. **7**, 125 (1979).
22. C.J. Wen, R.A. Huggins, J. Electrochem.Soc. **128**, 1181 (1981).
23. G.R. Goward, N.J. Taylor, D.C.S. Souza, L.F. Nazar, J. Alloys Compds. **329**, 82 (2001).
24. G.-T. Zhou, O. Palchik, I. Nowik, R. Herber, Y. Koltypin, A. Gedanken, J. Solid State Chem. 177 (2004) 3014.
25. I.A. Courtney, J.S. Tse, O. Mao, J. Hafner, J.R. Dahn, Phys. Rev. B **58**, 15583 (1998).
26. P.E. Lippens (unpublished results).
27. R.A. Dunlap, D.A Small, D.D MacNeil, M.N Obravac, J.R. Dahn, J. Alloys Compd. **289**, 135 (1999).
28. F. Robert, P.E. Lippens, R. Fourcade, J.C. Jumas, M. Morcrette, J.M. Tarascon, Hyp. Int. **167**, 797 (2006).
29. J. Chouvin, J. Olivier-Fourcade, J.C. Jumas, B. Simon, Ph. Biensan, F.J. Fernandez Madrigal, J.L. Tirado, C. Pérez Vicente, J. Electroanal. Chem. **494**, 136 (2000).
30. I.A. Courtney, W.R. McKinnon, J.R. Dahn, J. Electrochem. Soc. **146**, 597 (1999).
31. P. Blaha, K. Schwarz, J. Luitz, WIEN97, Vienna University of Technology, 1997 [improved and updated Unix version of the original copyright WIEN code, which was published by P. Blaha, K. Schwarz, P. Sorantin, S.B. Trickey, Comput. Phys. Commun. **59**, 399 (1990)].
32. P.E. Lippens, J.C. Jumas, J. Olivier-Fourcade, Hyp. Int. **156**, 327 (2004).
33. F. Robert, P.E. Lippens, J. Olivier-Fourcade, J.C. Jumas, M. Morcrette, J. Power Sources **146**, 492 (2005).
34. Y. Idota, T. Kubota, A. Matsufuji, Y. Maekawa, T. Miyasaka, Science **276**, 1395 (1997).
35. I.A. Courtney, J.R. Dahn, J. Electrochem. Soc. **144**, 2045 (1997).
36. J. Chouvin, C. Branci, J. Sarradin, J. Olivier-Fourcade, J.C. Jumas, B. Simon, Ph. Biensan, J. Power Sources **81–82**, 277 (1999).

37. H. Li, L. Shi, Q. Wang, L. Chen, X. Huang, Solid State Ionic 8263 (2002).
38. F. Robert, F. Morato, L. Aldon, P.E. Lippens, J. Olivier-Fourcade, J.C. Jumas, B. Simon, P. Biensan, J. Power Sources **119**, 581 (2003).
39. J.C. Jumas, P.E. Lippens, J. Olivier-Fourcade, F. Robert, P. Willmann, Matériau composite d'électrode négative, procédé de fabrication, électrode négative et accumulateur Lithium-ion, Patent n° 04 51742 30/07/2004.
40. A. Aboulaich, F. Robert, P.E. Lippens, L. Aldon, J. Olivier-Fourcade, P. Willmann, J.C. Jumas, Hyp. Int. **167**, 733 (2007).
41. A. Aboulaich, M. Mouyane, F. Robert, P.E. Lippens, J. Olivier-Fourcade, P. Willmann, J.C. Jumas, J. Power Source (in press).
42. O. Mao, R.L. Turner, I.A. Courtney, B.D. Fredericksen, M.I. Buckett, L.J. Krause, J.R. Dahn, Electrochem. Solid State Lett. **2**, 3–5 (1999).
43. R.A. Dunlap, O. Mao, J.R. Dahn, Phys. Rev. B **59**, 3494 (1999).
44. S. Sharma, L. Fransson, E. Sjöstedt, L. Nordström, B. Johansson, K. Edström, J. Electrochem. Soc. **150**, A330 (2003).
45. S. Naille, P.E. Lippens, F. Morato, J. Olivier-Fourcade, Hyp. Int. **167**, 785 (2006).
46. S. Naille, M.C. Ionica, P.E. Lippens, F. Robert, F. Morato, J. Olivier-Fourcade, J. Power Source (in press).
47. L. Aldon, C. Ionica, P.E. Lippens, M. Morcrette, D. Larcher, J.M. Tarascon, J.C. Jumas, Hyp. Int. **167**, 729 (2006).
48. R. Dedryvère, S. Leroy, S. Naille, H. Martinez, P.E. Lippens, J.C. Jumas, D. Gonbeau, J. Power Source (in press).
49. Sony, USA Patent 0053131 (2005).
50. C.M. Bousquet, P.E. Lippens, L. Aldon, J. Olivier-Fourcade, J.C. Jumas, Chem. Mater. **18**, 6442 (2006).
51. J. Sangster, A.D. Pelton, J. Phase Equilibria **14**, 4 (1993).
52. J.T. Vaughey, C.S. Johnson, A.J. Kropf, R. Benedek, M.M. Thackeray, H. Tostmann, T. Sarakonsri, S. Hackney, L. Fransson, K. Edström, J.O. Thomas, J. Power Sources **97**, 194 (2001).
53. L.M.L. Fransson, J.T. Vaughey, K. Edström, M.M. Thackeray, J. Electrochem. Soc. **150**, A86 (2003).
54. C.M. Ionica, L. Aldon, P.E. Lippens, F. Morato, J. Olivier-Fourcade and J.C. Jumas Hyperfine Interactions **156**, 555 (2004).
55. L. Aldon, P. Kubiak, A. Picard, P.E. Lippens, J.C. Jumas, J. Olivier-Fourcade, Hyp. Int. **156**, 497 (2004).
56. C.M. Ionica, P.E. Lippens, J. Olivier Fourcade, J.C. Jumas, J. Power Sources **146**, 481 (2005).
57. L. Häggström, C. Ionica, J.C. Jumas, L. Aldon, P.E. Lippens, K. Edström, Hyp. Int. **167**, 759 (2006).
58. C. Ionica, M. Womes, P.E. Lippens, J. Olivier-Fourcade, B. Ducourant, A. Chadwick, Hyp. Int. **167**, 773 (2006).
59. P. Kubiak, A. Garcia, M. Womes, L. Aldon, J. Olivier-Fourcade, P.E. Lippens, J.C. Jumas, J. Power Sources **119–121**, 626 (2003); T. Ozhuku, A. Ueda, N. Yamamoto, J. Electrochem. Soc. 142, 1431 (1995).
60. S. Panero, D. Satolli, M. Salomon, B. Scrosati, Electrochem. Commun. **2**, 810 (2000).
61. A. Wiebel, R. Bouchet S.L.P. Savin, A.V. Chadwick, P.E. Lippens, M. Womes, P. Knauth, ChemPhysChem **7**, 2377 (2006).
62. L. Aldon, P. Kubiak, A. Picard, J.C. Jumas, J. Olivier-Fourcade, Chem. Mater. 18, 1401 (2006).

Index

A
"ab initio" modeling, 27, 60–63, 177, 196, 209
2D exchange NMR, 227
Activation energy, 31, 33, 40, 62, 79, 80, 93, 111, 231, 240, 243, 244
Alumina, 3, 4, 12, 31, 64, 99, 127, 130, 217, 220, 236–238
Amorphous, 1, 2, 9, 10, 19–21, 26–28, 33, 35, 40, 45, 49
Anode, 1, 82, 83, 119, 140, 220, 261, 267, 269
Antimony, 248, 263–265, 269
Arrhenius law, 79
Atomic scale, 179, 206, 247

B
BaF_2, 228, 234, 235
Ball-milling, 231, 232, 234, 235, 244
Binary oxides, 72, 73, 88, 89, 91, 97, 98
BPP theory, 230, 231

C
CaF_2, 228, 234, 235, 244
Calix[6]pyrrole, 14, 16, 23–25
Casting, 3, 73, 93
Cation dynamics, 63, 186, 187
CEMS, 247, 259, 260, 269
Ceramic, 4, 5, 9, 40, 45, 215, 234, 255, 259
Clay-polymers nanocomposites, 9, 175, 176, 181, 182, 185–187, 192, 196–199
Composite, 1–14, 17, 19–21, 23, 25–29, 31, 32, 34–36
Conductivity, 2–6, 8–14, 17, 20, 21, 23, 25–31, 34–41
Conductor-insulator composites, 239
Co-ordination Number, 213, 216, 217
Correlation function, 230
$CuBr:TiO_2$, 229

D
Debye-Waller factor, 208
Density-Functional Theory (DFT), 60, 177–179, 196
Dielectric relaxation, 27, 28
Diffusion, 16, 28, 60, 63, 77–79, 82, 84, 85, 94, 98, 100, 109, 111
Diffusion induced NMR spin-lattice relaxation, 227, 229, 230, 231, 233
Diffusion pathway(s), 60, 94, 229, 232f, 238, 244
Dipolar relaxation, 229, 230
Dipole-dipole interactions, 231
Direct methanol fuel cells, 87, 87
Dispersed ion conductor, 228, 229

E
Effective Medium Theory (EMT), 35, 37–45, 49, 51–56, 58, 105
Electric field gradient, 230, 232, 250, 252, 265
Electrical conductivity, 46, 79, 122, 133–135, 139, 183
Electrode, 2–4, 12, 17, 28, 48, 65, 77, 82–84, 87, 88, 91
Electrode-electrolyte interface, 17
Energy conversion, 113
Enhanced diffusivity in nano-materials, 234
EXAFS, 206–208, 210–213, 215–221, 233
External magnetic field, 229, 230
Extreme narrowing, 230

F
Fick's law, 79
Field gradient NMR, 84, 227, 230, 232
Filler, 2–12, 13, 17, 18, 20–27, 31, 33, 35
Finite Element, 35, 36, 165, 170, 198
First-principles calculations, 269

Fluorine, 146
Fraction of mobile charge carriers, 31, 232, 236
Free induction decay, FID, 229, 239–241, 243
Fuel cell, 1, 72, 74, 76, 77, 82–84, 86–93, 95
Functional polymer, 60–62, 75, 84, 189, 198

G
Glass transition, 16, 27, 31, 40, 57, 79, 80, 87, 90, 99, 100, 108, 145, 164
Grain boundaries, 28, 36, 44, 45, 232, 234, 235, 237
Grotthuss mechanism, 77, 111–113

H
H_2SO_4, 7, 18
Heterocontacts, 228, 229, 238
Heterogeneous, 37, 44, 46, 228, 230, 232, 234, 235, 239, 241, 244
Heterogeneous materials, 44, 228, 230
Hetero-interfaces, 238
Heteropolyacids, 73, 77, 88, 93, 94, 99, 102
High-energy ball-milling, 231, 237, 243
Homocontacts
Homo-interfaces, 238
HR-TEM, 233, 234
Hybrid, 61, 71–74, 76, 77, 81, 84, 86, 91, 95, 97, 98, 102
Hydration, 76, 77, 81, 93, 96, 99, 108–112, 186, 187, 192
Hyperfine, 249, 257, 269

I
Impedance spectroscopy, 28, 135–137, 238
In situ, 17, 73, 74, 92, 94, 97, 137, 140, 145, 147, 148
Inert-gas condensation, 234
Insulator, 39, 41, 44, 45, 227–229, 236–240, 242–244
Interface, 2, 3, 4, 10, 17, 20, 36, 37, 39, 40, 59, 82, 89, 106
Interfacial regions, 104, 139, 228, 232, 233–235, 238, 239, 242–244
Intermetallic, 261, 263, 269
Ionic conductivity, 2–4, 6, 7, 10, 11, 13, 17, 20, 25, 30, 39, 43, 45, 53, 65, 95, 101, 106, 229
Ionic dopant, 2
Ionic transport, 1, 19, 23, 25, 28, 233, 235

L
Large-scale simulations, 185, 192, 199, 200
Larmor frequency, 230, 231
Lewis acid-base, 9–12, 21–23, 27
Li conductor, 234, 236
Li mobility, 64
Li NMR, 186, 229, 231–233, 236–240, 242–244
Li_2O, 228, 231–238, 242–244, 256–258
$Li_2O:Al_2O_3$, 228, 235–243
$Li_2O:B_2O_3$, 228, 229, 235, 236, 238, 240
$Li_4Ti_5O_{12}$, 228, 267, 268
$LiBF_4$, 13, 16, 24, 64
$LiCF_3SO_3$, 63, 13, 16, 24, 26, 63
$LiClO_4$, 5, 7, 8, 10–13, 22, 24, 42, 53, 122
$LiI:Al_2O_3$, 44, 229
Li-ion, 64, 139, 140, 248, 269
Li-ion conductivity, 139
$LiNbO_3$, 228, 232–235, 244
Line shapes, 186, 227, 229, 231–237, 244
$LiTaO_3$, 227
Lithium, 1–3, 10, 12–14, 16, 17, 23–25, 27, 59–63, 65
Li_xSn, 253–255, 258, 259, 269
Li_xTiS_2, 228
Local environment, 207, 208, 215, 218, 220, 258, 269
Local probe, 248, 267
Longitudinal relaxation, 240
Long-range, 175, 185, 227, 231
Low-temperature flank, 240

M
Magnetization transient, 231, 241–243
MEA, 77, 82–84, 91, 98
Mechanical properties, 4, 6, 10, 80, 86, 87, 89, 93, 94, 96, 97, 101, 102, 113
Mechanochemical reaction, 236
Membrane, 1, 10, 72–74, 76–78, 80–103
Microcrystalline materials, 228, 231, 232
Microstructure, 33, 36, 51, 63, 71, 72, 80, 84, 99, 113, 124, 206
Models, 3, 19, 21, 23, 27, 35–37, 44, 45, 55, 56, 58, 59, 61
Molecular dynamics (MD), 27, 59, 60, 62–65, 109, 113, 181–195, 197–199
Molecular scale, 27, 35, 59–65, 103
Monte Carlo, 181–184, 186, 192
Montmorillonite (MMT), 9, 88, 94, 95, 99, 146, 149, 185–198
Morphology, 20, 45, 72, 74, 77, 84, 87, 92, 95, 106, 108, 119, 121–123, 126, 139, 152, 171

Mössbauer, 176, 247–252, 254–269
Motional narrowing, 229, 234

N
Nafion, 71–74, 77, 78, 81, 84–100, 104–113
NaI, 4–6, 26, 29, 32–34, 41–43, 55, 56, 58
Nanocomposite (s), 9, 65, 71–74, 76–78, 81, 83, 85, 86, 89–95
Nanocrystalline materials, 227, 228, 232, 235, 244
Nanoparticle-Inorganic Matrix, 205, 213–216
Nanoparticle-Polymer Matrix, 171, 205, 206, 216, 217
Nanoparticle-Nanoparticle Composite, 205, 217–220
Nanostructured materials, 119, 143, 247
Nanostructures, 113, 215, 229, 247
NASICON, 4, 5, 26
Nernst-Einstein equation, 79
NEXAFS, 13, 176, 208
Nuclear magnetic resonance (NMR), 5, 6, 16, 26, 84, 183, 186, 187, 218, 219, 227–237
NMR line shapes
NMR line width, 229, 236
NMR relaxation model, 239
Number fraction of mobile ions, 244

O
Organic salt, 1
Organic-inorganic hybrids, 103
ORMOSIL, 74, 91, 92, 98, 102
Oxidation state, 206, 208, 209, 213, 220, 258
Oxygen, 10, 11, 22, 23, 27, 61–63, 80, 83, 89, 90, 92, 102, 110

P
PAAM, 10, 11, 41, 42, 51–54
PEGME, 8, 21–23
PEM, 82, 84, 88, 91, 102, 112, 113
Polyethylene oxide (PEO), 1–7, 9–14, 16, 19, 20, 24–26, 29, 32–34, 41–43, 51, 61, 65, 73, 103, 186, 191, 193
Percolation pathways, 4, 238, 239
Percolation theory, 27, 229
Phase scale model, 35
PMMA, 10
Polyacrylate, 193
Polybenzimidazole (PBI), 73, 96, 100–102
Polyethyleneglycol, 8, 21, 191, 193–195
Polymer electrolyte, 1, 2, 12, 13, 17, 20, 26, 27, 29, 30, 59, 60
Polymer electrolyte membrane fuel cells, 90
Potential, 36, 46, 48, 62, 63, 77–79, 83, 91, 94, 98, 119, 137
Pre-exponential factor, 40, 231, 243, 244
Preparation, 2, 3, 6, 7, 60, 73, 74, 91, 92, 98, 100–102, 145
Proton conductivity, 76–81, 85, 87–93, 95–98, 100, 102, 112, 113
Proton exchange membranes, 76, 90

Q
Quadrupolar interactions, 231, 234
Quadrupolar relaxation, 230
Quadrupole nuclei, 231
Quantum mechanics, 3, 27, 59, 60, 62, 63, 177–179, 198
Quick EXAFS, 210

R
Radial Distances, 208, 211, 212, 219, 221
Radio frequency pulse, 229, 230
Random Resistor Network (RRN), 20, 35, 36, 37, 44–58
Relative humidity, 72, 81, 88–90, 92–94, 101, 102, 151, 157, 158
Relaxation behaviour, 228, 230
Resonance frequency, 229, 230, 235, 236
Rigid lattice
  rotating reference frame, 229

S
SANS
  saturation recovery experiment, 240
SAXS
  separation of NMR relaxation rates, 227, 231, 239, 240, 243
  shape of NMR resonance lines, 186, 232
Short-range, 231, 239
Silylation, 75
Single-exponential behaviour, 241, 242
Single-phase materials, 227, 231–235, 237, 238, 244
SiO_2, 21, 41, 42, 55, 56, 73, 86, 88, 90, 91, 94, 97, 101
Slow Li motions, 236
Sol-gel, 74, 75, 86, 89–92, 94, 97, 103, 113, 213, 217, 234, 259
Space charge region, 21, 108, 229, 234, 235
Spectral density function, 230
Sulfonated polyetheretherketone (SPEEK), 73, 75, 84–86, 96–100, 102, 103

Spin-alignment echo NMR, 227
Spin-lattice relaxation, 227, 229–231, 233, 240–244
Spin-spin relaxation, 229–231, 239, 243, 244
Statistical ensembles, 182, 184
Structural disorder, 228, 231, 234
Structure-properties relationship, 77, 120
Sulfonation, 75, 95–99, 102, 103
Superacids, 12, 17, 23, 24, 90, 91, 112
Supramolecular, 3, 9, 12, 14, 23, 24, 60

T
TEOS, 74, 92, 97
Thin-film adhesion, 133, 134
Three-component NMR line shape, 233
Tin, 126, 218, 247, 248, 253, 255, 258, 259, 261, 269
Tin oxide, 218, 255–258
TiO_2, 60, 61, 63, 65, 73, 86, 88–91, 97, 98, 130
Titanate, 147, 148, 248, 267, 269
Titania, 73, 85, 90, 213
Transference number, 2, 3, 9, 12–14, 16, 17, 23–25, 28, 65
Transport mechanism, 12, 77, 111, 113
Transport properties, 28, 77, 84, 228, 244
Transverse magnetization, 229
Two-component FID, 243
Two-component NMR line shape, 232, 234, 235, 236, 243
Two-phase materials, 235–239, 244

V
Vehicle mechanism, 77, 111
Vogel-Tammann-Fulcher equation (VTF), 40, 79, 80

W
Water diffusion, 84, 85
Water uptake, 80–82, 84, 87, 90, 91, 93, 95, 96, 103, 110, 113

X
XAS, 206, 207, 210–213, 215–218, 220, 221
X-Ray Absorption, 13, 206, 208, 211, 215
XANES, 206–210, 213–219
XRD (line broadening), 236

Z
ZrO_2, 12, 13, 73, 86, 88, 90, 91, 97–99, 112, 113, 211, 212, 217, 218

Printed in the United States of America.